*FM 3-55.93 (FM 7-93)

Field Manual
No. 3-55.93 (FM 7-93)

Headquarters
Department of the Army
Washington, DC
23 June 2009

Long-Range Surveillance Unit Operations

Contents

	Page
Preface	xiii
Summary of Change	xv

Chapter 1 FULL-SPECTRUM OPERATIONS ... 1-1
    Section I. TYPES AND COMBINATIONS OF ARMY OPERATIONS ... 1-1
        Four Types of Army Operations ... 1-1
        Intelligence ... 1-2
    Section II. INFANTRY RECONNAISSANCE AND SURVEILLANCE UNITS ... 1-2
        Characteristics ... 1-2
        Missions and Organizations ... 1-3
    Section III. LONG-RANGE SURVEILLANCE COMPANY ... 1-3
        Primary Missions ... 1-4
        Secondary Missions ... 1-4
        Comparison to Special Operations Forces ... 1-4
        Organization ... 1-4
        Sustainment ... 1-5
        Subordinate Organizations and Key Personnel ... 1-5
        LRSD and LRS Teams ... 1-7
        Characteristics ... 1-7
        Capabilities ... 1-8
        Limitations ... 1-8

Chapter 2 COMMAND AND CONTROL ... 2-1
    Section I. OVERVIEW ... 2-1
        Standard LRSU Command and Control ... 2-1
        BFSB Staff ... 2-2
        Reconnaissance and Surveillance Squadron ... 2-6
        Nonstandard LRSU Command and Control ... 2-7
        Command Posts ... 2-7
        Tactical Operations Center ... 2-10
        Task Organization Outside NAMED Areas of Interest ... 2-13
        Liaison Duties, Employment, and Coordination ... 2-13

Distribution Restriction: Approved for public release; distribution is unlimited.

*This publication supersedes FM 7-93, 3 October 1995.

Contents

	Section II. COMMUNICATIONS, COMPUTERS, AND INTELLIGENCE	2-14
	Communications	2-14
	Computers and Intelligence	2-15
Chapter 3	**MISSION DEVELOPMENT**	3-1
	Section I. ISR OPERATIONS AND MISSION ORDERS	3-1
	ISR Operations	3-1
	Mission Orders	3-3
	Section II. MISSION PLANNING FOLDER	3-5
	Development	3-5
	Contents	3-5
	Target Folder Format	3-9
	Section III. OPERATIONS SECURITY	3-12
	Personnel Security	3-12
	Mission Classification	3-12
	Separation	3-13
	Security during Coordination	3-13
Chapter 4	**TEAM OPERATIONS**	4-1
	Section I. PHASES	4-1
	Planning Phase	4-1
	Infiltration Phase and Insertion Method	4-16
	Execution (Actions on Objective)	4-17
	Exfiltration Phase and Extraction Method	4-17
	Recovery	4-18
	Section II. RECONNAISSANCE OPERATIONS	4-21
	Area Reconnaissance	4-21
	Zone Reconnaissance	4-23
	Route Reconnaissance	4-25
	Bridge Classification	4-25
	Leader Reconnaissance	4-25
	Section III. SURVEILLANCE OPERATIONS	4-26
	Selection and Occupation of Sites	4-26
	Security and Reports	4-27
	Linkup and Dissemination of Information	4-29
	Contingencies	4-29
	Heavy Team and Platoon Operations	4-30
	Section IV. COMBAT ASSESSMENT	4-30
	Definition and Purpose	4-30
	Considerations	4-30
	Characteristics	4-31
	Damage Types and Levels	4-31
	Bridges	4-32
	Buildings	4-33
	Bunkers	4-35
	Dams and Locks	4-35
	Distillation Towers	4-36
	Military Equipment	4-36

	Ground Force Personnel	4-38
	Storage Tanks For Petroleum, Oil, Lubricants	4-39
	Power Plant Turbines and Generators	4-39
	Rail Lines and Rail Yards	4-40
	Roads	4-40
	Runways and Taxiways	4-41
	Satellite Dishes	4-41
	Ships	4-42
	Steel Towers	4-43
	Transformers	4-43
	Tunnel Entrances Or Portals	4-43
	Tunnel Facility Air Vents	4-44
	Section V. TARGET ACQUISITION	**4-44**
	Combat Patrol	4-45
	Mission Planning Factors	4-45
	Employment of Laser Designators	4-45
	Fire Support	4-46
	Fire Plans	4-47
	Section VI. URBAN TERRAIN	**4-47**
	Surveillance Operations	4-47
	Reconnaissance Operations	4-48
	Plans	4-48
	Hide and Surveillance Sites	4-52
	Section VII. IMAGERY COLLECTION AND TRANSMISSION	**4-53**
	Imagery Labels	4-53
	Image-Gathering Equipment	4-54
	Section VIII. STABILITY OPERATIONS	**4-61**
	Types	4-62
	Capabilities and Limitations	4-67
	Section IX. SPECIAL MISSIONS	**4-68**
	Chemical, Biological, Radiological, and Nuclear	4-68
	Pathfinder	4-68
	Personnel Recovery	4-68
Chapter 5	**INSERTION AND EXTRACTION METHODS**	**5-1**
	Section I. WATERBORNE OPERATIONS	**5-1**
	Considerations	5-1
	Combat Rubber Raiding Reconnaissance Craft	5-1
	Scout Swimmers	5-5
	Helocasting Operations	5-7
	Section II. HELICOPTER OPERATIONS	**5-9**
	Special Patrol Insertion/Extraction System	5-9
	Fast-Rope Insertion/Extraction System	5-20
	Army Aviation and Air Assault	5-25
	Pickup and Landing Zones	5-25
	UH-60 Loading Sequence	5-31

Contents

	Section III. VEHICLE OPERATIONS	5-31
	Mobility Platforms	5-32
	Planning Considerations	5-32
	Section IV. OTHER OPERATIONS	5-46
	Airborne Operations	5-47
	Stay-Behind Operations	5-47
	Foot Movement Operations	5-48
Chapter 6	COMMUNICATIONS	6-1
	Section I. NETWORKS	6-1
	Architecture and Frequency Management	6-1
	Operations Bases	6-1
	Teams	6-2
	Section II. RADIOS, COMPUTERS, AND THE BASE RADIO STATION	6-2
	Elements of Success	6-2
	HF, VHF, and UHF Radios	6-2
	Primary, Alternate, and Contingency Radios	6-3
	Fundamentals	6-4
	Beyond-Line-of-Sight Equipment	6-4
	Ruggedized COTS Laptop	6-6
	Communications Base Radio Station Platform	6-7
	Section III. OPERATIONS	6-7
	Tactical Employment	6-7
	Site Selection	6-8
	Tactical Satellite	6-9
	Section IV. REPORTS	6-9
	Messages and Report Formats	6-9
	Communications Security	6-16
	Section V. ELECTRONIC WARFARE	6-16
	Electronic attack	6-17
	Electronic warfare support	6-17
	Electronic protection	6-17
	Section VI. ANTENNAS	6-19
	Wavelength and Frequency	6-19
	Resonance	6-19
	Polarization	6-20
	Radio Wave Propagation	6-20
	Classification	6-23
	Construction and Selection	6-26
	Common Types of Antennas	6-27
	Field-Expedient Antennas	6-31
	High Frequency, Directional, Field-Expedient Antennas	6-34
	Section VII. UNUSUAL ENVIRONMENTS	6-36
	Desert Operations	6-36
	Jungle Operations	6-36
	Cold Weather Operations	6-37

	Mountain Operations	6-38
	Urban Operations	6-39
Chapter 7	**INTELLIGENCE PREPARATION OF THE BATTLEFIELD**	**7-1**
	SECTION I. DEFINE THE OPERATIONAL ENVIRONMENT	**7-1**
	Identify Significant Characteristics of the Environment	7-1
	Identify the Limits of the Command's Area of Operations	7-3
	Establish the Limits of the Area of Influence and the Area of Interest	7-3
	Evaluate Existing Databases and Identify Intelligence Gaps	7-4
	Initiate Collection of Information Required to Complete IPB	7-4
	SECTION II. DESCRIBE ENVIRONMENTAL EFFECTS ON OPERATIONS	**7-4**
	Analyze the Environment	7-4
	Describe Environmental Effects	7-9
	SECTION III. EVALUATE THE THREAT	**7-9**
	Analyze Threat Factors	7-9
	Update or Create Threat Models	7-10
	Identify Threat Capabilities	7-16
	SECTION IV. DETERMINE THREAT COURSES OF ACTION	**7-17**
	Identify the Threat's Likely Objectives and Desired Endstate	7-17
	Identify the Full Set of COAs Available to the Threat	7-17
	Evaluate and Prioritize Each Course of Action	7-18
	Develop Courses of Action	7-18
	Identify Initial ISR Requirements	7-21
	Prepare Decision Support Template	7-21
Chapter 8	**EVASION AND RECOVERY**	**8-1**
	Fundamentals	8-1
	Chain of Command	8-2
	Plans	8-2
	Types of Recovery	8-2
	Classifications of Evasion	8-3
	Movement	8-4
	Disguises	8-6
	Evasion Aids	8-7
	Evasion Areas	8-9
Appendix A	**RECRUITMENT, ASSESSMENT, AND SELECTION PROGRAM**	**A-1**
	Purpose and Organization	A-1
	Recruitment	A-1
	Assessment	A-2
	Selection	A-4
	Reassignment	A-5
Appendix B	**ORDERS AND BRIEFS**	**B-1**
	Section I. ORDERS	**B-1**
	Warning Order	B-1
	Operation Order	B-4
	Fragmentary Order	B-8

Contents

	Section II. BRIEFS	B-8
	Types	B-8
	Confirmation Brief	B-11
	Mission Analysis Brief	B-11
	Decision Brief	B-17
	Backbrief	B-17
	Mission Concept Brief	B-20
	Debrief	B-34
Appendix C	**PLANNING AREA FACILITIES AND SITES**	**C-1**
	Facilities	C-1
	Fixed Site	C-1
	Field Site	C-1
Appendix D	**GEOGRAPHIC ENVIRONMENTS**	**D-1**
	Jungle Operations	D-1
	Desert Operations	D-1
	Mountain Operations	D-2
	Cold Weather Operations	D-2
Appendix E	**CONTINGENCY PLANS**	**E-1**
	Branches and Sequels	E-1
	Contingency Plan Matrix	E-1
Appendix F	**COORDINATION FOR ARMY AVIATION**	**F-1**
Appendix G	**HIDE AND SURVEILLANCE SITES**	**G-1**
	Surface Sites	G-1
	Hasty Subsurface Sites	G-2
	Finished Subsurface Site	G-4
	Site Selection	G-7
	Leader Reconnaissance	G-7
	Occupation of Hide Site	G-7
	Actions in Hide Site	G-9
	Priority of Work	G-9
	Site Sterilization	G-9
Appendix H	**BATTLE DRILLS**	**H-1**
	Break Contact	H-1
	React to Air Attack	H-5
	React to Indirect Fire	H-6
	React to Flares	H-6
	Break from Hide or Surveillance Site	H-6
Appendix I	**TRACKING AND COUNTERTRACKING**	**I-1**
	Concepts of Tracking	I-1
	Organization of Tracking Team	I-5
	Tracker and Dog Team	I-5
Appendix J	**NIGHT OPERATIONS**	**J-1**
	Night Vision	J-1
	Hearing	J-4
	Smell	J-4

	Fatigue	J-5
	Selection of Route	J-5
	Night Walking	J-6
	Signals	J-7
	Target Detection	J-8
	Movement	J-9
Appendix K	EXAMPLE EVASION AND RECOVERY PLAN	K-1
GLOSSARY		Glossary-1
REFERENCES		References-1
INDEX		Index-1

Figures

Figure 1-1. Organization of long-range surveillance company. 1-5
Figure 1-2. Concept of intelligence, surveillance, and reconnaissance. 1-10
Figure 2-1. Company operations base. 2-8
Figure 2-2. Example C2 employment schematic for a LRSC. 2-9
Figure 2-3. Example C2 employment schematic for a LRSD in an MSS. 2-10
Figure 3-1. Intelligence, surveillance, and reconnaissance. 3-2
Figure 3-2. LRS mission development process. 3-4
Figure 3-3. Example format for target folder. 3-10
Figure 4-1. TLP and METT-TC. 4-7
Figure 4-2. Development of courses of action. 4-12
Figure 4-3. Reconnaissance and surveillance elements. 4-22
Figure 4-4. Fan method. 4-23
Figure 4-5. Converging routes method. 4-24
Figure 4-6. Successive sector method. 4-24
Figure 4-7. Example surveillance site. 4-27
Figure 4-8. Example hide site. 4-28
Figure 4-9. Example imagery labels. 4-53
Figure 4-10. Example imagery legend. 4-54
Figure 4-11. Example panoramic sketch. 4-55
Figure 4-12. Example topographic sketch. 4-56
Figure 4-13. Example objective sketchpad. 4-57
Figure 4-14. Example drawing technique: whole to part. 4-59
Figure 4-15. Example drawing technique: use of common shapes to show common objects. 4-60
Figure 4-16. Example drawing technique: use of perspective to represent depth. 4-60
Figure 4-17. Example drawing technique: use of vanishing points to indicate distance. 4-60
Figure 4-18. Example drawing technique: hatching. 4-61
Figure 5-1. Rubber boat. 5-2
Figure 5-2. Short count, long count. 5-4

Figure 5-3. SPIES rope rigging on UH-60 ... 5-17
Figure 5-4. Rigging of snap links. ... 5-17
Figure 5-5. Rigging of wood block. ... 5-18
Figure 5-6. Excess cargo straps secured. ... 5-18
Figure 5-7. Recovery line with Prusik knot. ... 5-19
Figure 5-8. SPIES rigging procedures for CH-46 or CH-47. ... 5-20
Figure 5-9. Fast-rope rigging procedures for UH-60. ... 5-21
Figure 5-10. UH-60 rigged for fast roping ... 5-22
Figure 5-11. Fast-rope rigging procedures for other aircraft. ... 5-22
Figure 5-12. Marking procedures for landing and pickup zones. ... 5-26
Figure 5-13. Example coordination checklist. ... 5-28
Figure 5-14. UH-60 loading sequence. ... 5-31
Figure 5-15. Example vehicle load configuration. ... 5-35
Figure 5-16. Procedures for loading HMMWV into CH-47 for infiltration. ... 5-38
Figure 5-17. HMMWVs in wedge formation. ... 5-41
Figure 5-18. Single camouflaged HMMWV. ... 5-43
Figure 5-19. Multiple camouflaged HMMWVs. ... 5-43
Figure 6-1. Automatic link sequence. ... 6-5
Figure 6-2. AN/PRC-150(C) in vehicular AN/VRC-104 (V)3 configuration. ... 6-7
Figure 6-3. Communications data wire diagram. ... 6-10
Figure 6-4. Report format. ... 6-11
Figure 6-5. Example message header. ... 6-12
Figure 6-6. Measurement of a wavelength. ... 6-19
Figure 6-7. Components of ground wave. ... 6-21
Figure 6-8. Structure of ionosphere. ... 6-22
Figure 6-9. HF skip zone and distance. ... 6-23
Figure 6-10. Unidirectional antenna pattern. ... 6-24
Figure 6-11. Bidirectional antenna pattern. ... 6-25
Figure 6-12. Omni-directional antenna pattern. ... 6-26
Figure 6-13. Half-wave dipole antenna. ... 6-27
Figure 6-14. Formula for calculating length of half-wave dipole antenna applied to example. ... 6-28
Figure 6-15. Inverted "V" antenna. ... 6-28
Figure 6-16. Long-wire antenna. ... 6-29
Figure 6-17. Sloping wire antenna. ... 6-30
Figure 6-18. Terminated sloping "V" antenna. ... 6-30
Figure 6-19. Repair procedure, whip antenna. ... 6-31
Figure 6-20. Expedient insulators. ... 6-32
Figure 6-21. High frequency antenna, long-wire type. ... 6-34
Figure 6-22. High frequency antenna, half-rhombic type. ... 6-34
Figure 6-23. High frequency antenna, "V" type. ... 6-35
Figure 6-24. Sloping antenna, "V" type. ... 6-35

Figure 7-1. Classification of urban area by size. 7-2
Figure 7-2. Example link diagram. 7-12
Figure 7-3. Example association matrix. 7-13
Figure 7-4. Example relationship matrix. 7-14
Figure 7-5. Example activities matrix. 7-15
Figure 7-6. Example time event chart. 7-16
Figure 7-7. Example pattern analysis plot sheet. 7-19
Figure 8-1. Types of recovery. 8-3
Figure B-1. Example warning order. B-3
Figure B-2. Example operation order. B-5
Figure B-3. Example fragmentary order. B-8
Figure B-4. Brief types. B-8
Figure B-5. Analysis of mission, intent, and priority intelligence requirements. B-12
Figure B-6. Analysis of specified and implied tasks. B-13
Figure B-7. Analysis of facts. B-14
Figure B-8. Analysis of assumptions. B-15
Figure B-9. Analysis of mission roadblocks, issues, and restated mission. B-16
Figure B-10. Example backbrief. B-18
Figure B-11. Slide 1, LRSC mission concept brief. B-21
Figure B-12. Slide 2, statement of purpose. B-21
Figure B-13. Slide 3, LRSC insertion conditions check. B-22
Figure B-14. Slide 4, agenda. B-22
Figure B-15. Slide 5, recommendation. B-23
Figure B-16. Slide 6, ISR fusion element. B-23
Figure B-17. Slide 7, S-2/ISR fusion element. B-24
Figure B-18. Slide 8, ATO slide. B-24
Figure B-19. Slide 9, team insertion and extraction. B-25
Figure B-20. Slide 10, LRSC insertion conditions check. B-25
Figure B-21. Slide 11, movement and maneuver. B-26
Figure B-22. Slide 12, fire support. B-26
Figure B-23. Slide 13, air protection. B-27
Figure B-24. Slide 14, sustainment. B-27
Figure B-25. Slide 15, command and control. B-28
Figure B-26. Slide 16, intelligence. B-28
Figure B-27. Slide 17, LRSC. B-29
Figure B-28. Slide 18, "Do we know what to look for?". B-29
Figure B-29. Slide 19, LRSC IPB. B-30
Figure B-30. Slide 20, LRSC maneuver. B-30
Figure B-31. Slide 21, LRSC fire support. B-31
Figure B-32. Slide 22, LRSC sustainment. B-31
Figure B-33. Slide 23, LRSC C2. .... B-32

Figure B-34. Slide 24, LRSC C2, communications. B-32
Figure B-35. Slide 25, LRSC abort criteria. B-33
Figure B-36. Slide 26, LRSC risk mitigation. B-33
Figure B-37. Slide 27, LRSC recommendation. B-34
Figure B-38. Slide 28, commander's decision. B-34
Figure B-39. Example debrief. B-35
Figure B-40. Example intelligence estimate annex. B-40
Figure B-41. Example communications annex. B-42
Figure B-42. Example fire support annex. B-43
Figure B-43. Example linkup annex. B-44
Figure B-44. Example vehicle movement annex. B-45
Figure B-45. Example air infiltration/exfiltration annex. B-47
Figure C-1. Example fixed site for planning. C-2
Figure C-2. Example planning area. C-3
Figure C-3. Use of intermediate staging area for planning. C-4
Figure E-1. Example completed contingency matrix. E-2
Figure F-1. Example OPORD. F-2
Figure F-2. Example fire support annex. F-4
Figure F-3. Example intelligence annex. F-5
Figure F-4. Example rehearsal area annex. F-6
Figure F-5. Example vehicular movement coordination annex. F-7
Figure G-1. Two-man surface site using ghillie suits. G-1
Figure G-2. Suspension line-weave site. G-3
Figure G-3. Polyvinyl chloride site. G-3
Figure G-4. Example subsurface site. G-6
Figure G-5. Fishhook and dog-leg methods. G-8
Figure G-6. Forcible occupation of site. G-8
Figure H-1. Break contact front (diamond or file). H-2
Figure H-2. Break contact front, left and right (Australian peel). H-3
Figure H-3. Break contact left, right (diamond or file). H-4
Figure H-4. React to enemy air attack. H-5
Figure H-5. React to indirect fire or air attack. H-6
Figure H-6A. Break contact from hide or surveillance site. H-8
Figure I-1. Areas surveyed for indicators by tracker. I-1
Figure I-2. Examples of displacement. I-2
Figure I-3. Types of footprints. I-3
Figure I-4. Box method for determination of number of footprints. I-4
Figure J-1. Typical scanning patterns. J-2
Figure J-2. Off-center viewing technique. J-2
Figure K-1. Example plan of action for an evasion. K-2
Figure K-2. Example of DD Form 1833 TEST (V2) (front). K-5

Figure K-3. Example of DD Form 1833 TEST (V2) (back). K-6

Tables

Table 4-1. Actions and responsibilities of LRSU personnel. 4-4
Table 4-2. LRS tasks by operation. 4-10
Table 4-3. Priority of actions for rehearsal. 4-14
Table 5-1. Minimum recommended landing point diameters. 5-26
Table 6-1. Radios that work with AN/PRC-150 in various security modes. 6-3
Table 6-2. Radio interoperability capabilities and characteristics. 6-6
Table 6-3. Procedure for use of duress codes. 6-11
Table 6-4. Report formats. 6-13
Table 6-5. Typical format for an Angus (Initial Entry) Report. 6-13
Table 6-6. Typical format for a Boris (Intelligence) Report. 6-14
Table 6-7. Typical format for a Cyril (Situation) Report. 6-15
Table 6-8. Typical format for an Under (Cache) Report. 6-15
Table 6-9. Typical format for a Crack (Battle Damage Assessment) Report. 6-16
Table 6-10. Priority for destruction of communications devices. 6-18
Table 6-11. Contents of a MIJI report. 6-18
Table 6-12. High frequency ranges in ionosphere. 6-21
Table 7-1. Identification of gaps in existing databases. 7-4
Table B-1. Comparison of brief types. B-9
Table J-1. Light sources and distances. J-3
Table J-2. Sounds and distances. J-4
Table J-3. Odor sources and distances. J-5

Preface

This manual is primarily written for US Army long-range surveillance units (LRSU) and other Infantry reconnaissance and surveillance (R&S) units. It is also provided for use by corps, division, brigade combat team (BCT); battlefield surveillance brigade (BFSB); and reconnaissance and surveillance squadron commanders and staffs; instructors of US Army corps, division, and BCT intelligence, surveillance and reconnaissance (ISR) operations. In addition, many of the subjects covered should be a ready and useful reference for other branches of the US Army and US military, and for multinational forces working in a joint environment.

This manual defines the organization, roles, operational requirements, mission tasks, battlefield functions, and command and control (C2) relationships of LRSCs organic to the R&S squadron of the BFSB. It also provides the doctrine for LRSU to use in combat training and combat. It establishes a common base of tactical knowledge from which leaders can develop specific solutions to LRSU tactical problems. It increases the effectiveness of LRSU operations by also providing doctrinal principles and selected battlefield-proven *tactics, techniques, and procedures* (TTPs). The Digital Training Management System (DTMS) contains the LRSC combined arms training strategies (CATS) and collective tasks for training the LRSU. Before leaders can use this manual to develop and execute training for, and to plan, coordinate, and execute LRS missions, they must first know FM 3-21.8, *Infantry Rifle Platoon and Squad,* and LRSC CATS.

This publication applies to the Active Army, the Army National Guard (ARNG)/Army National Guard of the United States (ARNGUS), and the United States Army Reserve (USAR) unless otherwise stated.

The *Summary of Changes* lists major changes from the previous edition. Changes include lessons learned.

The proponent for this publication is the US Army Training and Doctrine Command (TRADOC). The preparing agency is the US Army Infantry School. Send comments and recommendations by any means, US mail, e-mail, fax, or telephone, using the format of DA Form 2028, *Recommended Changes to Publications and Blank Forms.*

E-Mail	john.edmunds@conus.army.mil
Office/Fax	(706) 544-6448/-6421 (DSN 834)
US Mail	Commander, Ranger Training Brigade
	ATTN: ATSH-RB/Edmunds
	10850 Schneider Rd, Bldg 5024
	Ft Benning, GA 31905

Unless this publication states otherwise, masculine nouns and pronouns may refer to either men or women.

Some or all of the uniforms shown in this manual were drawn without camouflage to improve the clarity of the illustration.

Summary of Changes

This manual provides a major update of FM 7-93, *Long-Range Surveillance Unit Operations*. Most significantly, this edition--

- Introduces--
—Full-spectrum operations.
—Intelligence, Surveillance and Reconnaissance (ISR) Task Force concept.
—BFSB staff elements that provide products and services in support of LRS operations.
—R&S squadron units and staff elements that provide products and services in support of LRS operations.
—Transformation High-Frequency Radio System (THFRS) as the interim replacement LRS team and base station radio system.
—Automatic link establishment (ALE) in beyond line-of-sight LRSU communications.
- Details--
—The new Long-Range Surveillance Company (LRSC) organization organic to the Reconnaissance and Surveillance (R&S) Squadron of the Battlefield Surveillance Brigade (BFSB).
—The deliberate planning methodology in context with the BFSB and R&S squadron organizations.
—The requirements for the next generation of LRSU base radio stations.
- Updates the chapters on—
—Intelligence preparation of the battlefield (IPB, Chapter 7).
—Evasion and recovery (E&R, Chapter 8).
- Updates the chapters on—
—Intelligence preparation of the battlefield (IPB, Chapter 7).
—Evasion and recovery (E&R, Chapter 8).
- Adds sections or paragraphs on—
—Combat assessment (Chapter 4, Section IV).
—Target acquisition (Chapter 4, Section V).
—Urban operations (Chapter 6, Section VII).
—Stability operations (Chapter 4, Section VIII).
—Collection and transmission of imagery, to include digital photos (Chapter 4, Section VII).
—Special missions (Chapter 4, Section IX).
—Vehicle infiltration and exfiltration (Chapter 5, Section III).
—Multiple LRS team and LRS detachment tactical operations (Chapter 4, Section III).
- Adds appendixes on—
—Planning area facilities and sites.
—Army aviation coordination.
- Expands—
—The appendix on personnel recruitment, assessment, and selection procedures.
—The explanations and examples of briefings that LRS teams may have to perform to gain approval for mission execution.
—The purpose, contents and format for a LRSU mission planning folder (MPF), including the target folder.
—The appendix on LRS team contingency planning.
—The section on message formats.
- Deletes all references to "isolation" or "isolation facility (ISOFAC)" and replaces these terms with "planning" and "planning facility." Isolation and ISOFAC are special operations forces terms and are not used by LRSU.

Chapter 1
Full-Spectrum Operations

Section I of this chapter discusses full-spectrum operations as a basis for understanding the role of Infantry reconnaissance and surveillance (R&S) units in the modular force. Section II defines and discusses those units. It also overviews Infantry surveillance and reconnaissance units, details LRSU organizations and missions, and introduces R&S terms and concepts. Section III introduces the discussion of the long-range surveillance company (LRSC) that will continue throughout the remainder of the book. It also provides an example of how long-range surveillance units (LRSU) have been successfully used in stability operations.

The term "long-range surveillance unit" (LRSU) includes all LRSC and subordinate LRSD units, unless otherwise stated. The tactics, techniques, and procedures (TTP) in this manual are also useful to all US Army R&S units, to include Ranger reconnaissance units, Pathfinder companies, BCT scout platoons, reconnaissance squadrons and troops, and special operations units.

Many of the subjects covered in this manual are common to all Infantry surveillance and reconnaissance units and should be a ready and useful reference.

Section I. TYPES AND COMBINATIONS OF ARMY OPERATIONS

Full-spectrum operations involve the simultaneous conduct of any combination of the four types of Army operations, *offensive*, *defensive*, *stability*, and *civil support*, across the spectrum of conflict. Two critical components of each are the commander's understanding of the enemy and the operational environment (OE). Therefore, this section discusses how the commander can combine the types of Army operations based on the area of operations (AO). The mission dictates which type predominates. Finally, because Infantry R&S units are integral parts of the intelligence warfighting function (WFF), this section discusses how they directly support the commander's understanding of the enemy. (See FM 3-0 for more information about the four types of Army operations.)

FOUR TYPES OF ARMY OPERATIONS

1-1. This paragraph defines each type of Army operation and its relationship to the others in full-spectrum operations. The mission dictates which type predominates:

OFFENSIVE OPERATIONS

1-2. These operations carry the fight to the enemy by closing with and destroying enemy forces, seizing territory and vital resources, and imposing the commander's will on the enemy. To do this, the commander focuses on seizing, retaining, and exploiting the initiative.

DEFENSIVE OPERATIONS

1-3. These operations counter enemy offensive operations. They defeat attacks, destroying as many attackers as necessary. They control land, resources, and populations. They also retain terrain, guard populations, and protect key resources.

STABILITY OPERATIONS

1-4. These operations sustain or establish civil security and control over areas, populations, and resources. They use military capabilities to reconstruct or restore essential services and governance, and they support civilian agencies. Stability operations include both coercive and cooperative actions. They can occur before, during, or after offensive and defensive operations, or they may occur separately, usually at the low end of the spectrum of conflict. Stability operations lead to an environment in which, in cooperation with a legitimate government, the other instruments of national power can predominate.

Chapter 1

CIVIL SUPPORT OPERATIONS

1-5. These operations are conducted within the United States and its territories. They address the consequences of man-made or natural accidents and incidents beyond the capabilities of civilian authorities, or in support of homeland security. Homeland security provides the nation with strategic flexibility by protecting its citizens and infrastructure from conventional and unconventional threats. It has two components. The first is *homeland defense*. If the United States comes under direct attack, or is threatened by hostile armed forces, Army forces under joint command conduct offensive and defensive operations to defend the homeland. The other is *civil support*, which is the fourth type of Army operation. Civil support operations take the following forms:

- Support to civil authorities.
- Support to civil law enforcement.
- Protection of military and civilian critical assets.
- Response and recovery.

COMBINATIONS BASED ON AREA OF OPERATIONS

1-6. The four types of Army operations are combined in full-spectrum operations, depending on the area of operations (AO):

JOINT CAMPAIGNS ABROAD

- Offensive.
- Defensive.
- Stability.

HOMELAND SECURITY

- Offensive.
- Defensive.
- Civil support.

INTELLIGENCE

1-7. Key to fighting the Army's modular organizations in full-spectrum operations is moving to positions of advantage and acting before the enemy can respond. Essential to this operational concept is timely, relevant, and accurate intelligence. Infantry R&S units are organized, trained and equipped to provide the critical and timely information needed by the commander and his staff in order to develop intelligence.

Section II. INFANTRY RECONNAISSANCE AND SURVEILLANCE UNITS

Infantry R&S units primarily collect combat information. However, these units also have organic equipment, such as small-unmanned aircraft systems (UASs), long-range advanced scout surveillance system (LRAS3), low-light amplification systems, and digital cameras--that greatly assist in collecting information. The information these and other systems collect can be enhanced, compressed, and sent immediately to the commander or intelligence section of the supported unit. This section defines the characteristics, missions, and organizations of Infantry R&S units.

CHARACTERISTICS

1-8. Infantry R&S units share many characteristics, capabilities, limitations, organizational structures, and missions. Each unit is organized and equipped to address the specific information needs of the commander. These units are typically lightly armed, operate in squad size or smaller, and are foot mobile, but can conduct mobile reconnaissance. Their leaders and Soldiers must have special qualifications such as Airborne, Ranger, military diver, military freefall, or Pathfinder.

MISSIONS AND ORGANIZATIONS

1-9. Infantry R&S unit operations directly contribute to the collection of intelligence at the tactical, operational, and strategic levels of war. The information these units gather is critical to successful operations by combined arms, joint and multinational commanders, and units.

UNIT TYPES

1-10. Each of these six types of Infantry surveillance and reconnaissance units has a doctrinal manual that covers its missions, organizations, and equipment:

Scout Platoons organic to Infantry battalions of an Infantry brigade combat team (IBCT).

Infantry Reconnaissance Companies organic to reconnaissance squadrons of IBCTs.

Ranger Scout Platoons organic to Ranger battalions of the 75th Ranger Regiment.

Ranger Reconnaissance Company organic to the 75th Ranger Regiment.

Pathfinder Companies organic to combat aviation brigades medium (CAB(M)).

LRS Companies organic to R&S squadrons of battlefield surveillance brigades (BFSB). Each LRSC has three LRSDs.

SCOUT PLATOONS AND INFANTRY RECONNAISSANCE COMPANIES

1-11. Scout platoons and Infantry reconnaissance companies collect tactical combat information for Infantry battalions and IBCTs. For detailed mission, capabilities, and organizations of these units, see FM 3-21.20 and FM 7-92. These units can conduct the following missions:

- Conduct zone, area, and route reconnaissance.
- Screen.
- Conduct surveillance.
- Conduct linkup and liaison.
- Guide maneuver forces.
- Detect, survey, and monitor chemicals and radiation.

RANGER SCOUT PLATOONS AND THE RANGER RECONNAISSANCE COMPANY

1-12. These are special operations force (SOF) units. They primarily collect information for their parent units. For missions, capabilities, and organizations of Ranger scout platoons and the Ranger reconnaissance company, see FM 7-85.

PATHFINDER COMPANY

1-13. The Pathfinder company's primary mission is to provide navigational aid and advisory services to military aircraft in areas designated by the supported unit commander. Inherent in this mission is the ability to conduct R&S of these areas, and to report tactical combat information to the supported unit commander. After the R&S mission, the primary mission of the Pathfinder can be conducted. Pathfinder companies can also conduct R&S as a stand-alone mission. This mission supports the intelligence WFF needs of both operational and tactical level commanders. For detailed missions, capabilities, and organizations of Pathfinder companies, see FM 3-21.38.

Section III. LONG-RANGE SURVEILLANCE COMPANY

The LRSC serve the intelligence WFF needs of both operational and tactical level commanders.

Chapter 1

PRIMARY MISSIONS

1-14. LRSCs perform the following four primary missions:

- Surveillance.
- Zone and area reconnaissance.
- Target acquisition.
- Target interdiction.

1-15. Combat assessment, and its subcomponent battle damage assessment, is not a stand-alone LRSU mission. It is an inherent capability in *all* LRSU missions (Chapter 4, Section IV).

SECONDARY MISSIONS

1-16. In addition to these primary missions, LRSU can perform the following secondary missions, given time, training resources, additional personnel and equipment:

- Route reconnaissance.
- Emplacement and recovery of sensors.
- Pathfinder operations.
- Personnel recovery (PR) and combat search and rescue (CSAR).
- Chemical detection and radiological surveillance and monitoring operations.

Note: Providing security for other units is outside the range of a doctrinal or organizationally supported LRSU mission. LRS teams are lightly armed and lack organic automatic weapons necessary for defense and escort missions. In addition, LRS teams are limited-in-number and should only be used for the stated purpose of combat information-gathering.

COMPARISON TO SPECIAL OPERATIONS FORCES

1-17. LRSU are not SOF, although they share many of the same tactics, techniques, procedures, terms, equipment, and organizational structure. Similarly, scouts and cavalry units are not LRSU but also share many TTPs and equipment. The clearest distinction between these units is who they work for and where they operate on the battlefield:

Strategic Level--SOF, including Army Special Forces, when assigned a special reconnaissance mission, generally operate at the strategic level.

Operational Level--LRSU generally operate at the operational level.

Tactical Level--Scouts and cavalry units generally operate at the tactical level.

ORGANIZATION

1-18. The LRSC has one organizational structure. The total active force structure will consist of six LRS companies with 90 LRS teams. The reserve component also has six LRSC with 90 LRS teams (Figure 1-1).

Full-Spectrum Operations

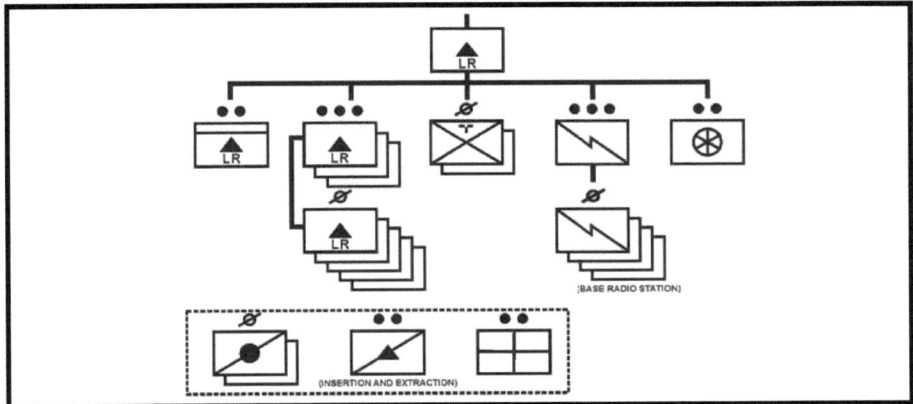

Figure 1-1. Organization of long-range surveillance company.

SUSTAINMENT

1-19. Sustainment for the LRSC is provided by the BFSB brigade support company (BSC). The LRSC depends on additional outside support for parachute rigging, and on aircraft for insertion and extraction operations.

SUBORDINATE ORGANIZATIONS AND KEY PERSONNEL

1-20. The LRSC is not a designed modular unit. However, it shares many of the characteristics of a modular unit. When supplemented with additional medical, fires, communications, rigging, logistics, and operations planning support, the LRSC or a LRSD can be deployed and employed as a separate unit.

HEADQUARTERS SECTION

1-21. The headquarters section serves both a planning and logistics role for the LRSC. Due to the austere nature of the headquarters section, the LRSC relies heavily on other organizations, such as the R&S squadron, the BFSB S-2 and S-3 (for LRS team mission planning and execution), and general logistics support. The section is composed of--

Commander

1-22. The LRSC headquarters section is led by the company commander. The LRSC commander is an Airborne, Ranger-qualified Infantry officer. He should be a graduate of the Long-Range Surveillance Leaders Course (LRSLC) or the Reconnaissance and Surveillance Leaders Course (RSLC). He should also be qualified or familiar with military free fall (MFF), underwater military diver, Pathfinder, and sniper (target interdiction) operations. The company commander is required in the course of his duties to regularly interact with senior commanders and staff officers. Minimum qualifications include previous very successful company command and staff experience. By table of organization and equipment, the company commander is a Captain. However, experience and required duties indicate that the more appropriate rank for the LRSC commander is that of Major.

Executive Officer

1-23. The executive officer (XO) is an Airborne, Ranger-qualified Infantry officer, who serves as second in command and provides logistics planning and control. He should be a graduate of the LRSLC or RSLC. The XO is required in the course of his duties to regularly interact with senior commanders and staff officers. Minimum qualifications include previous very successful company XO and staff experience. By table of organization and equipment, the XO is a Lieutenant. However, experience and required duties indicate that the more appropriate rank for the LRSC XO is that of Captain.

Chapter 1

First Sergeant

1-24. The 1SG is the most experienced NCO in LRS operations. As such, he helps plan, operationally control, and administer the unit. He should be a graduate of the LRSLC or RSLC, and both battle-planning staff and MFF qualified.

Master Sergeant (Operations Sergeant)

1-25. The master sergeant, who serves as operations sergeant, should be a graduate of the LRSLC or RSLC. He is both battle-planning staff and MFF qualified. He is the primary planner and is experienced in LRS operations.

Supply Sergeant and Armorer

1-26. The supply sergeant and assistant are both Pathfinder qualified.

COMMUNICATIONS PLATOON

1-27. The communication platoon has a headquarters section and four base radio stations (BRSs).

Headquarters Section

1-28. The platoon leader is an Airborne and Ranger-qualified Signal Officer. The platoon leader is required in the course of his duties to regularly interact with senior commanders and staff officers. At the least, his qualifications must include very successful signal platoon leader and staff experience. The platoon sergeant is a communications Master Sergeant who is Airborne, Ranger, and battle-planning staff qualified. The platoon leader and platoon sergeant should be graduates of the LRSLC or RSLC. In addition to the platoon leader and platoon sergeant, the headquarters section has two NCO radio operators/maintainers--both are Airborne and Ranger qualified.

1-29. The communications platoon headquarters section is responsible for planning and executing all aspects of the communications plan to link deployed LRS teams to the company operations base (COB) or mission support site (MSS) and the supported unit headquarters. This includes frequency spectrum planning, communications security (COMSEC) and unit-level equipment repair and evacuation. At least two of the members of the communications platoon headquarters section should be graduates of the Frequency Managers Course.

Base Radio Stations

1-30. Each of the four BRSs has a six-Soldier team, which is mainly responsible for maintaining communication between deployed LRS teams, the COB or MSS and the supported unit headquarters. Depending on the communications means used between LRS teams and the COB or MSS, the BRSs can support three MSS or relay sites and the COB simultaneously.

SNIPERS

1-31. The sniper teams are organized into two teams of two men each: senior sniper and sniper. The senior sniper is Airborne Ranger qualified and the sniper is Airborne qualified. Both the senior sniper and the sniper should be graduates of the LRSLC or RSLC and of either the US Army Special Forces or Marine sniper course. All the snipers should also be MFF qualified. The company sniper employment officer (SEO) is normally the company commander, the XO or 1SG. (FM 23.10 gives more information about sniper team training and employment.)

TRANSPORTATION SECTION

1-32. The transportation section is led by a staff sergeant, 88M. It is manned by five more transportation Soldiers. The section has three trucks and two trailers. The section's main duty is to transport personnel and material at the direction of the company XO and 1SG.

LRSD AND LRS TEAMS

LRSD

1-33. The three LRSD are each led by an Airborne, Ranger-qualified Infantry officer. Each detachment leader should be a graduate of the LRSLC or RSLC. Also, he should be MFF and Pathfinder qualified. He is responsible for the training, readiness, and employment of the LRS teams in the detachment. The detachment leader is required in the course of his duties to regularly interact with senior commanders and staff officers. He is also required to be prepared to operate independent of the LRSC utilizing a MSS. Minimum qualifications include: previous very successful platoon leader and staff experience. By table of organization and equipment, LRSD leaders are Lieutenants. However, experience and required duties indicate that the more appropriate rank for a detachment leader is that of Captain.

LRS TEAMS

1-34. Each LRS team is led by an Airborne, Ranger qualified Infantry Sergeant First Class or Staff Sergeant. There is one Sergeant First Class per detachment and he doubles as the detachment sergeant when his LRS team is not planning, on a mission or recovering. Team leaders should be graduates of the LRSLC or RSLC. Normally, the team leader is also battle-planning staff, MFF, underwater military diver, and Pathfinder qualified, or a combination of these qualifications.

1-35. LRS team members are, at a minimum, Airborne qualified. All team members should be graduates of the LRSLC or RSLC. They should also have a combination of MFF, underwater military diver and Pathfinder qualifications.

LEGACY LRSC AND LRSD

1-36. The legacy Army of Excellence LRSC was a unit normally organic to a Military Intelligence Brigade previously found at corps level. This type of LRSC is not addressed in this manual as all are currently scheduled for deactivation.

1-37. The legacy Army of Excellence LRSD was a unit normally organic to a Military Intelligence Battalion previously found at division level. While these LRSD are all scheduled for deactivation, some commanders have maintained them or they may be temporarily authorized while awaiting activation of the BFSB R&S Squadron LRSC. This type of LRSD is not specifically addressed in this manual.

CHARACTERISTICS

1-38. All LRS operations are characterized by the following:

- They follow strict operations security (OPSEC) procedures before, during, and after mission employment. This limits operational exposure of other LRS teams conducting missions.
- Units receive detailed intelligence preparation of the battlefield (IPB) from the BFSB S-2. They plan and rehearse in detail with internal and supporting assets. This helps ensure successful operations.
- Division- or corps-level assets, such as joint aviation, joint fire support, and communications, must support planning, infiltration and insertion, exfiltration and extraction, and contingencies.

1-39. LRS teams are also characterized by the following:

- Teams provide persistent surveillance on targets that can be covered only intermittently by most other systems.
- Teams avoid contact with enemy forces and local population.
- Team firepower is normally limited to small arms, grenades, and Claymore mines.
- Teams have organic ground mobility assets.
- Team members depend on an expert knowledge of digital high frequency (HF) and ultra high frequency (UHF) tactical satellite (TACSAT) communications equipment, enemy TTPs, order

Chapter 1

of battle (OB), and practiced skill in identifying equipment to successfully execute missions.

- Team members depend on the use of stealth, cover, concealment, and their Infantry and Ranger skills to remain undetected.
- Team equipment and supplies include only what the team can man-pack or cache, if dismounted.

CAPABILITIES

1-40. Modular force LRSCs are fully self-mobile once deployed to the AO. The LRSC is not a modular unit but can deploy detachments as independent units if provided additional support. LRSC and LRSD can do the following:

- Establish long-range digital communications using HF or UHF radio systems.
- Transmit both voice and data.
- Use planned or emergency resupply drops.

1-41. LRS teams have additional capabilities:

- Collect and transmit near-real-time digital imagery.
- Operate in inclement weather and over difficult terrain.
- Evade.
- Use special equipment cache sites emplaced by the LRSU or other friendly forces.
- Use captured supplies and equipment.
- Demonstrate the "art of camouflage" by—
    —Blending in with units and formations who have already established a presence.
    —Using clandestine cameras and sensors.
    —Shadowing blinds and false walls.
- Stay behind or infiltrate over land (dismounted or mounted), over water (small boat and underwater), or by air (rotary wing, static line parachute, or MFF).
- Exfiltrate over land, water, or air; or link up with advancing friendly forces.
- Operate with the use of organic ground mobility assets and, if resupply is available, conduct extended range and duration reconnaissance missions.
- Conduct dismounted operations for up to seven days with little or no external direction and support.
- Conduct operations in all types of terrain and environments.

LIMITATIONS

1-42. LRSU have the following limitations:

- Neither the LRSC nor the LRSD are modular. Both require additional support in the areas of medical, fires, communications, rigging, logistics, intelligence, and operations planning in order to provide a full range of operational capability.
- They lack organic medical capability. The LRSC depends on attached medics from the R&S squadron or on individual first aid. Ideally, LRS team members are certified emergency medical technicians (EMTs). At a minimum, each team member must be a qualified combat lifesaver (CLS).
- They are lightly armed and have limited self-defense capabilities.

- They require the support from the R&S squadron, BFSB S-2 and S-3, or division or corps staffs and units. This support includes—
    —Intelligence products.
    —Integrated area communications.
    —Access to a common-user telephone system.
    —Frequency management.
    —Satellite communications channel access.
    —Packing, rigging, and loading supplies.
    —Equipment for aerial resupply and parachute insertions.
    —Air transport to the area of operations.
    —Maintenance, supply, mess, administration, finance, personnel, and chaplain services.
- The tactical situation, equipment limitations, or enemy electronic surveillance or jamming may prevent LRS teams from maintaining continuous communications with the controlling headquarters.

Chapter 1

Example LRSU Operational Employment--ISR Task Force

Operational areas can be assigned, or remain unassigned, and thus the responsibility of the higher headquarters such as the division or corps. When the higher headquarters leaves areas unassigned, that portion of the AO requires a level of situational understanding (SU) sufficient to detect and classify threats. During past operations, an intelligence, surveillance and reconnaissance (ISR) task force, successfully provided SU to the division or corps headquarters in an unassigned AO (**Figure 1-2**).

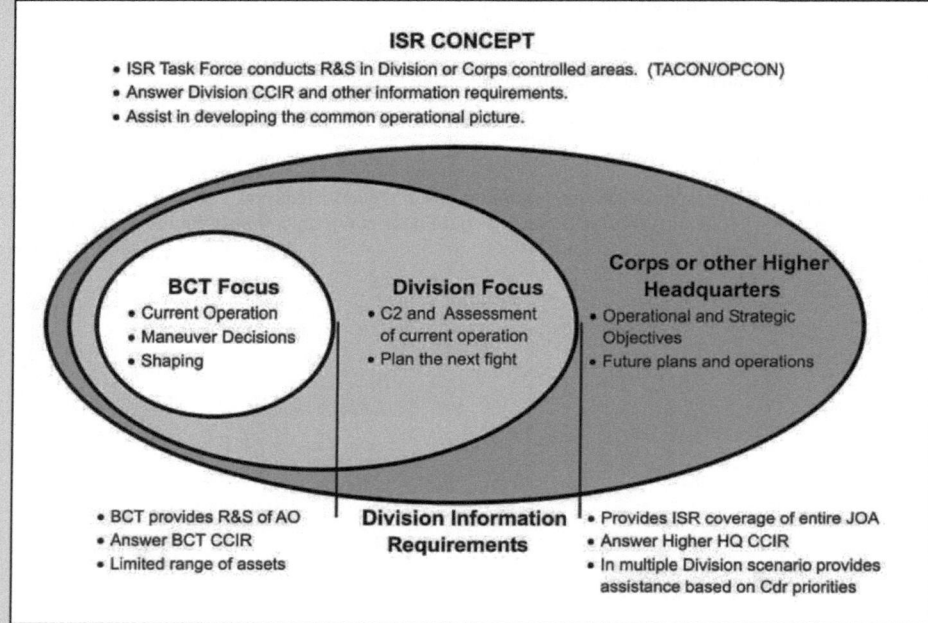

Figure 1-2. Concept of intelligence, surveillance, and reconnaissance.

A key enabler of future warfighting concepts is enhanced situational awareness (SA) that leads to actionable SU. This is achieved by fusing information obtained through a layered network of Soldiers, sensors, and collection platforms with information on friendly forces, enemy forces, and the environment to obtain a common operational picture that is shared across the force. The ISR task force integrated multiple intelligence-gathering systems and provided information about the enemy or potential enemy to the decision maker.

The most successful use of the ISR Task Force concept and organization has occurred during stability operations, due to the relative freedom of movement available to both friendly and enemy forces. The enemy force often hides among the population, making them difficult to track and target. The ISR Task Force integrated many sources to create actionable intelligence.

LRSU have typically been integrated into the ISR Task Force to provide combat information through persistent surveillance on named areas of interest (NAIs) or targets. In addition to providing combat information, the LRSU clandestinely emplaced, monitored, and recovered sensors for other units attached to the ISR Task Force. The LRSU also acquired targets for artillery and close air support (CAS), and interdicted targets as part of (sniper) operations against selected targets.

The BFSB is similar to an ISR Task Force in that it is organized to integrate intelligence, R&S and other combat units to provide immediate response to actionable intelligence. The BFBS headquarters element is sufficiently robust that it can accept attachment of combat forces.

Chapter 2
Command and Control

This chapter describes how LRSU use Infantry and Ranger skills, communications operators, and intelligence personnel to collect and report battlefield information. LRSU operations depend on effective command and control (Section I) and communications, computers, and intelligence (Section II).

Section I. OVERVIEW

Command and control is the exercise of authority and direction by a properly designated commander over assigned and attached forces in the accomplishment of a mission. Commanders perform C2 functions through a C2 system (FM 6-0). The LRSU C2 system is structured for rapid deployment, and for collecting and reporting information. This section discusses the LRSU internal C2 system and the functions and organizations that provide mission taskings and support. It also discusses—

- Standard and nonstandard command and control.
- BFSB sections.
- R&S squadron.
- Command posts.
- Tactical operations centers (TOCs).
- Task organizations outside named areas of interest.
- Liaisons.

STANDARD LRSU COMMAND AND CONTROL

2-1. The LRSU commander receives missions in the tradition manner of mission-type orders. Based on these orders, he assesses the situation, makes decisions, and directs actions like any other commander. However, he employs individual LRS teams IAW separate mission taskings that specify areas, objectives, or people to observe. LRSU can have multiple subunits spread out over hundreds of kilometers simultaneously planning, conducting, and recovering from operations. To C2 these operations, the commander depends on a robust communications system, detailed and well practiced SOPs, and subunit leaders who are highly skilled in all aspects of R&S.

SUPPORT

2-2. Support to LRS teams includes intelligence, operations, communications, and sustainment:

Intelligence Support includes target information and IPB products.

Operations Support includes planning, coordinating and developing mission planning folders (MPF), infiltration, exfiltration and fire support.

Communications Support includes frequency planning and positioning communications sites.

Sustainment Support includes planning facility, mess, medical, maintenance, and general supplies.

COMPANY

2-3. The LRSC depends on the BFSB staff, the R&S squadron, and the BSC for most intelligence, operations, and sustainment support. The LRSC deploys to an operational area as part of the BFSB. The BFSB S-2 ISR fusion element produces target information and IPB products. If the LRSC deploys as an attached element to a corps, division, JTF, or multinational headquarters without the BFSB, then the

G-2/J-2 and G-3/J-3 must perform these critical functions, along with all other operations and sustainment support.

DETACHMENT

2-4. A LRSD deploys to an operational area as part of a larger unit, normally the LRSC, because it lacks the ability to plan and execute missions, or to sustain itself. A LRSD can be attached or OPCON to almost any type of larger unit, and can be doctrinally employed, if the unit provides intelligence, planning, insertion, extraction, and sustainment support.

BFSB STAFF

2-5. The BFSB has three staff sections, which the LRSU primarily interacts with and depends on for planning and support: S-2, S-3 and the S-4.

S-2 SECTION

2-6. The BFSB S-2 section coordinates IPB for staff planning, decision making, and targeting. They support the commander's and staff's planning portion of the military decision-making process (MDMP). To answer the commanders' priority intelligence requirement (PIR) and other intelligence requirements, the S-2 section processes and analyzes all information it collects, and then produces and disseminates timely, relevant, and accurate intelligence. The S-2 also monitors the current threat situation and any environmental factors that might influence friendly or enemy courses of action. The following elements in the BFSB S-2 each provide specific products and services in support of LRS operations:

ISR Fusion Element

2-7. The BFSB ISR fusion element conducts all-source analysis and integrates (fuses) current intelligence. Working under the supervision of the S-2, the ISR fusion element develops and maintains an intelligence database consisting of all unprocessed information and other intelligence products. The ISR fusion element keeps the commander and staff situationally aware, and provides intelligence to exploit enemy weaknesses and vulnerabilities.

Functions

2-8. The ISR fusion element—

- Produces the IPB and mission support products.
- Develops intelligence estimates.
- Analyzes the battlefield area.
- Develops situational and targeting data to support maneuver and fires.
- Prepares combat assessments.

Products

2-9. The ISR fusion element produces—

- Target information as part of the LRS team MPF.
- Country studies.
- Imagery.
- Threat data (objective, capabilities, composition, and disposition).
- Threat templates.
- Situational templates.
- Event templates.
- Modified combined obstacle overlays (MCOO) in coordination with the geospatial information and services (GI&S) section.
- Environmental data, for example, edible plants and indigenous animals.
- Solar and lunar data in coordination with the combat weather team.

Geospatial Information and Services Section

2-10. The GI&S section supports the commander and staff with required geospatial information and maintains databases up-to-date to support the unit with terrain and weather effects products and analysis.

Functions

2-11. The GI&S section—

- Provides a terrain visualization mission folder to determine the terrain's effect on both friendly and enemy operations.
- Produces maps and terrain products.
- Helps the ISR fusion element prepare the IPB.

Products

2-12. The GI&S section produces—

- Hydrology overlays.
- Cover and concealment overlays.
- Soil composition overlays.
- Vegetation composition overlays.
- Obstacle overlays.
- Combined obstacle overlays.
- Special maps.

USAF Combat Weather Team

2-13. The USAF combat weather team provides weather forecasting support for organic and attached units of the BFSB. They can also provide weather effects analysis as it pertains to military operations. This greatly enhances the ability of the BFSB to conduct collection of information to support the commander.

Functions

2-14. The USAF combat weather team—

- Advises the task force commander on Air Force weather capabilities, support limitations, and how weather information can enhance combat operations.
- Evaluates and disseminates weather data such as—
  - Forecasts.
  - Warnings.
  - Advisories.
- Monitors the overall weather support mission for the commander.
- Serves as the commander's agent, identifying and resolving weather-support responsibilities.
- Determines weather-support data requirements.
- Advises the Air Force on the operational weather-support requirements of the supported Army command.
- Participates in targeting meetings.
- Prepares climatological studies, and analyzes them in support of planned exercises, operations, and commitments.

Chapter 2

- Coordinates weather support to subordinate units.
- Helps Army aircraft accident-investigation boards.

Products

2-15. USAF combat weather team produces—

- Weather forecasts.
- Climatological data.
- Solar, lunar, and darkness data.
- Weather effects on friendly and enemy operations and equipment.

BFSB S-3 SECTION

2-16. The following elements in the BFSB S-3 section each provide specific products and services in support of LRS operations:

S-3, Operations

2-17. The S-3 manages training, operations, plans, force development, and modernization. The BFSB S-3 section prepares and distributes orders to subordinate units. Normally it does not give specific orders to the LRSC. The BFSB commander gives mission type orders to the R&S squadron. The R&S squadron then conducts the MDMP and gives mission type orders to the LRSC. However, the BFSB S-3 section does assist in the LRSC in mission planning and execution for individual LRS team employment.

2-18. The S-3—

- Supervises the tactical command post (TACCP) when deployed.
- Develops courses of action.
- Prepares, coordinates, and disseminates standing operating procedures, orders, and directives (with other staff sections' input).
- Synchronizes all operations, to include reviewing and coordinating subordinate plans and actions.
- Coordinates with other brigades to place BFSB assets within their AOs as necessary.
- Requests and coordinates external assets for the R&S squadron's insertion and extraction of ground elements.
- Requests and coordinates joint personnel recovery assets as required.
- Coordinates and directs terrain and airspace management.
- Coordinates extended range multipurpose UAS support from the combat aviation brigade.
- Recommends priorities for allocating command resources and support.
- Directly assists the commander in controlling preparation for and execution of operations.
- Coordinates staff planning, execution, and supervision of information operations.
- Plans and coordinates stability operations.
- Conducts BFSB collection operations.
- Performs terrain management within the BFSB AO, when assigned.
- Supports sensitive site exploitation.
- Integrates risk management into operational planning.

Command and Control

2-19. The S-3 staff—

- Plans, executes, and supervises OPSEC.
- Conducts civil-military operations.
- Conducts BFSB collection operations.
- Performs terrain management within the BFSB AO, when assigned.
- Supervises the actions of the following special staff officers that have LRSU coordination and support responsibilities:
  — Air defense and airspace management/brigade aviation element (ADAM/BAE).
  — Air Force tactical air control party attached to the BFSB.
  — Fires cell.

Air Defense and Airspace Management/Brigade Aviation Element

2-20. The ADAM/BAE—

- Evaluates any air threat to BFSB operations.
- Analyzes aerial threat factors bearing on the OE.
- Analyzes the effects of weather and terrain on air operations.
- Determines threat aerial courses of action.
- Develops an air and missile defense (AMD) concept of support for each course of action (COA), wargames the AMD concept of support, and compares courses of action.
- Maintains the AMD running estimate.
- Prepares the AMD annex.
- Synchronizes AMD operations.
- Plans and coordinates airspace command and control, and deconflicts airspace.
- Integrates aviation into BFSB sustaining operations, such as medical evacuation and aerial resupply.
- Represents aviation during the MDMP and other planning processes such as IPB, ISR synchronization, and targeting.
- Maintains the aerial portion of the common operational picture.
- Assists the fires element to analyze airspace control orders and air tasking orders.
- Coordinates directly with the aviation brigade or the supporting aviation task force for detailed air mission planning.

Air Force Tactical Air Control Party

2-21. The TACP is the aligned US Air Force element that coordinates tactical air assets and operations such as CAS, air interdiction, joint suppression of enemy air defenses (J-SEAD), reconnaissance, and airlift. Specifically, the TACP—

- Advises the commander and staff on the employment of tactical air (TACAIR).
- Operates and maintains Air Force TACAIR direction radio net and air request net.
- Transmits requests for immediate close air and reconnaissance support.
- Transmits advance notification of impending immediate airlift requirements.

Chapter 2

- Coordinates tactical air support missions with the fire-support element and the appropriate Army airspace command and control (AC2) element.
- Recommends intelligence requirements to the S-2 through the S-3.
- Plans the simultaneous employment of air and surface fires.
- Supervises forward air controllers (FACs).
- Integrates air-support sorties with the BFSB's scheme of maneuver.
- Participates in targeting meetings.
- Serves as a member of the targeting cell.
- In the absence of a FAC, assists the fire support officer (FSO) in directing air strikes.
- Provides Air Force input into the AC2.

Fires Cell

2-22. The FC—
- Develops and recommends surface targets to attack.
- Employs fires to influence the will of and to destroy, neutralize, or suppress enemy forces.
- Plans, coordinates, disseminates, and maintains fire support coordination measures.
- Plans and requests CAS.
- Coordinates clearance of fires.
- Plans the employment of nonlethal fires.
- Synchronizes fires within the BFSB AO.

BFSB S-4 Section

2-23. The S-4 is responsible for planning brigade sustainment. The S-4 is the link between the BFSB's organic BSC and the theater support command for coordinating logistics support. Specific to LRSU operations, the S-4 is responsible for coordinating riggers in support of Airborne (static line, MFF and resupply) operations.

RECONNAISSANCE AND SURVEILLANCE SQUADRON

2-24. The R&S squadron is a multifunctional unit designed to collect and report information that answers the BFBS commander's critical information requirements (CCIR) and other information requirements (IR) using manned ground assets and small UAS. It can also acquire targets and, on a limited basis, interdict targets. The squadron has a headquarters and headquarters troop (HHT), one LRSC, and two mounted ground reconnaissance troops. It has long-range communications that allow it to operate dispersed and communicate throughout the division AO. The R&S squadron is designed to simultaneously employ the LRSC and the reconnaissance troops dispersed within the higher unit's AO. Except for its organic medical platoon, the R&S squadron depends on the BSC for all of its logistical support.

Headquarters and Headquarters Troop

2-25. The R&S squadron HHT has organic elements specifically designed to support the employment of the LRSC: medics, forward observers (FOs), and the insertion and extraction section organic to the S-3 section. These Soldiers are distinguished by special qualifications identifiers and additional skill identifiers (SQIs and ASIs) in the R&S squadron headquarters and headquarters troop's TOE.

2-26. The combat medic section of the medical platoon has six Soldiers identified for direct support (DS) to the LRSC: the staff sergeant section NCO and five specialist medics. Each of these Soldiers is Airborne and special operations combat medic qualified, so they can support LRS team missions.

2-27. Within the fire support platoon, two fire support staff sergeants, and two sergeant FOs are Airborne and Ranger qualified. This gives them the basic skills needed to participate in LRS team missions when target acquisition is required.

2-28. The insertion and extraction section helps plan and execute infiltration and exfiltration of LRS teams, and provides general planning support for the LRSC and ground reconnaissance troops. The section has five Soldiers: a Captain assistant operations officer, a sergeant first class assistant operations NCO, a staff sergeant assistant operations NCO, and two specialist vehicle drivers. The Captain must be Airborne, Ranger, and Pathfinder-qualified. The assistant operations NCOs are Airborne, Ranger, Pathfinder, and underwater military diver-qualified. The section--

- Helps plan, coordinate, and develop MPFs.
- Keeps a list, overlays of, and additional information about possible landing zones (LZs), drop zones (DZs), and pickup zones (PZs).
- Coordinates requests for any air operations with aviation support units.
- Posts the schedule of infiltration and exfiltration operations.

2-29. In addition to the R&S squadron, Soldiers and sections that directly support the LRSC, the S-2 and the S-4 have additional important support functions. The S-2 section collaborates with the BFSB ISR fusion element to produce target information for LRS team missions. The S-4 section coordinates with the BFSB S-4 section for riggers to support LRS team insertion operations through other sustainment units as needed.

NONSTANDARD LRSU COMMAND AND CONTROL

2-30. LRSU are in high demand and limited in number. Therefore, even a LRSC or a task-organized LRSD is often told to support units other than a BFSB or R&S squadron. However, support and functions provided by the BFSB, the R&S squadron, and the BSC are essential and necessary to the successful execution of a LRSU operation.

2-31. Without this support, the LRSU would be limited as to the spectrum of missions it could successfully perform. These limitations should be explained clearly to the gaining unit commander and considered during the conduct of the MDMP.

2-32. LRSDs can also be task-organized between LRSCs. For instance, if a LRSC requires additional assets to perform its mission, a LRSD can be attached from another LRSC. For rapid interoperability, communications SOPs (message formats, COMSEC, and reporting procedures) must be similar or, if possible, identical.

COMMAND POSTS

2-33. LRSUs normally employ three types of command posts: a company operations base (COB), an alternate operations base (AOB), and an MSS.

COMPANY OPERATIONS BASE

2-34. Figure 2-1 shows an example COB, which normally collocates with or near the R&S squadron tactical operations center (TOC) or the BFSB main command post (MCP). A key consideration for the location of the COB is the requirement for a secure location or facility for LRS team planning. Teams should be separated from each other to preserve operational security and to allow them to conduct detailed planning with minimal interruption. This location is called the planning facility. Most members of the LRSU locate in the COB.

Chapter 2

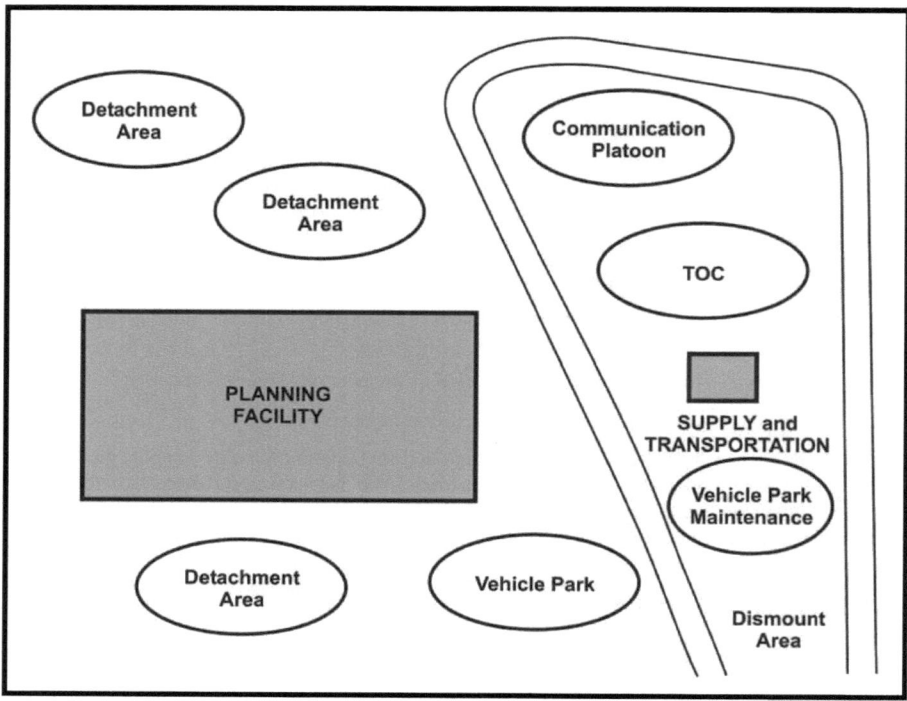

Figure 2-1. Company operations base.

2-35. In addition to the planning facility, the COB includes areas for a TOC, the company headquarters, the communications platoon or BRS, motor park, helicopter LZ, and LRSD and team living areas. METT-TC determines whether the site should be a fixed or a field site.

2-36. The company commander selects the general location of the LRSU COB. Based on the commander's requirements, the 1SG or XO determines the exact location. He oversees the setup of the base and the implementation of security for it.

2-37. Each LRSD or LRS team is assigned an area, within which it sets up a CP. When a team deploys, the platoon sergeant, or a designated team leader, provides for security in the team area and for any equipment that the team did not take on the mission.

Operations Section

2-38. The operations section sets up, secures, and restricts access to the company TOC. The section prepares and marks an LZ nearby. The operations NCO normally controls the LZ, however a LRSD or LRS team can also be given the mission to set up and control it.

Communications Platoon

2-39. The communications platoon is assigned an area. It sets up required HF, UHF TACSAT, and very high frequency (VHF) antenna systems; operates the company or detachment wire net; and provides communication equipment maintenance and logistical support. The communications platoon coordinates for the unit to be included into the local area network (LAN). This allows them access to the higher unit switching system.

Command and Control

Company Headquarters

2-40. The company headquarters is assigned an area from which it provides administrative and logistical support as required. The 1SG initiates and enforces the operations base security plan.

Base Radio Station

2-41. Intelligence reports (INTREP) received by the BRS go directly to the R&S squadron S-2 and the BFSB ISR fusion element. The LRSU operation section neither delays nor changes any INTREP. If a BRS receives a message at the AOB and not the COB, the operator sends it, exactly as received, by the fastest, most secure means to the COB. Unit SOPs largely determines how INTREPs are processed once received at the COB.

ALTERNATE OPERATIONS BASE

2-42. The primary mission of the AOB is to serve as a communications relay between the COB and deployed LRS teams. AOB planning considerations derive from the communication requirements of the COB and on the deployed LRS teams. Normally, the communication platoon leader controls the AOB but is not necessarily located there. Selected personnel from the operations section also normally locate at the AOB. The task organization for the LRSU AOB varies based upon METT-TC considerations and unit SOPs.

2-43. The COB and AOBs maintain communication with employed teams using HF or UHF TACSAT radios. For OPSEC purposes, each team has a separate frequency and cryptographic keys. The tactical switching system between the two bases allows them to communicate. Depending on METT-TC requirements, HF or line of sight (LOS) radio systems provide backup communications. (Figure 2-2 shows an example C2 employment schematic for a LRSC.)

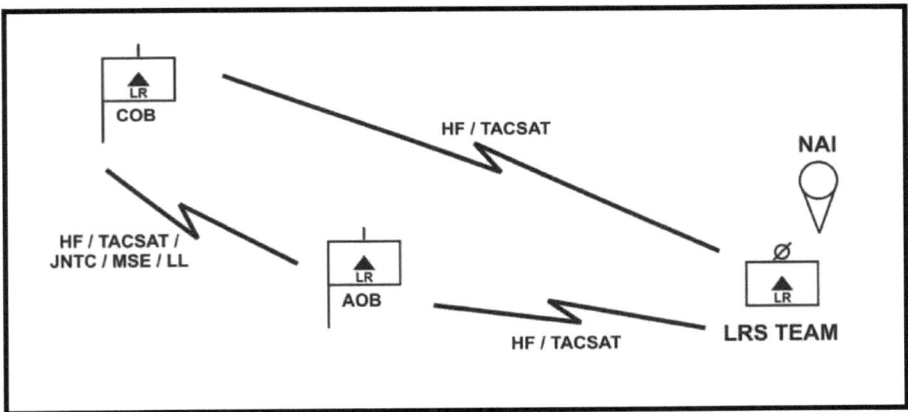

Figure 2-2. Example C2 employment schematic for a LRSC.

2-44. The AOB locates where it can best support communications with deployed LRS teams. Because of its limited ability to provide for its own security, the AOB normally collocates with another unit. For linkup operations, the LRSC can collocate a BRS-equipped liaison officer (LNO) team with the moving unit.

MISSION SUPPORT SITE

2-45. If required, the LRSU can also use an MSS. An MSS is a temporary grouping of operations, communications, and support personnel, formed to conduct a specific operation or mission. Normally, the MSS commander is a LRS detachment leader. The only difference between C2 for a LRSD in an MSS and a LRSC is that the LRSD may not use an AOB. (Figure 2-3 shows an example C2 employment schematic for a LRSD in an MSS.)

Chapter 2

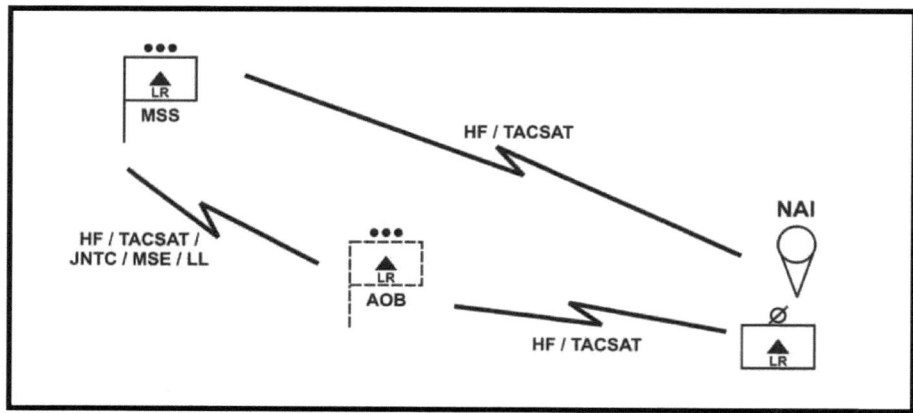

Figure 2-3. Example C2 employment schematic for a LRSD in an MSS.

TACTICAL OPERATIONS CENTER

2-46. The LRSU TOC sets up in the COB. The TOC gives commanders a C2 capability with higher headquarters and subordinate elements.

LRSC

2-47. In the LRSC TOC, personnel perform specific functions as follows:

Battle Captain or NCO

2-48. The battle captain or NCO supervises operations in the TOC. In the LRSC, the battle captain is normally a LRS detachment leader. This duty normally rotates between the three LRS detachment leaders. The battle captain or NCO plans and coordinates the company's tactical operations based on the commander's guidance and--

- Analyzes assigned missions.
- Plans employment of teams.
- Prepares or approves operations orders (OPORD) before they go to the commander.
- Updates the commander on current and projected tactical situations.
- Supervises the preparation of all operational and intelligence documents.
- Supervises coordination with higher and supporting headquarters.
- Reports the operational status of committed and uncommitted LRS teams.
- Ensures that the current situation is posted on all maps and charts.
- Collects and forwards combat information from the LRS teams to higher.
- Maintains the operations workbook.
- Approves the TOC personnel work schedule.
- Ensures preparation of the briefing area and maps.
- Posts the mission planning chart.
- Collects combat information from LRS team operations.
- Updates mission folders after receiving info from BFSB and R&S squadron staff elements.
- Posts and maintains the enemy situation overlay.

- Informs teams of critical information that could affect their missions.
- Conducts final security inspections of LRS teams before deployment.

Operations Sergeant

2-49. The operations sergeant supervises the TOC enlisted personnel. He assumes responsibility for the TOC in the absence of the battle captain. He also--

- Helps prepare and edit all tactical operations plans.
- Supervises the operation of the planning area.
- Posts the current situation on the friendly situation overlay.
- Posts current data from deployed teams on mission-status charts.
- Establishes the TOC personnel work schedule.
- Coordinates with the 1SG for TOC messengers and guards.
- Ensures that only authorized personnel enter the TOC.
- Posts the manning chart.
- Prepares the situation report (SITREP) for the period.
- Helps maintain the operation workbook.
- Acts as a shift leader to maintain a 24-hour capability in the TOC.

Chemical, Biological, Radiological, and Nuclear NCO

2-50. The CBRN NCO helps establish, administer, and apply defensive CBRN operations. He also--

- Supervises preparation of CBRN reports.
- Supervises maintenance of CBRN supply.
- Supervises unit and individual CBRN training records.
- Serves as principal NCO of the CBRN defense team.
- Collects CBRN information and data.
- Interprets and analyzes CBRN information and data.
- Disseminates CBRN information and data.
- Acts as a shift leader to maintain a 24-hour capability in the TOC.

Liaison NCO

2-51. The LNO coordinates test fire areas; rehearsal areas; communication equipment test areas; and supply, medical, transportation requirements.

Ongoing Actions

2-52. During the mission, the TOC personnel monitor the progress of surveillance teams and are prepared to coordinate for exfiltration, fire support, resupply, linkups or any other action that might arise. They do this by sending updated SITREPs and any changes to the LRS team's mission. They also receive, log, and disseminate combat and administrative information from the teams. In addition, they--

- Monitor the guard frequency 24 hours a day.
- Coordinate resupply for committed teams.
- Coordinate emergency extractions.
- Coordinate medical evacuations.

Chapter 2

- Coordinate other required support.
- Plan and coordinate additional missions as directed by the commander.
- Monitor scheduled communication times.

Debriefing

2-53. Immediately after exfiltration, personnel from the R&S squadron S-2 or the BFSB S-2, debrief each LRS team. In addition, the LRSC communications platoon leader or a BRS section sergeant debriefs the team on communications specific details.

Messages

2-54. The battle captain or NCO provides a receipt for all incoming messages. He forwards INTREPs from LRS teams to the BFSB S-2 ISR fusion element and R&S squadron S-2, as required. He also--

- Records in the staff journal the receipt of each message.
- Posts the information from each message to appropriate maps and charts.
- Files each message in the journal file by journal-entry number.
- Records in the journal all messages that go out of the TOC.

Journal

2-55. The Daily Staff Journal or Duty Officer's Log (DA form 1594) is a chronological record of events pertaining to the unit during a given period. The battle captain or NCO maintains the journal. He cross-references all items to the journal entries by journal item number. He posts all messages to the journal, and notes the following information about each:

- Sender.
- Message number.
- Message title or a description of the event.
- Time of receipt.
- Journal item number.
- Message center number (if applicable).
- Action taken.
- His initials.

Security

2-56. Normally the 1SG or the operations NCO restricts and controls personnel access to the TOC and to the planning facility. Unit SOPs establish procedures for controlling and identifying visitors. The TOC and the planning facility should have only one secure entrance. The entrance to the planning facility should have limited access and be secured at all times when LRS teams are conducting mission planning. Unit SOP details appropriate security measures in the safeguarding and handling of all classified material, to include preparing and rehearsing and emergency destruction.

Displacement

2-57. When directed to displace, the on-duty shift continues to operate; the off-duty shift breaks down all equipment and loads it on the vehicles. The COB notifies the AOB of the departure time, route and proposed relocation site. When the COB is ready to displace, it transfers control to the AOB. The AOB monitors committed teams and controls operations. When the COB is again operational, it reassumes control and the AOB sends an update of the situation.

LRSD

2-58. When employed at an MSS, the LRS detachment leader, detachment sergeant, and attached communications personnel perform all functions. The LRSC commander may supplement the LRSD personnel as needed to assist in planning and operation of the MSS.

TASK ORGANIZATION OUTSIDE NAMED AREAS OF INTEREST

2-59. All LRSU should use the same communication procedures. This facilitates task organizing LRSU as battlefield conditions change. The rapid pace of some operations may require the LRSU to coordinate C2 of deployed LRS teams and to exchange information to meet the commander's intelligence needs. Special situations could include the employment of the LRSU outside the BFSB's area of operations or area of interest.

2-60. During retrograde operations, the C2 of any LRS teams beyond the BFSB's area of interest is temporarily transferred to the unit now responsible for that area. This action requires an LNO with a BRS to locate with the new controlling headquarters. Once teams are extracted, they return to the parent LRSU.

TASK ORGANIZATION FOR BRIGADE, TASK FORCE, OR BOTH

2-61. For certain contingency operations, a LRSD, or portions of a LRSC, might be OPCON of a BCT. This most often occurs in a stability operation and before the BFSB S-2 ISR fusion element deploys to the AO. It also occurs when BCTs expand control of a sector and deployed LRS teams are operating in that sector. When the latter occurs, an LNO with a BRS locates with the BCT MCP. For mission planning, a member of the BFSB S-2 ISR fusion element or R&S squadron S-2 LNO can accompany the LRS control element.

LIAISON DUTIES, EMPLOYMENT, AND COORDINATION

2-62. The term "liaison" refers to the contact or intercommunication maintained between elements of military forces in order to ensure mutual understanding and unity of purpose and action. A trained, competent, trusted, and informed officer or noncommissioned officer is the key to effective liaison. Normally, the unit provides its own LNO.

DUTIES

2-63. The commander must trust the LNO completely. The LNO must hold the appropriate rank and have the appropriate experience for each particular mission. During LRS operations, he provides the critical link between the LRSU and external agencies. He coordinates the planning process, assists as needed, and generally supports the LRS operation.

EMPLOYMENT

2-64. The actual method of employing LNOs varies, depending on each unit's requirements and SOPs.

Corps, Division, or Both

2-65. The LRSU can be tasked to provide the corps or division headquarters with personnel to help coordinate C2 between the LRSU and its controlling headquarters. The LNO can assist in the coordination and tracking of LRS team operations, and provide other assistance as required.

Brigade or Task Force

2-66. When required, the LRSU might need to provide an LNO to work with a BCT or task force staff. His duties would be similar to the duties provided to the corps and or division staff.

Headquarters or Operations Section

2-67. The headquarters or operations section of the LRSC normally employs internal LNOs to work with LRS teams preparing for an operation. These LNOs help the LRS team plan and coordinate upcoming missions. They can also help prepare and set up the planning facility and enforce security procedures. To ensure the effectiveness of the LNOs, they should be senior in rank, and experienced in LRS

Chapter 2

operations. Each should work with his assigned team throughout the planning process and infiltration phase. Because he knows that teams' mission, he can serve in place of any injured or incapacitated team member. After infiltration, if he is no longer needed on the team, he can help the operations section track the battle and exfiltrate, extract, debrief, and recover the LRS team. However, until the supported team's mission is complete, he should not be allowed to deploy on another LRS team mission until the supported LRS team's mission is complete.

COORDINATION

2-68.   Specific LNO duties include--

- Internally, the LNO coordinates test fire areas; rehearsal areas; communication equipment test areas; and supply, medical, transportation requirements. He also helps destroy excess LRS team planning materials, and inventories and secures team equipment not needed for the mission.
- Externally, the LNO coordinates with any outside unit or element. He normally has the authority to coordinate with all relevant staff agencies and sister service units.

Section II. COMMUNICATIONS, COMPUTERS, AND INTELLIGENCE

The LRS team is a valuable combat information collection asset assigned to the BFSB. Rapid and efficient information flow between the BFSB and the LRS TOC requires an efficient and timely dissemination link. This section first discusses communications in general as it applies to LRSU, and then discusses computers and intelligence.

COMMUNICATIONS

2-69.   The rapid flow of information to the BFSB S-2 ISR fusion element is vital for decision-making, and directly affects the successful execution of operations. Once the BRS receives a message from a deployed LRS team, it is forwarded to the COB for decryption and analysis. The BRS at the COB uses organic wire lines or runners (when collocated) to relay team reports. If the BRS operator cannot collocate with the COB, he must send messages by secure UHF TACSAT, VHF or HF radio.

2-70.   The BRS at the COB must stay in constant communication with the AOB. If the BRS at the COB is destroyed or otherwise incapacitated, then the AOB assumes the mission of the COB, or collocate with them. The AOB battle tracks the entire mission. BRS-to-BRS communication systems include—

- HF radio.
- UHF TACSAT.
- DSVT with facsimile.
- VHF (needed if AOB must be collocated with COB).
- Local and wide area networks (LAN and WAN).

2-71.   If a BRS collocates with a support unit that employs mobile subscriber equipment (MSE), the AOB uses the digital secure-voice terminal (DSVT) with facsimile.

2-72.   The COB and the AOB should also use a digitized, lightweight video-reconnaissance system that can receive imagery from the deployed teams via HF, UHF TACSAT, and VHF.

2-73.   When the COB receives a message from either BRS, it decrypts and analyzes it, and then, as quickly as possible, sends it to the BFSB S-2 ISR fusion element. They also send it to the R&S squadron S-2. As a minimum, the link from the COB to the BFSB S-2 ISR fusion element should include—

- Secure VHF (computer, voice, or both).
- DSVT with facsimile.
- LAN.
- Runner.

COMPUTERS AND INTELLIGENCE

2-74. The LRSC TOC connects to the BFSB S-2 ISR fusion element and the R&S squadron S-2 LAN via both secure and nonsecure links.

ISR COMPUTER WORKSTATION

2-75. The LRSC TOC requires access to an ISR computer workstation. This workstation is normally linked to other intelligence network processors, which—

- Lets the LRSC report all ISR tasks and requests for information (RFIs) in real time. Sending information directly to the BFSB S-2 ISR fusion element updates the intelligence database and promotes immediate analyzing and processing of the information.
- Allows the LRSC access to current information in the friendly database, including imagery, enemy OB, situation maps, and enemy templates relevant to the LRSU mission. It also gives the unit access to current information from other intelligence databases.

LOCAL AREA NETWORK LINE

2-76. The secondary link to the BFSB S-2 ISR fusion element is a secure LAN line. This normally requires a DSVT or an MSE tactical facsimile machine.

HUMAN COURIER

2-77. This is normally the final contingency means of disseminating information between the LRSU TOC and the BFSB S-2 ISR fusion element. However, if the LRSU TOC is located away from the BFSB S-2 ISR fusion element, this takes more time than any other method.

Chapter 3
Mission Development

Successful LRS missions require detailed planning and coordination. The BFSB commanders' collection plans describe the desired PIR and intelligence requirements. The PIR drives the conduct of ISR operations. After the LRSU collects information, the BFSB S-2 ISR fusion element evaluates it during the continuous IPB process. This chapter answers the questions: Why and how do the BFSB, R&S squadron and the LRSC develop the LRS team mission? This chapter answers this question by discussing ISR operations and mission orders (Section I), the mission planning folder (Section II), and OPSEC (Section III).

Section I. ISR OPERATIONS AND MISSION ORDERS

ISR is defined as an activity that synchronizes and integrates the planning and operation of sensors, assets, and processing, exploitation, and dissemination systems in direct support of current and future operations. This is an integrated intelligence and operations function. (JP 1-02). Mission orders is a technique for completing combat orders that allows subordinates maximum freedom of planning and action to accomplish missions (FM 1-02). It leaves the "how" of mission accomplishment to the subordinate. How these two concepts work together is critical to understanding the mission development process for LRSU.

ISR OPERATIONS

3-1. The BFSB conducts ISR operations to enable the division commander to precisely focus joint elements of combat power and simultaneously execute current operations while preparing for future operations. Key to successful accomplishment of the BFSB mission is the targeted collection, analysis, and dissemination of intelligence that satisfies the needs of the supported unit commander. All elements of the BFSB are organized and trained for this purpose.

3-2. The BFSB fills two roles in conducting ISR operations. First, it executes the ISR plan for the supported unit (division, corps, JTF, or multinational). Second, it augments BCTs and other supporting brigades with ISR capabilities.

3-3. Collaborative planning is conducted between organizations so that at each level ISR plans are synchronized and integrated early in the MDMP. This also allows subordinate units to task collection assets as early as possible to collect the needed information.

ORGANIZATIONS ABOVE THE BFSB

3-4. As the BFSB's higher unit conducts the MDMP, the commander and staff identify information that is needed to make informed decisions. This information is PIR, which along with friendly force information requirements comprise the CCIR. As soon as the G-2/J-2 and the G-3/J-3 know the CCIR, they start formulating an ISR plan. This plan includes seeking answers from higher-level organizations, and tasking subordinate units. The MDMP process produces an OPORD or OPLAN. In addition to the main body of the order, the process produces Annex B Intelligence and Annex L Intelligence, Surveillance and Reconnaissance. Included in Annex L is the ISR tasking matrix, which assigns to subordinate units specific tasks associated with collecting information—the ISR tasks for which the unit is responsible.

BFSB PLANNING

3-5. As the BFSB receives its mission orders, it conducts its own MDMP. The BFSB follows the same process as the higher unit, which results in the production of mission orders for its subordinate units. However, unlike the ISR efforts of BCTs, the BFSB does not focus on internal requirements. It focuses on gathering and disseminating information that meets the requirements and priorities of the higher-level commander and staff.

Chapter 3

3-6. As part of the MDMP, the staff performs ISR synchronization and ISR integration to develop its ISR plan (Figure 3-1). ISR synchronization determines the intelligence requirements that must be met, compares them to the units or assets available and capable of collecting in the time and location required, and balances them with the higher unit's priorities. The S-2 leads this coordinated staff effort. The product is the ISR synchronization plan. The S-3 leads the ISR integration effort. The S-3 uses the ISR synchronization plan to develop ISR tasks. These tasks are assigned to either the staff or subordinate units. The ISR tasks—in conjunction with task organization, graphics, fire support plans, coordinating instructions, and sustainment information—are merged together into the ISR plan. The BFSB ISR plan is the basis for the BFSB operation order.

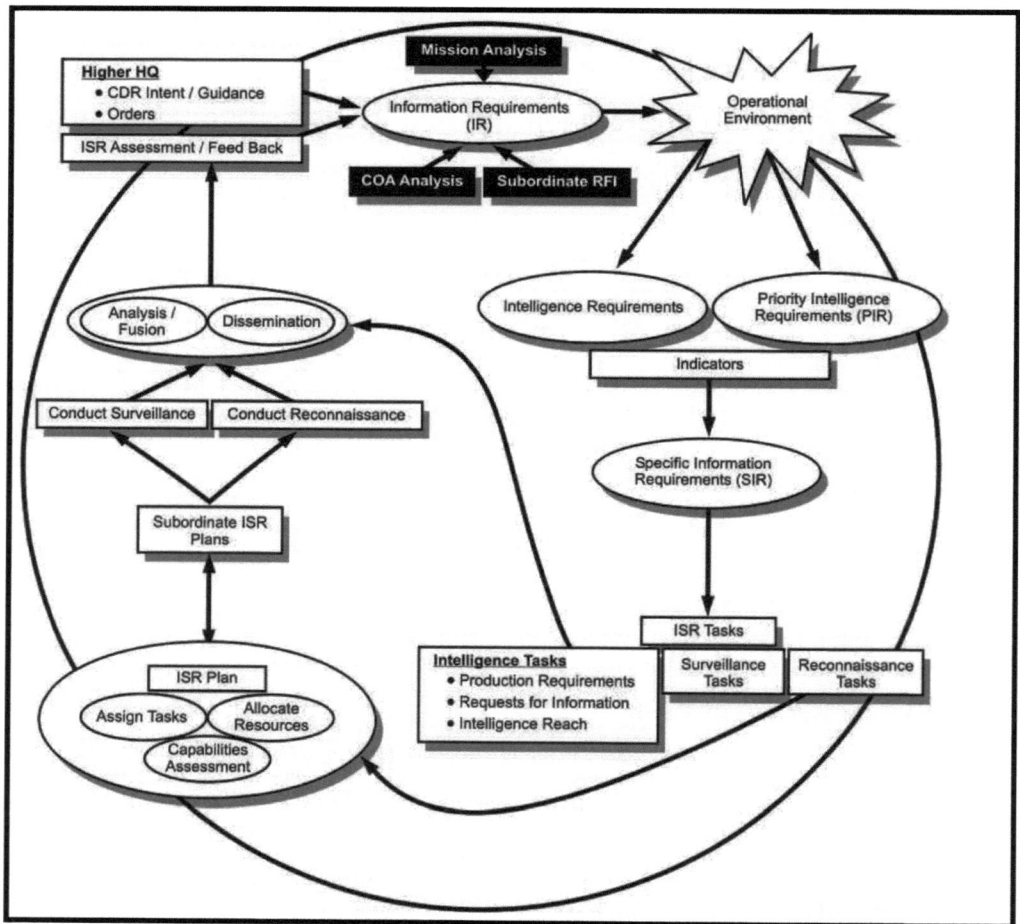

Figure 3-1. Intelligence, surveillance, and reconnaissance.

TARGET IDENTIFICATION

3-7. Potential LRS targets are first identified by the BFSB staff during the wargaming step of the MDMP. The BFSB S-2 ISR fusion element OIC participates in the MDMP, and during war-gaming, assesses which potential targets best suit the capabilities of LRS teams. He normally seeks the assistance of the R&S squadron S-3 insertion and extraction section or the LRSC commander, in evaluating the suitability and feasibility of potential targets. He then gains the concurrence of the BFSB S-2 and S-3.

3-8. The BFSB S-3 notifies the R&S squadron S-3, who consults with the LRSC commander again regarding the suitability and feasibility of the targets. Notification of the LRS commander is recommended before the LRS commander receives the OPORD from the R&S Squadron because this serves as a warning order. This allows the LRSC to start troop leading procedures (TLP).

3-9. Simultaneously, the BFSB S-2 ISR fusion element begins developing information on each potential target for inclusion into target folders. The target information is an essential part of the LRS team MPF.

R&S SQUADRON PLANNING

3-10. The same process is followed when the R&S squadron receives the BFSB OPORD, with one exception. The R&S squadron S-2 in most cases does not have the capability to provide the target information needed for LRS teams to plan missions. As a result, most LRS target information is developed and disseminated by the BFSB S-2 ISR fusion element.

MISSION ORDERS

3-11. The result of the ISR process is the need, for example, for a target to be surveiled. The higher-level unit avoids dictating how the LRS team should conduct the mission.

LRSC PLANNING

3-12. The LRSC receives the target information from the BFSB S-2 ISR fusion element as it becomes available. This allows the LRSC commander to issue warning orders (WARNO) and begin to identify assets required to conduct the anticipated mission.

3-13. The HQ section, communications platoon leader, LRS detachment leaders, the R&S squadron S-3 insertion and extraction section, and a LRS team LNO form the planning cell and assist the LRSC commander in the conduct of TLP. The commander ensures the OPORD provides detailed information on the friendly and enemy situation, communications and sustainment. The commander normally writes the company mission statement and the intent, task and purpose for each LRS team with the assistance of the LRS detachment leaders. This information is essential to answer the question "why" the individual LRS team is deploying to a particular target to collect specific information for the supported unit commander.

3-14. Normally, LRS detachment leaders do not write detachment or individual LRS team OPORDs. They act as members of the planning cell writing the company OPORD and assist in the production of individual team MPFs. If a LRSD is deployed to an MSS, the detachment leader normally does write an order.

3-15. The result of the company planning efforts is an OPORD and individual LRS team MPFs. The MPF is mission order based. It does not dictate how the mission will be performed. It does provide the who, what, when, where and why of the mission. (Figure 3-2 summarizes the LRS mission-development process.)

Chapter 3

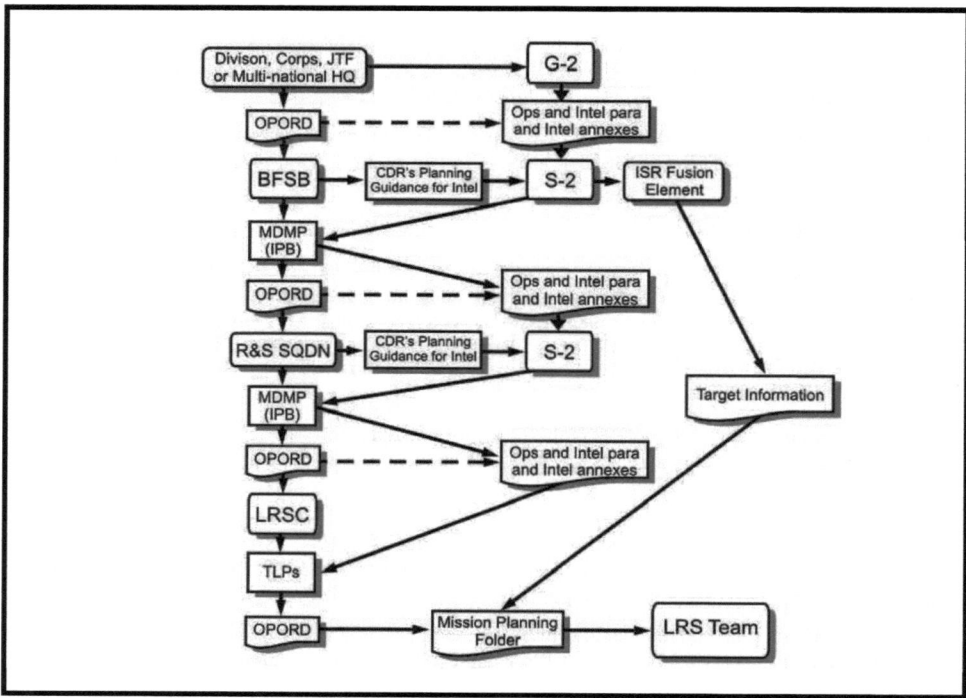

Figure 3-2. LRS mission development process.

ADDITIONAL PLANNING CONSIDERATIONS

3-16. Although LRS teams should not be held in reserve, the BFSB and R&S squadron should consider the need to conduct continuous operations. LRS teams need to recover from missions and staffs must anticipate needs for future operations. General guidelines for operational tempo are—1/3 of the teams conducting missions, 1/3 preparing for employment and 1/3 recovering, training and preparing to receive a new mission.

3-17. As information is collected and reported from deployed teams, each organization above the LRSC analyzes it to determine if it met any intelligence requirements. This allows each organization to make an initial determination if it can or must task new LRS teams or reallocate other assets to collect required information or begin collection on other priorities.

3-18. LRS teams generally require 24 to 48 hours for deliberate planning. Planning time can be reduced by well-written, understood, and rehearsed unit SOPs.

HASTY PLANNING

3-19. Sometimes, less than the recommended LRS team planning time of 24 to 48 hours is available. Although LRS teams are capable of employment on short notice, the potential benefit must be weighed against the risk associated with hasty planning and execution. At a minimum, the following information is needed for hasty planning and execution:

- Mission statement, to include area or object to surveil, latest time information of value (LTIOV), and anticipated length of mission.
- PIR, intelligence requirements, associated specific information requirements (SIR), ISR tasks, and RFIs.

- Enemy situation in the target area.
- Commander's intent for intelligence (can be stated by the BFSB or R&S squadron S-2 or S-3).
- Method of insertion, with abort criteria. Coordination time and place are included, if applicable.
- Fire support plan, to include assets available.
- Exfiltration plan.
- Communications plan (provided by the LRSC headquarters).
- Linkup, if applicable.
- Contingency planning guidance.

Section II. MISSION PLANNING FOLDER

The BFSB and R&S squadron S-2 and S-3 provide information and intelligence products to support the LRS mission. The LRSC commander, the communications platoon leader, the R&S squadron insertion and extraction section, and the LRS team LNO request information from the staffs. After developing, collating and packaging the information and products, the LRSC headquarters section assembles the MPF. The MPF contains primarily three products: the LRSC operations order, the team's mission specific information, and the target folder.

DEVELOPMENT

3-20. The MPF is a stand-alone document prepared by the LRSC headquarters to help the LRS team leader plan and execute his mission. It provides the LRS team leader with detailed information about his AO and mission, including maps, photographs, sketches, climatology, area geography, and recent enemy activity. It also provides coordination information, such as details about infiltration and exfiltration, insertion and extraction means and corridors.

3-21. Development of the MPF begins as soon as the LRSC is notified of a potential mission. The LRS headquarters section, communications platoon leader, LRS detachment leaders, the insertion and extraction section, and the LRS team LNO coordinate the mission and assemble the MPF. Requesting information from the BFSB and R&S squadron staff elements is a critical step for MPF development. As with any order, the LRSC company commander is ultimately responsible for the completeness and accuracy of the document.

3-22. Simultaneously, in writing the company order, the BFSB S-2 ISR fusion element develops specific information about the target and compiles it into the target folder for inclusion into the MPF.

CONTENTS

3-23. The MPF normally contains, at a minimum, base LRSC OPORD, team mission-specific information, and target information. The MPF can also contain additional information or resources to assist the LRS team in mission planning. This information includes: excepts from R&S squadron and BFSB order (as required), unit SOP formats for briefings, orders, RFIs, reports, supply requests, logs, planning area setup, and so on.

LRSC OPERATIONS ORDER

3-24. The LRSC operations order follows the standard five-paragraph order format with annexes. In general, the order includes the enemy road to war/enemy situation, and the mission, intent, and concept of operations for the BFSB, R&S squadron, and LRSC commanders. If a LRSD is operating independently of the LRSC in an MSS, the detachment leader will substitute his mission, intent, and concept of the operation for the LRSC commander.

Chapter 3

3-25. The LRSC operations order does not contain information that is specific to individual LRS team missions. For example, paragraph 5, command and signal, will provide information on the location of COB, AOB, frequency ranges used (which radios to carry) and when reports are expected. It will not contain the specific frequencies or times for communications windows the individual teams will use.

LRS TEAM-SPECIFIC OPERATIONS INFORMATION

3-26. Operational security considerations require that each team knows only the specific information that is pertinent to their mission. While commanders can make exceptions to this general rule, each team is only provided the information they need to conduct their specific mission.

3-27. Normally, the LRS team specific operations information also follows the standard five-paragraph order format with annexes. The information contained in this document will not repeat information contained within the LRSC operations order but is formatted to supplement the company order. For example, paragraph 5, command and signal, will specify the frequencies, communications windows and code words the individual team will use.

3-28. The planning for a LRS team is largely a bottom-up process. The LRSC provides information and packages the contents to assist the LRS team in analyzing and completing their plan. The information should most importantly be complete and through. Also, the information should be presented in a form that makes it easy for the team to reference and use. A common method of packaging is to provide annexes and appendices to the base order that group information by the five phases of a LRS team operation. Examples appendixes might cover insertion, extraction, evasion, and communications. Below is a list of items normally covered in the LRS team base order, appendices or annexes:

Critical Times

- Time schedule during planning.
- Event times during operations.

Fire Support

- Task and purpose.
- Supporting units.
- Unit locations.
- Frequencies and call signs.
- Type and size of fire support.
- Target numbers.
- No-fire areas (NFA) or restrictive fire areas (RFA) numbers (method of control).
- High-payoff target (HPT) list.
- CAS (A- or C-type ordnance).
- Gun target lines.
- Danger close.
- Planning ranges.

Available Air Insertion and Extraction Platforms

- Air assault, Airborne, MFF insertion.
- Unit supporting and point of contact.
- Type and number of supporting aircraft.
- Aircraft capabilities such as Special Purpose Insertion and Extraction System (SPIES) and Fast Rope Insertion and Extraction System (FRIES).

- Date-time group for aircraft availability.
- Air mission commander.
- Location of pickup zone.
- Tentative flight routes.
- Date-time groups for the initial planning conference and the air mission brief.
- Suppression of enemy air defenses (SEAD) plan (lethal and nonlethal).

Vehicle or Ground Insertion

- Supporting unit and point of contact.
- Type and number of vehicles.
- Date-time group (DTG) for vehicle availability.
- Pickup location.
- Tentative routes.
- Passage point date-time group and location.
- Frequencies and call signs.
- Recognition signals.
- Fire support.

Host Nation or Partisan Forces

- Supporting personnel.
- Type of insertion platform (time available and capabilities).
- Coordination for linkup.
- Linkup.
- Recognition signals.
- Routes.

Waterborne Operational Information Requirements

- Supporting unit.
- Insertion platform.
- Tables showing currents and tides (blue water).
- Tables showing currents and depths (brown water).
- Terrain at beach-landing site.
- Fire support.

Communications Data

- Frequency modulations (signal operation instructions).
- UHF TACSAT frequencies and availability.
- HF propagation charts.
- HF list.
- VHF frequency list.

- Reporting procedures from LRSC or LRSD to higher headquarters.
- IP address.

Evasion Plan

- Personnel recovery procedure and evasion corridors.
- DD Form 1833 TEST (V2).
- Location of selected area for evasion (SAFE) or designated area of recovery (DAR).
- Documents that describe SAFEs.
- Blood chits.
- General survival information.
- Civilian population information.
- Cache and air resupply.
- Medical information.
- Border information.
- Food sources.
- Water sources.
- Plant and wildlife data.
- Epidemic diseases.
- Air-tasking order, special instructions, and airspace-control order.

Maps, Products, and Imagery

3-29. These supplement target folder information.

- Gridded satellite imagery of possible LZs or PZs, recovery points, and linkup sites.
- Line of sight and field of view from the proposed surveillance site, 360 degrees from the objective.
- Maps.
    - 1: 250,000.
    - 1: 100,000.
    - 1: 50,000.
    - 1: 25,000.

- Gridded satellite imagery.
- Multispace imagery products.
- Elevation tint.
- Slope tint.
- Surface drainage.
- Panoramic graphs.
- MCOO.
- Hydrology overlay and charts.
- Overlay of landing or pickup zone.
- Cover and concealment overlay.
- Operational graphics (friendly unit locations).
- Overlay of DARs (if used).
- Flight corridors and air control points.
- Enemy situational template (most dangerous and probable courses of action).
- Enemy event template.
- Photos or pictures of enemy weapon systems and uniforms.

TARGET FOLDER FORMAT

3-30. The target folder is created by and the responsibility of the BFSB S-2 fusion element. The R&S squadron S-3 insertion and extraction section, the S-2 section, the LRSC operations section, and the LRS team LNO assist in the development of the target folder. The target folder format is derived from FM 3-05.102 (Appendix C). The advantage of using this format is the standardization of the information the BFSB S-2 fusion element is required to produce (Figure 3-3).

```
                        (Classification)
Section 1: Target Identification and Description
   A. Target identification data.
   B. Description and significance.
   C. Detailed target description.
   D. Target vulnerability assessment.

Section 2: Natural Environment
   A. Geographic data (including terrain and hazards to movement).
   B. Meteorological data (climatologically overview and tables and
illumination data).
   C. Hydrographic data (coastal, waterway, lakes, luminescence, and
so on).

Section 3: Threat
   A. Ground forces.
   B. Paramilitary and indigenous forces (including intelligence and
security and police services).
   C. Naval forces (including Coast Guard and maritime border guard).
   D. Air forces.
   E. Air defense forces (including radars, passive detectors, and
C2).
   F. Electronic order of battle.
   G. Space-based assets.
   H. Counter Intelligence environment (efforts of indigenous forces
to collect against R&S forces).
   I. Other.

                        (Classification)
```

Figure 3-3. Example format for target folder.

```
                        (Classification)
Section 4: Demographics and Cultural Features
   A. Area population characteristics.
   B. Languages, dialects, and ethnic composition.
   C. Social conditions.
   D. Religious factors.
   E. Political characteristics.
   F. Economic conditions.
   G. Miscellaneous (for example, currency, holidays, dress, and
customs).

Section 5: Lines of Communications and Information Systems
   A. Airfields.
   B. Railways.
   C. Roadways.
   D. Waterways.
   E. Ports.
   F. Petroleum, oils, and lubricants (POL).
   G. Power grid.
   H. Public information media and telecommunications (print, radio,
television, telephone, and so on).

Section 6: Infiltration and Exfiltration. This includes potential DZs,
LZs, recover zones, seaward launch and recovery points, and
beach landing zones.
   A. Potential zones.
   B. Choke points between insertion point(s) and objective.

Section 7: Survival, Evasion, Resistance, Escape, Recovery (SERER)
Data. See JP 3-50.3, Joint Doctrine for Evasion and Recovery.
   A. SAFE data.
   B. SAID data.
   C. Survival data.

Section 8: Unique Intelligence (mission-specific requirements not
covered above)

Section 9: Intelligence Shortfalls
                        (Classification)
```

Figure 3-3. Example format for target folder (continued).

Chapter 3

```
                         (Classification)
Appendix A: Bibliography

Appendix B: Glossary

Appendix C: Imagery

Appendix D: Maps and Charts

Appendix E: Sensitive Compartmented Information (if applicable)
                         (Classification)
```

Figure 3-3. Example format for target folder (continued).

Section III. OPERATIONS SECURITY

LRS units' OPSEC measures are important planning considerations. LRS mission classification seldom falls below SECRET during war or stability conditions. This section discusses security classifications; mission classification; the need and procedures for separating teams during planning; and security during coordination.

PERSONNEL SECURITY

LEADERS

3-31. Leaders require TOP SECRET clearance and inclusion on the BFSB S-2 ISR fusion element access roster as well as TOP SECRET SCI access to national-level assets so that they can plan missions in detail:

- LRSC commander.
- LRSC executive officer.
- LRS detachment leaders.
- LRSC communications platoon leader.
- First Sergeant.
- Operations NCO.
- Any LRSU Soldier acting as an LNO.
- The LRS team leader.

LRS TEAM MEMBERS

3-32. Each LRS team member must hold a current SECRET clearance and access level, or higher. The information that a LRS team needs for planning purposes is sometimes classified above the individual team member's access level. When a LRSU Soldier fails the mandatory investigation for a required security clearance, he receives a transfer to a position or unit with lower clearance requirements. Members of a LRSU who frequently engage in alcohol- or drug-related incidents, demonstrate financial or mental instability, or violate the law are reassigned or separated from military service. The BFSB commander must entrust a LRSU Soldier with mission-sensitive and classified information. Mishandling of information, no matter how innocent the compromise, constitutes a serious incident.

MISSION CLASSIFICATION

3-33. LRS missions receive SECRET-level classification due to the vulnerability of a six-man element, which might be operating deep behind enemy lines. LRS team locations are seldom posted or mentioned on a computer LAN, graphic or written OPORD, intelligence summary (INTSUM), or INTREP, regardless of the classification of the dissemination vehicle. The BFSB S-2 ISR fusion element OIC, S-2s, S-3s, G-2,

G-3, R&S squadron, BFSB, division and corps commanders are normally the only personnel outside the LRS operations cell that "need to know" LRS team locations. Teams working close together might need collateral support and, therefore, *might* need to know. At a minimum, they publish NFAs or RFAs on the FC's system. The system then automatically disseminates information to all units on the system. The BFSB generally needs to coordinate with the special operations coordinator (SOCOORD) LNO at division or corps as special forces and LRS teams might be operating in the same AO. Information provided to the SOCOORD LNO prevents fratricide among LRS and SOF operating in the same AO.

SEPARATION

3-34. LRS teams plan independently. What they need to know (critical information) generally depends on mission proximity. That is, LRS teams may require support from another team in the form of a communications relay or contact team (Joint Pub 3-50.12). The planning facility separates teams and insolates them from distractions and mission operational pace. Each team member keeps all information about his mission from the other teams. He only acknowledges, mentions, or discusses it with other members of his own team. Keeping mission information internal to the team ensures that, if one team is compromised and later captured, the enemy can only obtain information about that team and its mission.

3-35. Teams on the ground-conducting missions avoid communicating with each other. Radio communications are vulnerable to direction-finding (DF) equipment. Threat forces might target areas where LRS teams could be templated to operate. The less that it knows about other teams' missions, the more secure they and the LRSU as a whole remain.

SECURITY DURING COORDINATION

3-36. Coordination, such as air mission briefs, can also compromise a LRS team mission. The LRS LNO must consider OPSEC when coordinating. During an air mission brief, everyone avoids discussing the mission location, NAI, or duration. They only discuss details such as the infiltration route, checkpoints, call signs, and logger area.

Chapter 4
Team Operations

The success of LRS operations depends on thorough planning, acquisition of ISR tasks and RFIs, rapid and timely reporting, and avoidance of detection.

LRS teams collect critical information for the BFSB commander in support of division commanders' PIR. Answers to the PIR directly affect the commander's decision-making and dictate the successful execution of military operations. This chapter discusses the following aspects of LRS team operations:

Phases (Section I)
Reconnaissance operations (Section II)
Surveillance operations (Section III)
Combat assessment (Section IV)
Target acquisition (Section V)
Urban terrain (Section VI)
Imagery collection and transmission (Section VII)
Stability operations (Section VIII)
Special missions (Section IX)

Section I. PHASES

LRS team operations have five distinct phases--planning, infiltrating/inserting, executing (actions on the objective), exfiltrating/extracting, and recovering. Controlling or supporting deployed teams can overlap into more than one phase.

PLANNING PHASE

4-1. Detailed planning at all levels helps ensure mission success and team survival. The planning phase starts when the commander receives the mission folder from the BFSB S-2 ISR fusion element. It extends throughout the final inspection of the LRS team.

SEQUENCE

4-2. In most cases, the commander, XO, first sergeant, and operations personnel participate in the initial S-2 planning. To ensure the LRS team completes each of its planning tasks, the team follows a detailed timeline.

Type of Planning

4-3. The length of time available determines whether the team conducts deliberate or hasty planning.

Deliberate Planning (24 to 48 hours available for planning).

Hasty Planning (less than 24 hours available for planning).

Required Planning Events

4-4. The LRS team should cover the following events in each planning sequence, regardless of time available:

- Alert notification.
- N-Hour planning sequence (the sequence followed before deployment).
- H-Hour sequence (the deployment sequence).

Chapter 4

ACTIVITIES

4-5. To make the best use of time for planning at company, detachment, and team levels, all leaders use TLP. Table 4-1 (page 4-4 and 4-5) lists what each member of the company and detachment must do during the planning phase.

LRSU Commander

- Receives WARNOs and OPORDs from the R&S squadron and BFSB headquarters as required.
- Reviews target information from the BFSB S-2 ISR fusion element.
- Conducts TLP and mission analysis using METT-TC.
- Prepares and issues WARNOs as required.
- Prepares and issues OPORDs as required.
- Supervises the finalization of the MPF.
- Coordinates with BFSB and R&S squadron staff members.
- Approves the location of the COB, AOBs, and MSSs.
- Receives LRS team confirmation briefs.
- Receives LRS team decision briefs.
- Receives LRS team backbriefs.

LRSC Headquarters

- Receives WARNOs and OPORDs from the R&S squadron and the BFSB headquarters.
- Receives target information from the BFSB S-2 ISR fusion element.
- Finalizes MPFs and issues to LRS team leaders as required.
- Helps commander conduct TLP to include developing and issuing WARNOs and OPORDs.
- Establishes COB (including the planning facility), AOBs and MSSs.
- Helps coordinate and finalize team mission support requirements
  — Aviation.
  — Fire support.
  — Imagery.
  — Sustainment.
- Prepares briefings as required.

Communications Platoon

- Receives company WARNOs and OPORDs.
- Helps commander conduct TLP to include developing and issuing WARNOs and OPORDs.
- Conducts TLP for the communications platoon.
- Prepares and issues WARNOs as required.
- Prepares and issues OPORDs as required.
- Supervises the finalization of the communications plan for each MPF.
- Establishes BRSs.
- Confirms all radio frequencies with R&S squadron S-6 or BFSB S-6.

- Develops communications plan for company WARNOs, OPORDs and communication annexes.
- Coordinates and conducts communication rehearsals.

LRS Detachment Leader
- Receives WARNOs and OPORDs from the R&S squadron and BFSB headquarters as required.
- Receives WARNOs and OPORDs from the LRSU headquarters as required.
- Receives and reviews target information from the BFSB S-2 ISR fusion element as required.
- Helps establish the COB (including the planning facility), AOB and MSSs as required.
- Helps commander conduct TLP to include developing and issuing WARNOs and OPORDs.
- Conducts TLP for his detachment.
- Prepares and issues WARNOs as required.
- Prepares and issues OPORDs as required.
- Supervises the finalization of the MPF.
- Coordinates with BFSB and squadron staff members.
- Approves the location of MSSs in support of his LRSD.
- Receives LRS team-confirmation briefs.
- Receives LRS team-decision briefs.
- Receives LRS team backbriefs.

LRS Team Leader and Team Members
- Receives the initial mission analysis from LRSC headquarters.
- Receives WARNO from LRSC headquarters or LRS detachment leader.
  — The team leader and radio telephone operator (RTO) normally attend.
  — The RTO serves as the recorder.
- Conducts TLP.
- Issues team WARNOs.
- Conducts initial inspections.
- Moves to and establishes team area in the planning facility.
- Receives the OPORD with mission folder from LRSC headquarters or LRS detachment leader.
- Issues confirmation brief.
- Receives decision brief.
- Issues COA or decision brief to commander (after mission analysis brief).
- Directs the activities of the team LNO.
- Delivers OPORD to the LRS team (LRS team leader).
- Coordinates as needed with operations personnel.
- Memorizes plan and prepares for backbrief.
- Backbriefs the commander or his representative.
- Conducts final inspection.

Chapter 4

Note: The entire team normally attends the OPORD.

Table 4-1. Actions and responsibilities of LRSU personnel.

Commander	Operations	Communications Platoon	LRS detachment Leader	Team	LNO
Receives OPORD and MPF from R&S Squadron S-3 or BFSB S-3	[Select personnel] accompanies commander	Platoon leader accompanies commander	Accompanies commander		
Initiates alert procedures	Receives or relays the alert	Receives or relays the alert	Receives or relays the alert	Receives or relays alert	Receives or relays alert
Reviews mission folder Conducts TLP	Establishes the COB, to include planning facility	Begins mission preparation	Begins preparations for movement to planning facility	Begins preparations for movement to planning facility	Team LNO: Starts preparing to move to planning facility Company LNO: Deploys to higher headquarters
Develops company WARNO	Helps develop the WARNO	Helps develop the WARNO	Helps develop the WARNO (as required)	Continues preparations for movement to planning facility	Team LNO: continues preparing to move to planning facility
Issues company WARNO	Receives the WARNO	Receives WARNO	Receives WARNO	Receives WARNO	Team LNO: Receives WARNO Company LNO: Establishes contacts
Conducts mission analysis (METT-TC)	Helps the commander conduct METT-TC	Issues WARNO Establishes base stations Helps the commander conduct METT-TC Conducts mission analysis (METT-TC) for the PLT	Issues WARNO Moves to and establishes detachment area in the COB Helps the commander conduct METT-TC Conducts mission analysis (METT-TC) for the PLT	Issues WARNO Begins initial inspections Moves to and establishes team area in the planning facility	Team LNO: Helps establish team area in planning facility
Develops company OPORD	Helps the commander develop the OPORD Finalizes MPF	Helps the commander develop the OPORD Develop PLT OPORD	Helps the commander develop the OPORD Develops PLT OPORD	Continues TLP	Company LNO: Begins coordination
Issues company OPORD	Issues intelligence annex	Issues communications annex, paragraph 5	Develops detachment OPORD (as required)	Receives company OPORD	Team LNO: Receives company OPORD
Receives confirmation brief Provides guidance	Begins mission development and coordination	Begins BRS operations	Receives confirmation brief Provides guidance	Issues confirmation brief Leader--Issues updated WARNO to team	Receives guidance

Table 4-1. Actions and responsibilities of LRSU personnel (continued).

Commander	Operations	Communications Platoon	LRS detachment Leader	Team	LNO
Supervises mission preparation	Supervises mission preparation	Continues BRS operation	Supervises mission preparation	Conducts mission analysis (METT-TC)	Assists team with analysis and conducts coordination
Receives COA decision brief Gives guidance	Supervises mission preparation Receives COA decision brief	Continues BRS operation Receives COA decision brief	Receives COA decision brief Gives guidance	Conducts COA decision brief	Assists team with COA decision brief
Supervises mission preparation	Supervises mission preparation	Continues BRS operation	Supervises mission preparation	Develops OPORD	Assists team with analysis and conducts coordination Assists team with OPORD development
Supervises mission preparation	Supervises mission preparation	Continues BRS operation	Supervises mission preparation	Issues team OPORD	Receives OPORD
Receives team backbrief Provides guidance	Receives team briefback	Continues BRS operation Receives team backbrief	Receives team briefback Provides guidance	Conducts backbrief	Helps team conduct backbrief
Briefs R&S Squadron or BFSB commanders on team mission and execution (as required)	Helps commander conduct briefing	Helps commander conduct briefing Continues BRS operation	Helps commander conduct briefing Supervises mission preparation	Helps commander conduct briefing	Helps team prepare and coordinate mission
Supervises mission preparation	Supervises mission preparation	Continues BRS operation	Supervises mission preparation	Continues mission preparation Checks communications Conducts rehearsals	Helps team prepare and coordinate mission
Supervises mission preparation	Supervises mission preparation	Continues BRS operation	Supervises mission preparation	Checks communications Conducts rehearsals	Helps team prepare and coordinate mission
Supervises mission preparation	Supervises mission preparation	Continues BRS operation	Supervises mission preparation	Sterilizes planning facility Conducts final inspection Briefs back as needed	Helps team prepare and coordinate mission

ORDERS

4-6. LRSU OPORDs are very detailed and specific to LRS operations. To ensure mission success, LRSU orders must contain all planning coordination conducted for the pending mission. For this reason, OPORD format may deviate from the standard five-paragraph Army order format. Appendix B shows some example orders formats used in LRSU operations.

Movement Order--This stand-alone order facilitates an uncommitted, typically administrative, unit movement. While the arrangement of the troops and vehicles should expedite their movement and conserve time and energy, they should still maintain 360-degree security.

Warning Order--This notifies of an upcoming order or action. The WARNO can, but need not, follow the standard five-paragraph OPORD format.

Confirmation Brief--No later than 30 minutes after receiving the unit OPORD, each subordinate leader gives his higher commander a confirmation brief. This demonstrates his understanding of the higher unit OPORD. It also ensures the correct focus during mission planning.

COA or Decision Brief--The team conducts a METT-TC analysis, analyzes possible COAs, develops and compares war games. The team leader then briefs the commander on his chosen COA, explaining the criteria he used to choose it over other possible COAs. When the team leader finishes his briefing, the commander approves, disapproves, or modifies the COA and gives guidance on finalizing the team plan.

Operations Order--The unit leader uses the OPORD to coordinate the actions of subordinates in the execution of an operation. Sometimes called the five-paragraph field order, this order describes, at a minimum, the task organization, situation, mission, execution, administrative and logistics support, and command and signal requirements for the specific operation.

Fragmentary Order--An abbreviated OPORD, used to update the original OPORD. Fragmentary orders (FRAGO) can be issued anytime. Although the FRAGO follows the five-paragraph OPORD format, it normally only includes the paragraphs or items that have changed. New missions or significant mission changes require a new OPORD rather than just a FRAGO.

Backbrief--A briefing by subordinates to the commander to review how subordinates intend to accomplish their mission (FM 1-02). The backbrief can be a formal or informal presentation. During the backbrief, each leader conveys detailed information about the planned mission as he understands it. This gives the recipient of the briefing an idea of the leader and unit's ability to conduct a planned mission. After listening to the subordinate leader's backbrief, each leader briefs back his own understanding of the mission to his higher as required. The unit SOP usually identifies who briefs whom and what must be briefed.

PLANNING PROCEDURES

4-7. The LRSU uses detailed planning procedures to enhance mission planning and OPSEC. Unit SOPs and leaders cover planning procedures in detail. Each LRSU should have a planning facility, with a separate planning area for each team planning for a mission. In the planning area, the team only sees mission-essential personnel such as the LNO, members of the planning cell, or others on the access roster. The unit maintains OPSEC at all times to avoid compromising the teams and the mission. "Need to know" applies: The less each Soldier knows about the other teams' missions, the less can be divulged in case of capture.

4-8. The planning facility location can vary, but the unit can locate it wherever they can achieve the objective of separating a LRS team during planning from the remainder of the unit. A planning facility can be located at an intermediate staging base, depending on the unit's deployment plan. The planning facility can use tents, buildings, hangers, or navy vessels. The exact structures used matter less than the ability to achieve and maintain security and separation. A planning facility should only have one controlled entrance, and the unit must maintain an access roster.

4-9. An ideal planning facility has separate sleeping quarters, showers, electricity, heat or air-conditioning, and a latrine for each team. If possible, the planning facility should be structured to make the teams' accommodations comfortable.

TROOP LEADING PROCEDURES

4-10. The TLP provides leaders at company level and below with a framework to develop plans and orders, and to prepare for operations. TLP is a dynamic process used by small unit leaders to analyze a mission, develop a plan, and prepare for an operation. These procedures enable leaders to maximize available planning time while developing effective plans and adequately preparing their unit(s) for an operation. The TLP have eight steps, shown in Figure 4-1. The sequence is not rigid. They can be modified to meet the mission, situation, and available time. Some steps are performed concurrently, while others may continue throughout the operation (FM 5-0).

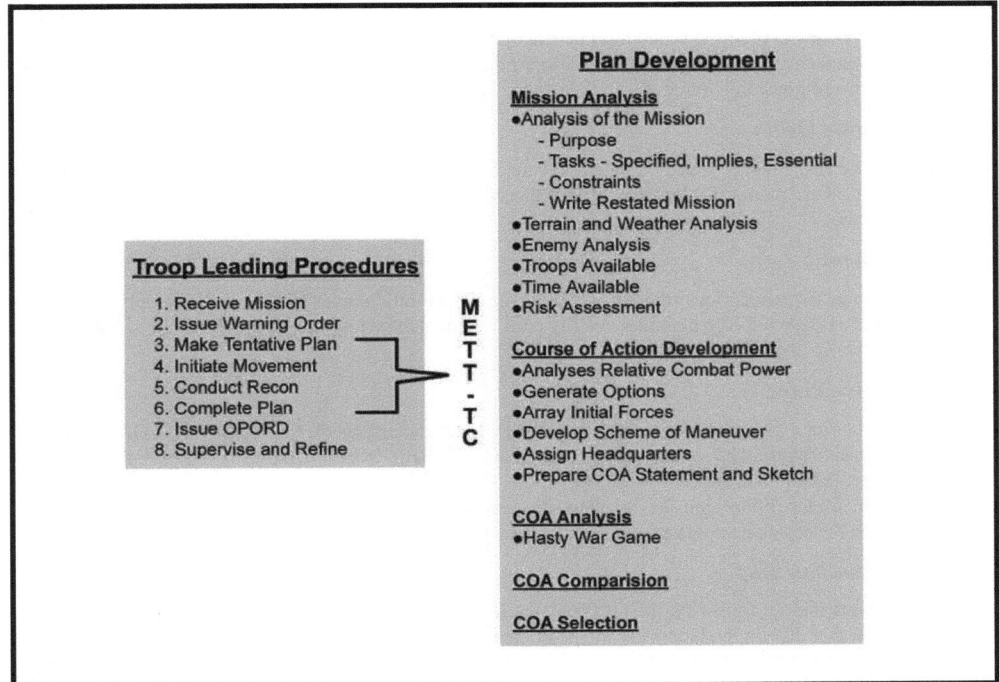

Figure 4-1. TLP and METT-TC.

4-11. LRS teams use TLP slightly differently than do other units. Normally, Steps 3 and 4 reverse, because the team moves to the planning facility before receiving the company OPORD and making a tentative plan (mission analysis). After Step 7, the team usually must backbrief to gain approval to conduct the mission. The following description of how TLP are executed uses the LRS team as an example. However, other LRSU leaders use TLP the same way:

Receive the Mission

4-12. Detailed planning ensures mission success and team survival. On receipt of the WARNO, the team begins an intensive preparatory phase at the unit area. The team leader has the team start initial inspections and prepare for movement to the planning facility. Occasionally, the team may need to receive the initial mission briefing (OPORD with MPF) at this time. However, it is preferred this briefing occur after the team is in the planning facility.

Chapter 4

Perform an Initial Assessment--During the initial assessment, the team leader conducts a METT-TC analysis with the available information. The team leader shares the results of this analysis with the team that are necessary to complete details of the WARNO. Details of the analysis may be withheld from the team for OPSEC reasons until arrival at the planning facility.

Allocate the Available Time--The LRS team leader produces a time schedule based on the information received in the company or detachment WARNO. Using the reverse planning methodology, knowing critical times and experienced judgment on time estimates for events are essential in creating a realistic time schedule. The team leader should also consider potential changes to the time schedule. For example, if the team were ordered to execute the mission six hours early, the team leader must determine what remaining critical must still be completed to ensure mission success.

Issue the Warning Order

4-13. The team leader issues a WARNO to the team. This WARNO generally provides the same information as the company WARNO, with the addition of both general and team-specific instructions (Appendix B).

Initiate Movement

4-14. After the inspections, the team compiles a list of deficiencies and an initial supply request. The team and LNO move to the planning facility area, where they set up the planning facility IAW unit SOP (Appendix D).

Make a Tentative Plan

4-15. Making a plan combines mission analysis, COA development, COA analysis, COA comparison, and COA approval. The team leader and RTO normally receive the mission briefing from the commander, detachment leader, or the operations section. At the same briefing, the team leader receives the mission folder. Following the briefing, the team leader conducts a confirmation briefing. The team leader, with the assistance of the team, conducts mission analysis using METT-TC.

Conduct a Confirmation Brief

4-16. Immediately following the OPORD, the team leader briefs the commander or detachment leader to confirm that he understands the mission and the commander's intent. (Appendix B provides an example confirmation brief.) Before mission planning begins, the team leader may give the team an updated WARNO based on material in the company OPORD.

Begin Team-Planning Process

4-17. The team begins the planning process. The team leader might conduct a visual reconnaissance of the AO. If feasible, the senior scout accompanies him. While he reconnoiters, the assistant team leader supervises equipment and personnel preparation. TOC personnel make themselves available for coordination throughout the planning phase.

Note: The OPORD, *not* the backbrief, drives planning. The backbrief is just a form of rehearsal.

Analyze the Mission

4-18. The team leader reviews and studies the mission, first identifying the specified and implied tasks. From the specified and implied tasks, the team leader identifies the essential tasks. These tasks must be completed to accomplish the mission. Failure to complete an essential task results in mission failure. He also identifies any constraints. The product of this part of the mission analysis is the restated mission.

Mission Statement

4-19. The restated mission is a simple, concise expression of the essential tasks the unit must accomplish, and the purpose to be achieved. The mission statement says who (the unit), what (the task), when (either the critical time or on order), where (location), and why (the purpose of the operation). Each mission statement has three distinct elements: operation, task, and purpose:

Operation

4-20. This is a military action. It consists of all the processes involved with combat: movement, supply, attack, defense, and maneuvers to gain objectives. LRSU conduct the following four operations:

Surveillance--This is a systematic observation of airspace or surface area by visual, aural (hearing), electronic, photographic, or other means.

Reconnaissance--This is any action taken to obtain information about the operational area. It includes any visual or other detection methods taken to learn the enemy's or potential enemy's activities and resources and the area's meteorological, hydrographic, or geographic characteristics.

Target Acquisition--This is the detection, identification, and location of a target in sufficient detail to permit the effective employment of weapons against it.

Target Interdiction--This is any action taken to divert, disrupt, delay, or destroy the enemy's surface military potential before it can be used effectively against friendly forces.

Task

4-21. A task is a clearly defined, measurable activity accomplished by individuals and organizations. (Table 4-2 identifies LRS tasks by operation.) A task includes specific actions that contribute to mission accomplishment or other requirements. A task is definable, obtainable, and decisive; and it is either specified or implied. A specified task is stated in the WARNO, mission order, annex, or overlay, or the commander directs it. An implied task is neither specified, routine, nor included in the SOP, yet it is inherent and mission specific. It is deduced from the order and its products:

Observe--This is the visual; audible; or mechanical, electrical, or photographic monitoring of enemy activities. It applies to missions in which the enemy's location is known or strongly suspected, such as an NAI where the enemy must pass.

Locate--This is to search or examine an area to find an enemy (or his equipment) known to be present in the AO, but whose specific location is unknown.

Detect--This is to discover or discern the existence or presence of enemy activity. This task applies to missions in which little, if anything is known about the enemy. The enemy may or may not be present.

Determine--This is to decide or settle conclusively that this task applies to missions in which much information is known about the enemy. However, some questions still exist about his exact disposition, location, or content. This mission is to gather one or two specified intelligence requirements.

Identify--This is to positively recognize enemy units, formations, equipment, and so on.

Evaluate--This is to examine and judge carefully and place a value or worth on the condition and state of specified structures or enemy capabilities. Structures or enemy location are known.

Confirm--This is to support or establish certainty or validity.

Deny--This is to prove untrue or invalid.

Report--This is to communicate collected information accurately and in a timely manner.

Mark--This is to designate a target by using lasers or other marking devices.

Pinpoint--This is to locate and precisely identify a target for engagement without using lasers.

Chapter 4

Measure--This is to estimate by comparing two distances.

Attack by fire--This is to use direct fires, supported by indirect fires, to engage an enemy without closing with him to destroy, suppress, fix, or deceive him.

Suppress--This results in the temporary degradation of the performance of a force or weapons system below the level needed to accomplish the mission.

Disrupt--This is to integrate direct and indirect fires, terrain, and obstacles to upset an enemy's formation or tempo, interrupt his timetable, or cause his forces to commit prematurely or attack in piecemeal fashion.

Table 4-2. LRS tasks by operation.

Operation	Surveillance	Reconnaissance	Target Acquisition	Target Interdiction
TASKS:	• Observe	• Report	• Mark	• Attack-by-fire
	• Locate	• Locate	• Locate	• Suppress
	• Detect	• Detect	• Detect	• Disrupt
	• Determine	• Identify	• Identify	
	• Identify	• Confirm	• Pinpoint	
	• Evaluate	• Deny		
	• Report	• Pinpoint		
	• Confirm			
	• Deny			

Purpose

4-22. The purpose generally supports the BFSB's purpose and is the same as other Army operations. LRSU study strengths, dispositions, composition, and capabilities of the friendly and enemy forces that could affect the team's mission.

• Plan use of time and prepare a written schedule for required actions. Include--

— Time.

— Event.

— Place.

— Uniform.

— Personnel who will attend.

• Use the reverse-planning technique.

• Select and request equipment (routine and special).

Enemy

4-23. The MPF provides information about the enemy's composition, disposition, strength, recent activities, ability to reinforce, and possible COAs. The team also determines what they do not know about the enemy and requests that information. It is likely the enemy situation will remain uncertain and the experience and training of the LRS team becomes essential in determining a realistic enemy situation.

Terrain

4-24. This aspect of mission analysis addresses the military aspects of terrain: observation and fields of fire; avenues of approach; key terrain; obstacles; and cover and concealment (OAKOC). The MPF provides significant details about the operations area including hydrology, LOS data, and locations of potential drop zones, landing and pick-up zones.

Weather

4-25. The five military aspects of weather include visibility, winds, precipitation, cloud cover, temperature, and humidity (FM 34-130). The consideration of their effects is an important part of the mission analysis. The team reviews the forecasts and conclusions available in the MPF and develops their own conclusions on the effects of weather on the mission. The analysis considers the effects on personnel, equipment, and supporting forces, such as air and artillery support. The team identifies the aspects of weather that can affect the mission. They focus on factors whose effects they can mitigate. For example, the team leader may modify SOP uniform and carrying loads based on the temperature. The team leader checks for compliance during preparation, especially during rehearsals (FM 5-0).

Troops and Support Available

4-26. The team leader knows the status of the teams' morale, their experience and training, and the strengths and weaknesses. The team leader realistically determines all available resources. This includes possible attachments like a sniper team, target acquisition or a medic. The assessment includes knowing the strength and status of team members and their equipment. It also includes understanding the full array of assets in support of the team. The team knows, for example, how much indirect fire, by type, is available, when it will become available and the time it takes to employ. They consider any new limitations based on level of training or recent fighting.

Time Available

4-27. The team leader continues to refine the time schedule. The team views its own tasks and enemy actions in relation to time. Most importantly, the team leader monitors the time available. As the situation changes, the team leader uses his experience to adjust the time schedule to ensure the team is best prepared to accomplish the mission.

Civil Considerations

4-28. Civil considerations are how the man-made infrastructure, civilian institutions, and attitudes and activities of the civilian leaders, populations, and organizations within an AO influence the conduct of military operations (FM 6-0). Rarely are military operations conducted in uninhabited areas. Most of the time, units are surrounded by noncombatants. LRS team operations are in fact even more complicated because they often depend on effectively hiding among the civilian population. These noncombatants include residents of the AO, local officials, and governmental and nongovernmental organizations (NGOs). Based on information from higher headquarters and their own knowledge and judgment, the team leader identifies civil considerations that affect their mission. Civil considerations are analyzed in terms of six factors, known by the memory aid ASCOPE (FM 6-0):

- Areas.
- Structures.
- Capabilities.
- Organizations.
- People.
- Events.

Chapter 4

Develop Courses of Action

4-29. The purpose of COA development (Figure 4-2) is simple: to determine one or more ways to accomplish the mission. Most missions and tasks can be accomplished in more than one way. However, in a time-constrained environment, the team leader may develop only one COA. Normally, the team will develop two or more. Usable COAs are suitable, feasible, acceptable, distinguishable, and complete. To develop them, the team focuses on the actions the team takes at the objective and conducts a reverse plan to the point infiltration starts.

COA DEVELOPMENT

1. Analyze relative combat power.
2. Generate options.
3. Array forces.
4. Develop the concept of operations.
5. Assign responsibilities.
6. Prepare COA statement and sketch.

Figure 4-2. Development of courses of action.

Analyze (War-Game)

4-30. For each COA, the team thinks through the operation from beginning to end. They compare each COA with the most likely enemy COA, given what the LRS team is doing at that instant. Normally, small unit leaders visualize a set of actions, reactions and counteractions. The LRS team leader does this also, but because the LRS team's objective is to not make enemy contact, the result should minimize the chance of contact. The team records the results of all wargames.

Compare

4-31. The team leader compares results of all the wargames and chooses the COA that has the best chance of mission accomplishment and preserves the team for future operations. Criteria normally include-- mission accomplishment, time to execute the mission, risk, and posture of the team for future operations.

Approve

4-32. The team leader picks the COA that best supports successful mission accomplishment. Normally, before the team continues plan development, the team leader must obtain approval for the chosen COA. This briefing is given to the detachment leader or company commander. Once approval is obtained and guidance given, the team begins development of the OPORD.

Conduct Reconnaissance

4-33. Ideally, the team leader reconnoiters the area visually. If this cannot be done, then the team leader continues to study aerial reconnaissance photos, UAS video and satellite imagery (if available). He confirms, clarifies, and supplements information gleaned from maps and other sources.

Complete the Plan

4-34. Complete detailed planning to include necessary coordination with all support elements required for the mission. (Appendix G provides an example list of the necessary coordination.)

Issue the OPORD

4-35. Include all necessary annexes. Use a detailed OPORD format. (Appendix B provides an example.) Use visual aids (terrain models, objective sketches, charts, photos, maps, and overlays) to reinforce information and to help ensure the team fully understands all aspects of the mission.

Supervise and Inspect Soldiers

4-36. The team leader and the assistant team leader supervise the team throughout the preparation to ensure timely completion of all required tasks. They inspect to ensure that--

- The team takes only the equipment required for the mission.
- All equipment is functional, complete, secured, and evenly distributed.
- Resupply bundles and packages meet SOP criteria.
- Cache meets SOP criteria.
- Insertion vehicle and aircraft preparation meet SOP criteria.

Check the Communications Equipment

4-37. Under the guidance of the team leader, the RTO checks all communications equipment on a distant BRS (HF and UHF TACSAT). They also check--

- Internal communications with all VHF frequency modulation (FM) communications equipment. Unserviceable and inoperable equipment is reported to the communications NCOIC.
- For all radios, frequencies are confirmed with the frequency manager, and, if needed, more frequencies are requested.

Rehearse Mission

4-38. After briefing the OPORD, the team leader conducts rehearsals. This includes inspecting personnel and equipment. During the rehearsal, the team wears the full uniform and carries all mission-essential equipment. The unit conducts detailed, full force rehearsals, because the team needs them to reinforce complex procedures. Terrain and conditions should, as much as possible, replicate those expected for the actual operation. Detailed rehearsals can include transportation and OPFOR, and as many contingencies as can be anticipated. The team simulates casualties among key personnel, with other team members assuming their duties. Leaders continually ask team members to answer mission-specific questions. Sand table briefings, a map study, and photograph examinations should complement all rehearsals. At a minimum, during hasty planning, the team should rehearse actions in the objective area (entering; maintaining; and sterilizing the hide, surveillance, and communication sites). Otherwise, rehearsals should cover as much as possible the following, also shown in Table 4-3.

- Off-loading and assembly procedures at points of insertion.
- Movement formations.
- "Lost-man" drill.
- Security halt procedures.
- Actions at possible danger areas.
- Reaction drill for aircraft flyover (friendly or enemy).
- Countertracking techniques.
- Actions on enemy contact such as chance, near/far ambush, sniper, air attack, indirect fire, flares.
- Loading procedures at the extraction site.
- Special actions (as required), and use of new or unfamiliar equipment.
- Procedures for emplacement and recovery of a cache.
- Actions at recovery points or contact points.
- Actions in the absence of a communications plan.
- Communications during scheduled windows and initial entry.

Table 4-3. Priority of actions for rehearsal.

Rehearsal Actions	Rehearsal Type		
	Hasty/ Critical	Minimal/ Important	Detailed/ Useful
Actions in the objective area (entering, maintaining, and sterilizing the hide, surveillance, and communication sites)	X	X	X
Off-loading and assembly procedures at points of insertion		X	X
Movement formations		X	X
"Lost-man" drill		X	X
Security halt procedures		X	X
Actions at possible danger areas		X	X
Reaction drill for aircraft flyover (friendly or enemy)		X	X
Countertracking techniques		X	X
Actions on enemy contact (chance, near and far ambush, sniper or air attack, indirect fire, flares)		X	X
Loading procedures at the extraction site		X	X
Special actions (as required) and use of new or unfamiliar equipment		X	X
Procedures for emplacing and recovering cache		X	X
Actions at recovery points or contact points		X	X
Actions in the absence of a communications plan		X	X
Communications during scheduled windows and initial entry		X	X
Transportation contingencies			X
Sand table briefings			X
Map study			X
Photographic examinations			X

Conduct Backbrief

4-39. When mission planning is complete, the team briefs back the entire mission to the commander or to the commander's designated representative. The backbrief ensures the commander that the team understands and is prepared for the mission. They can shorten the backbrief to accommodate condensed planning time or as the commander requests, based on his knowledge of the team's experience and on who attends the backbrief. The team rehearses the backbrief to ensure that all team members understand all aspects of the operation. (Appendix B provides an example backbrief format.)

Conduct Final Inspection

4-40. The team always conducts a final inspection before the team leaves the planning area. The team leader inspects personnel, personal equipment, and mission equipment, especially those items identified during previous inspections or during rehearsals as needing correction. The team leader questions team members to reinforce critical aspects of the mission.

Receive Intelligence Updates

4-41. The team leader receives intelligence updates from higher and adapts his plans accordingly.

CONTROL MEASURES

4-42. A control measure is a graphic or oral directive assigning responsibility, coordinating fires and maneuver, and controlling combat operations. Each measure is shown graphically and easily identified on the ground. Examples include boundaries, objectives, coordinating points, and contact points. Some control measures that help the team leader control team actions during the mission are--

- Times of departure and return.
- Points of departure and reentry.
- Checkpoints.
- Routes.
- Forward line of own troops (FLOT).
- Phase lines (PLs).
- Restrictive fire lines (RFLs) and NFAs or RFAs.
- Decision points (DPs).
- Designated areas for recovery (DARs).
- Limits of advance (LOA).
- Recovery points.
- Contact points.
- SAFEs.
- Airspace-recovery activation signals.
- No-fly areas.
- Identification friend or foe (IFF).

FIELD PLANNING ACTIVITIES

4-43. Leaders can conduct field planning one of two ways:

Field Planning

4-44. This generally means conducting deliberate planning in a tent. Special considerations include--

- The number of tents required for all elements to have sufficient planning and sleeping space.
- Type and quantity of lighting required.
- Heating requirements.
- Latrine facilities.
- Dining facilities. Ideally, teams should receive at least two hot meals a day.
- Security considerations such as wire, field phones, and guards.
- Support required to construct the site and maintain it for the period required.
- Communications to supporting units during planning and conduct of the mission.
- Rehearsal areas.

Chapter 4

Patrol-Base Planning

4-45. This refers to planning that occurs when the team receives a change of mission during the conduct of an operation. LRS patrol-base principles apply. The team--

- Maintains security.
- Plans and issues a FRAGO.
- Plans resupply.
- Plans for continuous communications.

CONTINGENCY PLANS

4-46. The contingency plan covers alternate, anticipated major events that could occur before, during, and after an operation, for example--

- Replacement of team members as needed to fulfill mission requirements.
- Transportation to the planning facility.
- The planning facilities (garrison or field).
- Direct support for unserviceable equipment.
- Acquisition of mission-essential equipment.
- Security during the mission planning process.
- Reposition to COB, AOB or MSS.

INFILTRATION PHASE AND INSERTION METHOD

4-47. The infiltration phase extends from the point of embarkation to arrival in the objective area. Insertion method options include--

- High altitude, high opening (HAHO).
- High altitude, low opening (HALO).
- FRIES.
- SCUBA.
- SPIES.
- Air.
- Vehicle.
- Rubber boat.
- Stay behind.
- Rollover.
- Foot movement.

4-48. Any requested SEAD and CAS start when the insertion platform crosses the FLOT and starts moving towards the insertion point. Insertion normally ends after the team caches nonmission-essential equipment and the insertion platform leaves the loiter area. Generally, the infiltration phase continues with team movement from the point of insertion to the security halt, and ends before the objective rally point (ORP). Ideally, infiltration occurs during times of limited visibility. If the team must halt during periods of increased visibility, they establish a clandestine patrol base. During infiltration, the team leader records the team's movements in the patrol log. In the log, the team leader records in detail--

- The general direction of movement.
- Deviations from planned infiltration route.
- Information about terrain and weather.
- Enemy sightings en route.
- Signs of activity.
- Grid locations.
- Any peculiarities.
- Time of initial entry report.

EXECUTION (ACTIONS ON OBJECTIVE)

4-49. The LRS team establishes a security halt and the leader reconnoiters to identify an ORP. After the team establishes the ORP, the leader moves out to pinpoint the objective or NAI, then to establish surveillance. Once the reconnaissance is complete, other team members move to and establish the tentative hide site and to set up communications with the COB or AOB. The surveillance team keeps their "eyes on" the objective and maintains continuous communications with the hide site. The surveillance site reports ISR tasks to the hide site.

4-50. If the LRS team cannot establish communications, the team and the LRS COB or AOB execute the "no communications" contingency plan (developed during the planning phase).

4-51. The hide site sends information to the COB or AOB via HF or UHF TACSAT communications. Once communications is established, the COB or AOB forwards the information to the BFBS S-2 ISR fusion element over a LAN. The R&S squadron is copied on all reports.

4-52. The LRS team keeps sending reports to the COB or AOB until the team meets the mission completion criteria or until the LTIOV. The LRS team reports during designated communication windows or, if the report is information answering a PIR, out of those windows. After the team pulls back from the objective they disseminate information, collect all surveillance logs and objective sketches.

EXFILTRATION PHASE AND EXTRACTION METHOD

4-53. This phase starts after the team links up and disseminates information at the ORP and ends when the team arrives at the debriefing location. Exfiltration routes normally differ from infiltration routes. The team leader (again) keeps a patrol log, which details--

- The general direction of movement.
- Deviations from the planned exfiltration route.
- The terrain and weather.
- Enemy sightings.
- Signs of activity.
- Peculiarities.

4-54. Extraction method options include--

- Linkup (friendly or partisan).
- Air.
- SPIES.
- Vehicle.
- Watercraft.
- Rollover.

Chapter 4

RECOVERY

4-55. This phase starts when the LRS team returns to the debriefing site, which is normally located at the planning facility, and starts the multiphase debriefing process. It ends with the after action review (AAR).

DEBRIEFING PROCESS

4-56. The whole LRS team attends all phases of the debriefing. The debriefing covers the team's actions and all related details, chronologically, from the start of the infiltration phase, through arrival at the debriefing site. The debriefing process normally includes the following in sequence: initial debriefing, post initial debriefing, review of documents and other materials, follow-up debriefing, AAR and team report of lessons learned.

Debrief Team

4-57. A trusted and knowledgeable intelligence representative, or LRS operations personnel conducts the debriefing, which starts NLT two hours after the team returns. The timely collection, analysis, and dissemination of information recovered in the debrief can provide many answers needed for follow-on missions in the objective area.

Follow-up Actions

4-58. After the debriefing, the LRS team starts equipment maintenance, refit operations, and training for follow-on missions.

Timing

4-59. Every LRS team is debriefed immediately after returning from a mission. The debriefing is important in the intelligence collection effort, especially when used to clarify and expound on information received via reporting.

Advantages

4-60. A debriefing can--

- Bring out unreported details that when collated with other information could alter the picture of the enemy situation.
- Update terrain information.
- Reveal the location of downed or missing friendly aircraft.
- Highlight shortcomings in pre-mission planning in unit SOPs.
- Help reconstruct a mission in which casualties were sustained, either to recover remains or to determine KIA, MIA, or POW status.
- Provide historical record of the mission for post-hostilities analysis.

Site

4-61. The LRS operations and the supporting intelligence section provide a facility for the debriefing. They must choose a quiet, secure site with few physical, visual, or audible distractions. For example, if they set up in a tent, they should do so away from generators. Before the LRS team arrives, the debriefing team sets up everything they need such as maps, overlays and other planning materials. They arrange for soup and coffee to keep team members alert and active during the debriefing process. If the debriefing team plans to record the debriefing on audio or videotape, they should set up and test the equipment before the team arrives.

Written or Taped Record of Information

4-62. Rather than recording the mission himself, the debriefer should delegate this task to two people (written record) or one person (electronic record that is, audio or videotape). The recorder(s) should concentrate on locations, times, direction of movement, and any other information the debriefer identifies as important. For best results, the recorders should use two electronic recording devices so that they can capture everything accurately. Voice and video recorders capture the facts of a debriefing and serve as excellent historical records. When a patrol is in progress or the pace of an operation increases, a unit can record the action and send copies of the tapes in place of formal patrol reports. Knowing that he can hear or see a tape of the proceedings later keeps most requesters from asking to attend the live debriefing. Operations provide the overlays produced during planning. These include the infiltration, exfiltration, evasion and recovery (E&R), enemy situation map, and enemy SITEMP overlays. On a clean overlay, the team records the actual routes taken. They also mark locations of key events such as halts, enemy sightings, and signs of enemy presence or passage. On the overlay, they can add short narratives of the events.

Protocol

4-63. Everyone attending must understand the rules. Although informal, the initial debriefing still follows a strict protocol. Regardless of rank, the debriefer is in charge of the debriefing at all times. This keeps the process orderly and prevents conflict. Interested parties may submit questions or comments for the debriefer to address during the session. This ensures that all personnel receive as much information as possible. "Strangers" may not interrogate the team. Only the debriefer may address them during the debriefing. This allows rapport to build between the debriefer and the team. The LRS team sits together in the center of the room or tent, in front of a map of their operational area. The debriefer sits or stands near the team, where they can all see and reach the maps. Other participants, such as recorders, interested staff members, and LNOs, sit in the second or third row of chairs from the map. This reinforces the fact that the team members are the center of the debrief. This encourages them to speak out, especially when attendance includes only a small group of people whom the team knows.

Attendees

4-64. The entire team, the debriefer(s), two recorders or one with an electronic recording device, the LRSU commander, the operations sergeant, and a communications representative attend the debriefing. A few interested parties, such as the R&S squadron or BFSB commanders, may also attend. If unable to attend, the commanders may give their questions to the debrief team beforehand. Keeping the group small helps the LRS and debriefing teams establish rapport. It also reduces the chance of hostilities.

Debriefer

4-65. The debriefer focuses on helping team members reconstruct their mission. Ideally, the debriefer is an enemy OB analyst or technician who worked with the team before they infiltrated. An enemy OB analyst will know the team's mission, the enemy situation in general, and how to deal with information from other sources.

Raw Data

4-66. Before the debriefing, the team gives the intelligence section representative all team maps and notes; patrol, surveillance, communications, and photo logs; film and sketches; captured material; and any other relevant materials.

Time-Sensitive Information

4-67. The intelligence staff immediately exploits time-sensitive information critical to their commander's, or higher level, decision-making process. For example, the debriefer might ask first about the location of insurgent safe houses, so that friendly direct-action teams can target them.

Chapter 4

TECHNIQUES

4-68. The debriefer can use several methods to gather information. A good technique is to use a map of the team's AO. He starts at the team's point of infiltration and follows their routes and actions through exfiltration. He monitors the flow of information to ensure the team covers all events, sightings, and activities up to the point of arrival at the debriefing site. After going over the initial mission information, he segments route information. He asks specific questions, emphasizing the specific WFF that affected the team's mission. On a clean map overlay, the team leader and other members of the team annotate route deviations, enemy sightings, or mission-sensitive information. Rather than leading participants through the reconstruction of the mission, he keeps them focused and asks them questions for clarification. The team leader is the key, but not the only, speaker. He helps ensure that each member gets the chance to say what he saw.

Prepared Questions

4-69. The debriefer asks about observed target types.

Spot Reports

4-70. The team elaborates on their earlier reports.

Map

4-71. The team talks through the mission as executed.

INITIAL DEBRIEFING

4-72. The LRS operations section, a member of the BFSB S-2 ISR fusion element and a representative of the R&S squadron S-2 section conduct this portion of the debriefing, but other staff elements may also attend. This debriefing should help answer PIR, intelligence requirements, SIR, and ISR tasks and RFIs. When the team arrives, the debriefer escorts them to the site. The initial debriefing is quick and to the point. The debriefer chooses what format and line of questioning he will use.

POST-INITIAL DEBRIEFING

4-73. At the end of the formal debrief, a communications representative debriefs the RTO. He covers communications-specific information when certain antennas worked best, which frequencies were best, and other communications-specific issues. After the initial intelligence debriefing, the intelligence staff gathers all maps, logs, notebooks, papers, exposed film, video tapes, photographs, recovered equipment, and other material. If necessary, they inventory all rucksacks, map cases, and uniform pockets to ensure that they have collected all items of intelligence interest. Then, they thoroughly review all of the collected items for data and formulate more detailed questions for the next stage of debriefing. The team remains separated from outside contact until after the follow-up debrief, but can start recovery operations (showering, eating, sleeping, and conducting post-mission maintenance).

REVIEW OF DOCUMENTS AND OTHER MATERIAL

4-74. The BFBS and R&S squadron S-2 sections process all of the information the team collected. Individual members of the team can be requisitioned to clarify a sketch or log entry. The LRS team also begins to complete a debrief report and to collate their AAR notes. The R&S squadron and LRS operations section collects information on the adequacy of MPF, intelligence and operations support. The original overlays and maps should not be destroyed or discarded. After properly classifying and annotating them, the squadron S-2 files them. Later, they will go into the unit historian's archives, for reference in case of future missions into the same AO. These artifacts, consisting of actual operations overlays, maps, orders, and debriefing records, should go into an historical database. Though considered of little immediate value, the information in these items assumes increased importance over time, not only for historical reasons, but also for reference for future operations.

FOLLOW-UP DEBRIEF AND AFTER-ACTION REVIEW

4-75. After individual debriefings, and not later than six hours after the team has recovered to the debriefing site, the LRS operations section assembles the LRS team and the staff for a follow-up debriefing and AAR. The commander may also attend. At this debriefing, the team leader summarizes the operation, focusing on the team's stated and implied missions. He also briefs unanticipated team or member activities such as actions to exploit a high-value source of information. Each staff section takes a turn questioning the team members. This debriefing gives the team members the chance to raise issues of support, communication, and coordination as well as any other perceived deficiencies in planning or execution. The commander provides any further guidance and releases the team to prepare their AAR and Report of Lessons Learned.

AFTER-ACTION REVIEW AND REPORT OF LESSONS LEARNED

4-76. The AAR identifies who, what, when, why, where, and how of the operation. It permanently records the team's major activities from planning to debriefing. As such, it serves as an extremely important template for comparison with past missions and planning of future missions. In his report, the team leader reflects on the operation and makes recommendations for the future. He organizes them IAW the WFF. He states what did and did not work, and identifies how the unit's existing TTP need to change.

Section II. RECONNAISSANCE OPERATIONS

The three forms of reconnaissance operations are area reconnaissance, zone reconnaissance and route reconnaissance. Reconnaissance missions greatly increase a team's vulnerability and chances of compromise. The team's mobility is generally limited to foot movement and the amount of equipment carried reduces the size of the area they can reconnoiter. Appendix J provides example formats for reconnaissance of bridges and routes.

AREA RECONNAISSANCE

4-77. Area reconnaissance is a form of reconnaissance operation. It is a directed effort to obtain detailed information concerning the terrain or enemy activity within a prescribed area (FM 1-02). The location may be given as a grid coordinate or an objective on an overlay. The team leader organizes the team to conduct the reconnaissance in one of two ways. Depending on the terrain and time, the team may either use single or multiple separate reconnaissance and security elements (Figure 4-3). Reconnaissance and security teams are normally used in any size reconnaissance patrol. When conducting reconnaissance missions in team-sized units, the leader can organize the team in any of several ways:

- One two- to three-Soldier reconnaissance and security team conducts the reconnaissance. The rest of the team stays at the release point and establishes a hide site.
- Two reconnaissance and security teams reconnoiter a separate portion of the objective, and then meet at a designated linkup point.
- One reconnaissance and security team, followed closely by a security team, acting as a quick-reaction force.

4-78. One or two Soldiers in a reconnaissance and security team can reconnoiter, while the rest of the element provides security. However, the number of Soldiers in a reconnaissance and security team varies, depending on the mission. Usually, three can provide both an adequate reconnaissance and the required security. The information used may vary according to the terrain. The most important planning consideration is that each member of the reconnaissance and security team *knows* the sector or area for which he is responsible.

Chapter 4

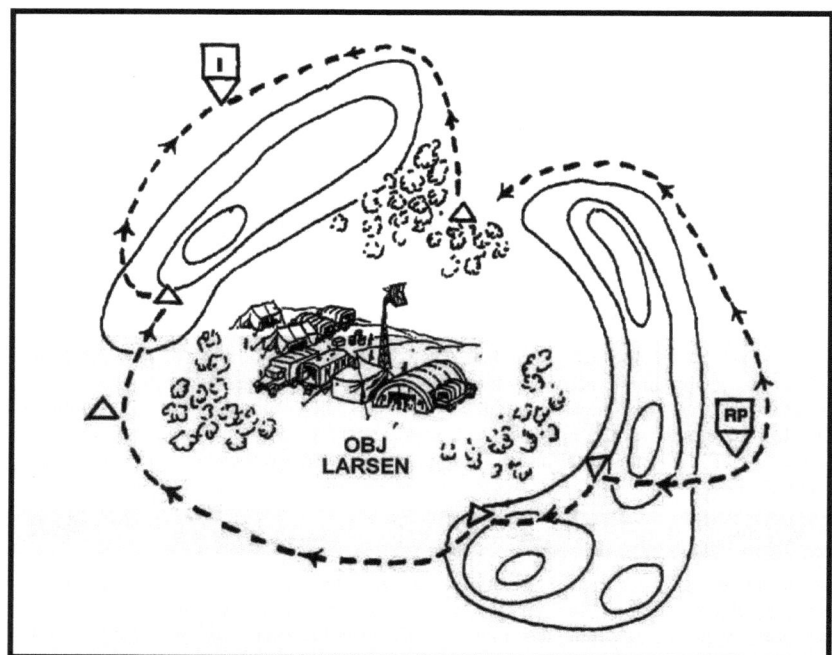

Figure 4-3. Reconnaissance and surveillance elements.

4-79. Before executing an area reconnaissance, the team develops a detailed plan using all available information. They conduct this mission the same as they would a surveillance mission. First, the leader pinpoints the reconnaissance objective or NAI, and second, he locates the best vantage points for a static surveillance. The best static surveillance techniques include—

Long-Range Observation and Surveillance

4-80. Long-range observation and surveillance means "the observation of an objective from a point" (an observation post [OP]). The OP is normally outside enemy small-arms weapons range and local security measures. The LRS team can use this technique whenever METT-TC lets them gather information from a distance. It is the best way to conduct an area reconnaissance, because the team remains far enough away to avoid detection. Using this technique also keeps the team's no-fire or restricted fire area from overlapping with the objective area. When the reconnaissance team cannot gather information from one OP, they can move to a series of OPs until they gather the required information. Observation posts require adequate cover and concealment and a good view of the objective. Routes between and from OPs to the hide site or RP also require cover and concealment.

Short-Range Observation and Surveillance

4-81. Short-range observation and surveillance is "the observation of an objective from a place that is within the range of enemy small-arms weapons fire and local security measures."

- Short-range observation works best when METT-TC requires a close approach to the objective to gain information.

- The reconnaissance teams can conduct short-range observation and surveillance from OPs, but they must usually move near the objective before they can find a place where they can observe. In some cases, the teams may gather information by listening, even though they cannot see the enemy.

- Observing at short ranges increases the chance of detection. The enemy might use anti-intrusion devices and patrols near key installations. Inclement weather can reduce the sounds of the reconnaissance team's movement, and limited visibility favors short-range observation. When the team must observe at short ranges, they use every means available to avoid detection.

4-82. To reconnoiter a road, the team leader selects multiple vantage points (OPs) along the road. The reconnaissance element reconnoiters bridges, defiles, bends in the road, and urban areas. The reconnaissance element reports the condition, trafficability, and width of the road; evidence of the enemy, obstacles; bridge and ford locations and conditions; and tunnel or underpass locations and dimensions.

4-83. To reconnoiter a wood line, the reconnaissance element uses concealed routes and stealth to reach the wood line and avoids contact. The reconnaissance element checks for evidence of enemy activity such as tracks, litter, old fighting positions, mines, booby traps, and obstacles. The reconnaissance element determines if the woods are trafficable. The element checks all positions from which the enemy could observe and fire on friendly elements in open areas, then reports.

ZONE RECONNAISSANCE

4-84. Zone reconnaissance is a directed effort to obtain detailed information on all routes, obstacles, terrain, and enemy forces within a zone defined by boundaries (FM 1-02). The team obtains detailed information about routes, obstacles, key terrain, and enemy activities in a zone established by lateral boundaries. The team can use the fan, converging-routes, or successive-sectors method.

FAN METHOD

4-85. The team leader selects a series of ORPs throughout the zone. When the team arrives at the first ORP, it halts and establishes security. The team leader confirms the team's location. He then selects reconnaissance routes to and from the ORP. The routes form a fan-shaped pattern around the ORP (Figure 4-4). The routes must overlap to ensure that the team reconnoiters the entire area. Once the routes are selected, the team leader sends out reconnaissance elements. He keeps a small reserve in the ORP. For example, if the team has three reconnaissance elements, he sends two, keeping the third in reserve. The team leader also sends the elements out on adjacent routes. This keeps the teams from making contact in two different directions. After the team has reconnoitered the area (fan), the leader reports the information. The team then moves to the next ORP and repeats these actions.

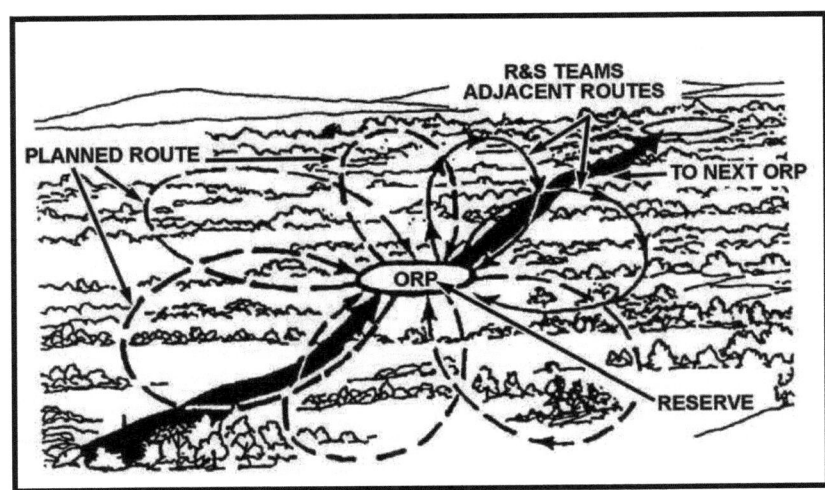

Figure 4-4. Fan method.

CONVERGING-ROUTES METHOD

4-86. The team leader selects an ORP, reconnaissance routes through the zone, and then a linkup point. He sends out a sub element on each route. He normally moves with the center element. The subunits normally reconnoiter their routes by using the fan method. The entire team meets at the linkup point at the designated time (Figure 4-5).

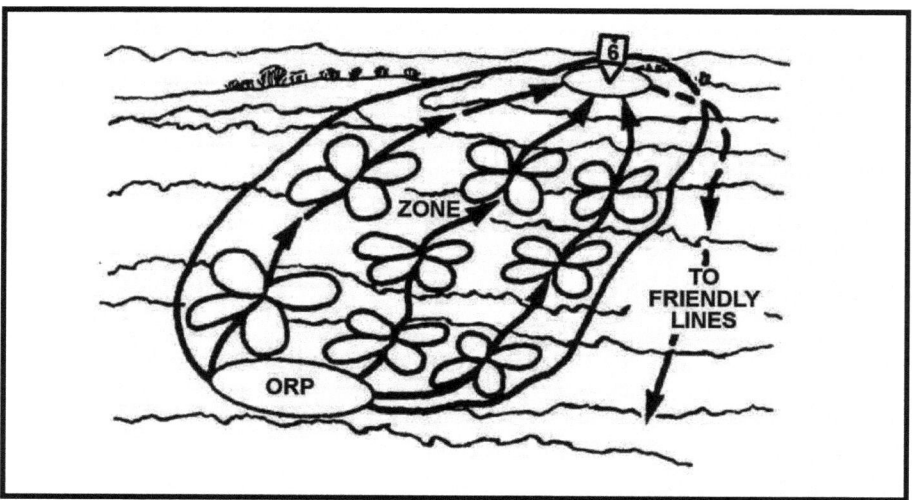

Figure 4-5. Converging routes method.

SUCCESSIVE-SECTOR METHOD

4-87. This method is a continuation of the converging-routes method. The team leader selects an ORP, a series of reconnaissance routes, and linkup points. The actions of the team from each ORP to each linkup point are the same as in the converging-routes method, that is, each linkup point becomes the ORP for the next phase. When the team meets, the team leader again designates reconnaissance routes, a linkup time, and the next linkup point. This action continues until the team has reconnoitered the entire zone (Figure 4-6).

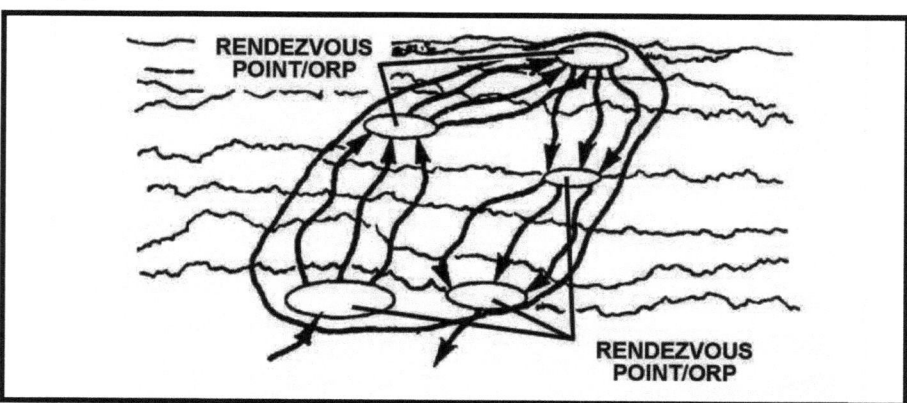

Figure 4-6. Successive sector method.

ROUTE RECONNAISSANCE

4-88. Route reconnaissance is a directed effort to obtain detailed information of a specified route and all terrain from which the enemy could influence movement along that route (FM 1-02).

4-89. Route reconnaissance obtains information about enemy activity, obstacles, route conditions, and critical terrain features along a specific route. It is generally beyond the capability of a LRS team to obtain precise data for road curves, widths and heights of underpasses, and dimensions of tunnels. However, they can report the types of vehicles that use the roads, or that enter or exit the tunnels. From this, intelligence personnel can estimate the weight limits, widths, and other information about the roads, tunnels, and underpasses. The LRS team follows the example report formats provided in FM 5-170. Example intelligence requirements for a LRS route reconnaissance include--

- The available space (in meters) in which a force can maneuver without having to bunch up to avoid obstacles. The size of trees and the density of forests are reported due to the effect on vehicle movement.
- The locations of all obstacles and of any available bypass(es).
- Any enemy forces that can influence movement along the route.
- The observation and fields of fire along the route and adjacent terrain.
- The locations along the route that provide good cover and concealment.
- Trafficability along the route.
- Landing and pickup zones along the route.
- Any bridges by construction and type, estimated dimensions of each, and any vehicles crossing the bridge. This helps intelligence personnel estimate its load classification.

BRIDGE CLASSIFICATION

4-90. This is not a separate category of reconnaissance, but it might be a necessary part of an area, zone, or route reconnaissance. The team follows procedures to ensure that they provide the dimensional data needed to analyze the bridge's structure for repairs, demolition, or military-load classification. Seldom can a team obtain precise measurements. However, if possible, they report the type and number of vehicles that cross the bridge. Intelligence can then estimate the weight, height, and weight limit of the bridge (FM 5-170).

LEADER RECONNAISSANCE

4-91. The leader plans and reconnoiters all primary and alternate surveillance and hide sites. The criteria for selecting these sites is similar, except that the hide site must be sited to allow for long-range communications, and the surveillance site must allow round-the-clock surveillance. Leaders evaluate all primary and alternate sites based on the following criteria, then they establish and disseminate rally points and break out plans for all sites:

HIDE SITE

- Does the site facilitate long-range (HF or UHF TACSAT) communications?
- Does the area provide concealment as well as routes of ingress and egress?
- Are dominant or unusual terrain features located nearby?
- Is the area wet, does it have adequate drainage, or is it prone to flooding?
- Is the area a place that the enemy would want to occupy?
- Is the site silhouetted against the skyline or a contrasting background?
- Are roads or trails located nearby?

- Are other natural lines of drift located nearby?
- Could the hide personnel become trapped easily in the site?
- Do obstacles, such as a ditch, fence, wall, stream, or river, prevent vehicle movement nearby?
- Are any inhabited areas located in the prevailing downwind area?
- Are any suitable communication sites located nearby?
- Is the site in the normal line of vision of enemy personnel in the area?
- Is there a source of water in the area?

SURVEILLANCE SITE

- Can the team place the designated surveillance target(s) under constant and effective observation and within the range of surveillance devices to be used?
- Would the surveillance site have to move if weather and light conditions change?
- Does the area provide concealment?
- Does the area provide adequate egress routes?
- Are dominant or unusual terrain features located nearby?
- Is the area wet, does it have adequate drainage, or is the area prone to flooding?
- Would the enemy want to occupy this area?
- Is the site silhouetted against the skyline or a contrasting background?
- Are any roads or trails located nearby?
- Are any other natural lines of drift located nearby such as gullies, draws, or any terrain easy for foot movement?
- Could the surveillance team become trapped easily in the site?
- Do any obstacles, such as a roadside ditch, fence, wall, stream, or river, prevent vehicle movement nearby?
- Are any inhabited areas located in the prevailing downwind area?
- Is the site in the normal line of vision of enemy personnel in the area?
- Is there a source of water in the area?
- Does the site facilitate communications?

Section III. SURVEILLANCE OPERATIONS

This section discusses selection and occupation of sites; security and reports; linkup and dissemination of information; contingencies and heavy team and platoon operations.

SELECTION AND OCCUPATION OF SITES

4-92. After he completes his reconnaissance (described in Section II), the leader selects the hide and surveillance sites. While selecting the sites, he determines whether each site will have nonmission-essential equipment. His decision determines the size of sites to construct and the type of breakout drills that the team must perform, if required.

HIDE SITE

4-93. The team should test communications from the tentative site before they start constructing it. Otherwise, they might have to start over with a new site.

SURVEILLANCE SITE

4-94. The leader selects each site based on the quality of its observation and communications with the other site. To determine if the planned surveillance site (Figure 4-7) will work well for surveillance, the team places all optics at the same levels they will use when they actually occupy the site, such as at ground level. Seldom do circumstances allow for rebuilding of sites. Finally, try to minimize digging signatures (sound and dust).

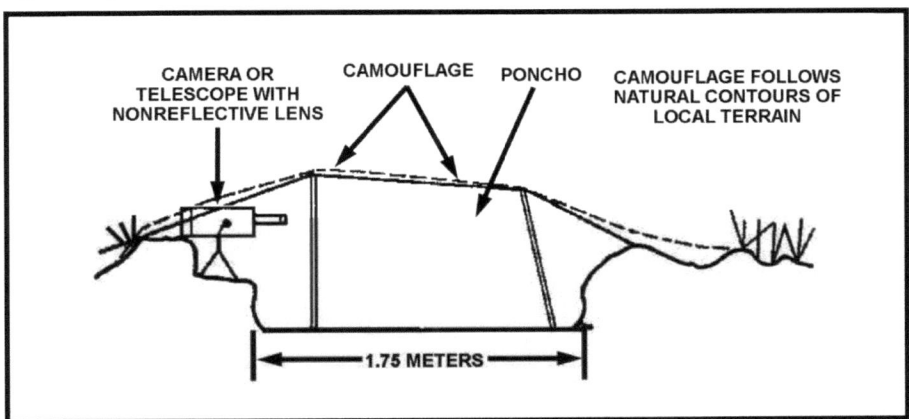

Figure 4-7. Example surveillance site.

OCCUPATION

4-95. Teams maintain security at all time. Security personnel must emplace far enough out to detect intruders, but must remain close enough to alert the team. A technique to alert the site of an intruder is to simply tug on a cord tied from the security position to a Soldier at the site.

4-96. The team should maintain all-round security, and pay close attention to the most likely avenues of approach. Each site needs all-round observation and a view of anyone or anything in the area. This helps prevent compromise, as it allows team members to detect movement and gives them the advantage if they must break out of the site.

4-97. Appendix H discusses how to construct hide and surveillance sites. In most cases, the team should only work on a surveillance site in limited visibility, but they can work on a hide site during the day, if needed. Generally, the team must camouflage all sites well enough to escape detection from greater than 10 meters. The sites should blend well with the surrounding terrain in texture and form.

4-98. The LRS team prepares and rehearses a complete breakout plan. Team members prepare equipment for executing a breakout drill immediately. Pre-positioned M18 Claymore mines and smoke grenades can facilitate a quick breakout of the site.

SECURITY AND REPORTS

4-99. The team prepares and sends reports during actions on the objective from on the start of observation activities until the dissemination of information. The reports the team members at each site must make depend on that site's specific tasking.

Chapter 4

HIDE SITE

4-100. The primary mission of the hide site is to facilitate team internal and long-range communications. On long missions, team members can rotate between the surveillance and hide sites (Figure 4-8). It is usually more secure than the surveillance site, because it is farther from the objective and designed for hiding. Antennas, a hide site's largest signature, should remain up only when in use. Otherwise, they are lowered or removed. One team member serves as lookout while another one constructs or adjusts the antennas. The lookout must be able to see the enemy before the enemy sees the Soldier working on the antennas. To prevent detection, team members minimize movement around the site. They use countertracking whenever someone moves around the site. They also minimize noise and light. Although the site is well camouflaged, noise and light can easily compromise the team's location.

4-101. The team keeps detailed communications and patrol logs throughout the mission. The communications log includes—

- Exact messages sent.
- Exact messages received.
- Antennas used or tried, and their configuration time(s) and location(s).

4-102. Before any element performs a shift change in the site, all information collected at the site by the outgoing shift is disseminated, to include—

- Message traffic.
- The status of the surveillance site.
- Any enemy activity on the objective.

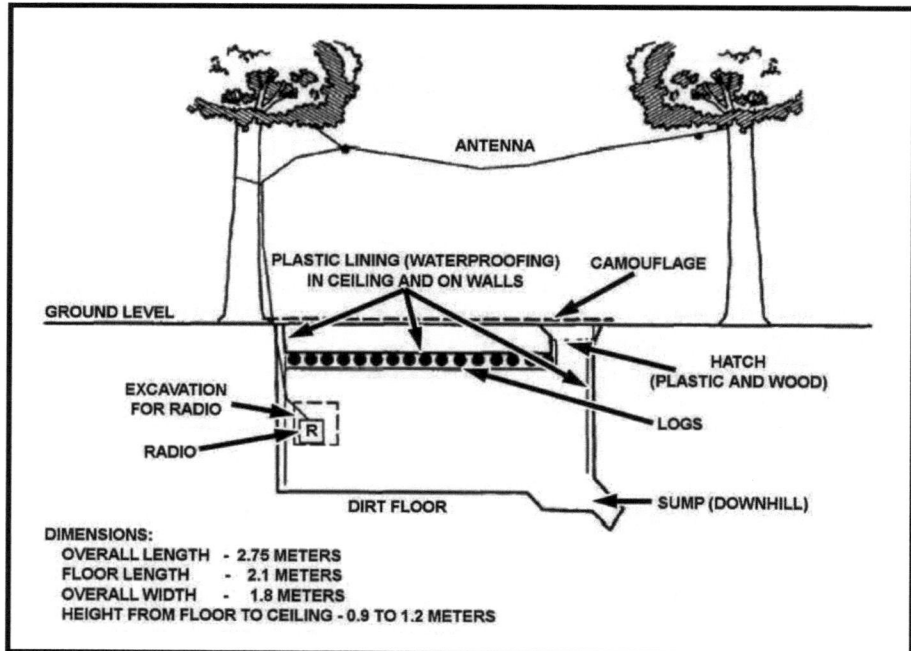

Figure 4-8. Example hide site.

SURVEILLANCE SITE

4-103. During periods of activity on the objective, all personnel should be awake and alert to aid in security and recording. One person cannot record and conduct surveillance at the same time during periods of increased activity. The team keeps detailed surveillance logs, to include sketches of the objective. The more detailed the information and sketches are, the more successful the mission will be. When the team recognizes PIR, intelligence requirements, SIR, ISR tasks and RFIs, they send it to the hide site promptly for transmission to the COB or AOB either immediately (PIR) or during a scheduled communications window. They can use either secure or nonsecure means, depending upon the equipment or time available. The messages must be complete and well written to reduce time needed for corrections or clarification. (Chapter 6 provides examples of communication orders formats.) The team pays particular attention to the amount of movement, noise and light discipline, and the reduction of any reflection that may come from using optics. The team must pack any equipment not in use so that they can evacuate the site quickly. As with the hide site, the team must completely disseminate all information collected before they change shifts. If the team plans to perform a breakout drill, they send a message to the hide site first, if possible.

LINKUP AND DISSEMINATION OF INFORMATION

4-104. When the decision is made to leave the objective, the team must perform a number of actions. The plan outlines the routes they will take, site breakdown procedures, and security during breakdown. After they break down the site, they thoroughly sterilize it to prevent the enemy from detecting their presence and knowing where to release tracker teams. The team must leave nothing behind that can give the enemy any information. They take extra care with their security. The team is in great danger from this point until the extraction is complete.

4-105. While the surveillance team moves out, the hide site team breaks down and sterilizes the site. They follow the same procedures and precautions as described for the surveillance site.

4-106. The team rehearses linkup procedures before the mission begins. During linkup, the team is at the greatest risk of fratricide. Everyone must understand all signals. Only one element may move into the linkup site at a time. Who moves first is decided during rehearsals. Communications and night vision capability assist in conducting linkups. After the linkup, the team disseminates all information gained in case they become separated again before extraction.

CONTINGENCIES

4-107. Due to the uncertainty of the situation, contingencies require plans, rapid response, and special procedures to ensure safety and readiness of personnel and equipment. The team must consider the following contingencies for the execution phase:

- Actions on enemy contact during insertion.
- Break in contact.
- Actions taken by the team if separated during insertion.
- Plan for priority of destruction of equipment.
- Rally or rendezvous plan to cover team during foot infiltration to objective, while on the objective, and during exfiltration.
- Plan for avoiding all known or suspected enemy forces, danger areas, or civilian concentrations.
- Security during movement, halts, cache, communications, and during hide or surveillance site construction.
- Cross-loading of equipment, sensitive items, and construction material.
- Lack of a communications plan (team internal and external to the COB).

- Actions on enemy contact.
- Linkup plan for both teams internally, and with other friendly forces.

HEAVY TEAM AND PLATOON OPERATIONS

4-108. Some METT-TC conditions may require a LRS team to have additional Soldiers attached or in support. A LRS team may require additional personnel to carry surveillance site construction materials and water. These additional Soldiers (three to six) are known as a "mule team." They accompany the team to the objective area, construct the subsurface surveillance site, and depart the objective area during limited visibility. The advantage to using these TTP is that the LRS team is rested when starting the surveillance mission.

4-109. METT-TC conditions may also allow for the use of a LRSD employed as a tactical platoon. The LRSD establishes a patrol base, and then emplaces multiple surveillance sites. The LRSD acts both as the sustainment base and as a quick reaction force in case a surveillance site is compromised. This technique has been very successfully employed in desert environments when the objective area is far from any support base and multiple surveillance sites are needed. Sustainment and rotation of surveillance site personnel occurs during the hours of limited visibility.

Section IV. COMBAT ASSESSMENT

Combat assessment is the determination of the overall effectiveness of force employment during military operations. Combat assessment has three major components: battle damage assessment (BDA), munitions effects assessment, and reattack recommendation (JP 3-60 and FM 1-02). However, LRS teams only participate in the BDA component. LRS teams are critical sensors for all fire-support platforms. At times, they may be the only source of targeting information available for high-value targets (HVTs). Keys to a successful combat assessment operation are using standardized report formats and sending the best possible information and images of the damage observed for trained analysts to evaluate.

DEFINITION AND PURPOSE

4-110. BDA is the timely and accurate estimate of damage resulting from the application of military force, either lethal or nonlethal, against a predetermined objective. BDA can be applied to the employment of all types of weapon systems (air, ground, naval, and special forces weapons systems) throughout the range of military operations. BDA is primarily an intelligence responsibility with required inputs and coordination from the operators (FM 1-02). BDA, in its most basic form, answers four questions quickly and accurately:

- Was the target hit?
- What was the extent of physical and functional damage?
- Were the commander's objectives achieved in full, in part, or not at all?
- Should the target be reattacked?

4-111. The LRS team can make a recommendation to reattack, but it is the targeting planners, not the LRS team, who decide whether the target should be reattacked.

CONSIDERATIONS

4-112. When assessing battle damage, the team should consider—

CONDITION OF THE TARGET

4-113. This includes the overall level of destruction or serviceability of the target and whether it was occupied or unoccupied, and whether the enemy was withdrawing or reinforcing.

CASUALTIES

4-114. This includes the number of wounded or dead.

EQUIPMENT DAMAGE OR SERVICEABILITY

4-115. This lists destroyed and neutralized vehicles and weapon systems, and identifies which need repair or recovery.

INSTALLATION DAMAGE OR SERVICEABILITY

4-116. This identifies damage levels and relative serviceability of runways, roads, buildings, lighting, bridges, power and phone lines, and any repair and damage-control activities.

MISSION PLANNING FACTORS

4-117. In addition to normal planning factors, BDA missions include—

- Task organization (forward observers or combat controllers).
- Personnel markings.
- Position markings.
- Communications equipment and frequencies.
- BDA criteria.
- Special equipment such as lasers and thermals.
- Pulse repetition codes.
- Authentication codes.
- Training in the use of special equipment.
- Rehearsals and precombat inspections.
- Withdrawal.

CHARACTERISTICS

4-118. Initial post-strike BDA reports are sent to the COB or AOB, who sends them to the controlling headquarters. The team is responsible for obtaining and reporting raw data about collateral, physical, and functional damage.

WEAPON AND ORDNANCE MALFUNCTIONS

4-119. Because the team observes and reports any observed battle damage, the team can also report the number and type of dud ordinance and any weapon malfunctions.

REPORTS

4-120. The team reports their BDA observations from the target area in either the SALUTER or INTREP format. They should include digital imagery, if available.

DAMAGE TYPES AND LEVELS

4-121. The team reports the extent of physical and functional damage sustained by each target. Physical damage assessments estimate the extent of physical damage resulting from the application of military force, that is, from munitions blasts, fragmentation, and the effects of fire. The team bases their assessment upon observed or interpreted damage.

Chapter 4

COLLATERAL DAMAGE

4-122. The team reports any collateral damage that occurs during BDA missions. Collateral damage is defined as unintentional or incidental damage to facilities or equipment, or casualties, that occur due to military actions directed against a targeted enemy force or facility.

PHYSICAL DAMAGE

4-123. Assessing physical damage is a judgment call. The key factors are the type and size of the target and warhead used, and the location of the detonation. The team also considers whether the enemy's use of camouflage, concealment, and deception reduced or increased the physical damage, because these factors could distort the assessment.

FUNCTIONAL DAMAGE

4-124. The team assesses the extent to which military force degraded or destroyed the ability of the targeted facility or objective to perform its intended mission. The level of success is based upon the operational objectives established against the target. Because assessment of functional damage is subjective, the team need not associate a confidence level with it. However, they can include an estimate of the time required for the recuperation or replacement of the target function.

- **No Functional Damage**--Target is undamaged or has sustained little or no damage to critical element(s), leaving normal functional capability intact. The target is fully operational or able to act; it is mission capable. This damage level does not require recuperation or replacement times.
- **Light Functional Damage**--The critical element(s) of the target has sustained damage, causing less than a 15 percent decrease in normal operational capability. This damage level requires an estimation of the time required for recuperation or replacement of the target function.
- **Moderate Functional Damage**--The critical element(s) of the target has sustained damage, causing a 15 to 45 percent decrease in normal operational capability. This damage level requires an estimation of the time required for recuperation or replacement of the target function.
- **Severe Functional Damage**--The critical element(s) of the target has sustained damage, causing greater than a 45 percent decrease in normal operational capability. This damage level requires an estimation of the time required for recuperation or replacement of the target function.
- **Functional Destruction**--The critical element(s) of the target has sustained damage rendering the entire target unusable for its original purpose. Target cannot support combat or production operations without repairing or replacing critical elements. This damage level requires an estimation of the time required for recuperation or replacement of the target function.
- **Abandoned**--Regardless of physical damage, this facility or equipment is not being used for its intended purpose. Target cannot support combat or production operations without being reoccupied, re-equipped, or both.
- **Unknown Functional Damage**--Although the critical element(s) of the target has been attacked, insufficient data exist to assess whether functional damage occurred.

BRIDGES

4-125. A bridge is designed to allow movement of personnel and equipment across an obstacle. Destruction of the bridge might not be required. For example, degrading the ability of the bridge to allow movement of vehicles might be enough to accomplish the mission.

PHYSICAL DAMAGE

4-126. When reporting physical damage, report the number of spans that are damaged and destroyed out of the total number of spans on the bridge. Note that the deck or floor of a railroad bridge can be solid or open track.

No Damage--Military action has caused no apparent damage.

Light Damage--The bridge has sustained superficial damage, but the roadway remains undamaged.

Moderate Damage--All spans remain intact, but one or more spans has sustained holes in the deck or floor. For pontoon bridges, one pontoon section has been sunk.

Severe Damage--All spans remain attached, but one or more spans has sustained 50 percent destruction to the deck or floor width. For pontoon bridges, two or more nonadjacent pontoon sections has been sunk.

Destruction--At least one span has been dropped. Piers or abutments might have sustained damage or they might have been destroyed. For pontoon bridges, two or more adjacent pontoon sections have been sunk.

FUNCTIONAL DAMAGE

4-127. Using the assessment of physical damage can help determine the level of functional damage.

Highway Bridge--When you report that a bridge has sustained moderate to severe damage, identify the number of lanes that remain open (on a highway bridge) and what how much traffic can still use the bridge.

Railroad Bridge--Moderate or greater physical damage to a railroad bridge generally renders it unusable.

Pontoon Bridge--Recuperation time may be short in duration for a pontoon bridge that has been destroyed due to the destruction of a few sections. Several factors such as the presence or availability of spare sections, repair capability, or both, can rapidly reverse the effects of an attack. Some might be repaired before the combat assessment report is completed.

Permanent Spanned Bridge--When a permanent bridge span sustains moderate to severe physical damage, assessing the extent of functional damage is difficult. The combat assessment analyst can seldom see under the bridge, but the damage there often exceeds the damage he can see on the deck. This is due to the delayed fuses used on most bridges.

BUILDINGS

4-128. A building is designed to environmentally shelter an enclosed function or equipment. Destruction of the building is not required. The point is to destroy the critical element(s) it houses.

TYPES

4-129. These include framed buildings, buildings with load-bearing walls, high multistory buildings, and buildings with multiple wings.

Framed Buildings--Framed structures (such as military headquarters, office buildings, and aircraft hangars) rarely collapse totally in an attack with conventional weapons. However, regardless of external and overall damage, the building's frame tends to remain intact. On the other hand, a steel or concrete frame need not collapse for the building to sustain damage at the levels previously described.

Buildings with Load Bearing Walls--In contrast to framed buildings, those with load-bearing walls, that is, walls that carry the weight of the floor and roof, generally sustain damage levels equal to the amount of building collapse, and the damaged elements generally include the load-bearing and non-load-bearing structural elements.

Chapter 4

High, Multistory Buildings--For buildings with more than four stories or with multiple sections (or wings), the team should report the level of damage for each story, section, or wing individually and that of the structure as a whole. For example, if a ten-story building receives severe damage to the upper three stories, they report the level of damage to those three floors plus the level of damage to the structure as a whole, which in this case might be assessed as moderate.

Buildings with Multiple Wings--For buildings with multiple wings, the team reports the destroyed wings and the damage to the remainder of the structure. For example, they might report that the North and South wings of a headquarters building have been destroyed, while the center section sustained only moderate damage.

PHYSICAL DAMAGE

4-130. Target-element-area damage includes damage to non-load-bearing elements such as facades and external sheathing, as well as broken windows and glass, blown-out curtain walls, and blown-out roof panels.

No Damage--Military action has caused no apparent damage.

Light Damage--At this level of damage, the target-element area has sustained up to 15 percent damage.

Moderate Damage--At this level of damage, the target-element area has sustained from 15 to 45 percent damage.

Severe Damage--At this level of damage, the target-element area has sustained from 45 to 75 percent damage.

Destruction--At this level of damage, the target-element area has sustained from 75 to 100 percent damage.

FUNCTIONAL DAMAGE

4-131. The greater the extent of physical damage to the building, the greater the likelihood that the critical element(s) within is damaged, hence the building requires a longer recuperation time to restore the function.

Location and Hardness--Although the level of physical damage to a building and functional damage to its critical element(s) correlate somewhat, the location and hardness of a building's contents are the keys to a meaningful functional assessment. For example, in an industrial building, the machinery may be less vulnerable than the structure in which it is contained. The structure might be moderately damaged, while the machinery it houses sustains little or no damage. On the other hand, fragile computer or other electronic equipment might be destroyed while the building that houses it sustains far less functional damage.

Contents--Framed structures tend to show less apparent physical damage and is less likely to collapse than other types of buildings. Thus, determining the functional damage to the contents of a framed building is more difficult than assessing those of a wall-bearing structure. These tend to show more physical damage, and they collapse more readily, causing greater functional damage to their contents than do framed structures.

Recuperation--The team reports recuperation for both the structure and for the critical elements.

Structural Damage as Unusability Criteria--General weaponeering guidance considers a building unusable (functionally destroyed) when it has sustained 50 percent structural damage. Depending on the type and location of critical elements, a lesser percentage of damage may be adequate to achieve the desired level of functional degradation.

Landmarks and Symbols--A building may also serve as an important landmark or other symbol of national unity and resolve; in these cases, the entire building may be the critical element.

BUNKERS

4-132. Bunkers were formerly called hardened facilities. It can be very difficult for LRS teams to provide accurate BDA on bunkers unless they are able to conduct a physical inspection.

PHYSICAL DAMAGE

4-133. Target-element-area damage includes damage to non-load-bearing elements such as facades and external sheathing, as well as to broken windows and glass, blown-out curtain walls, and blown-out roof panels.

No Damage--Military action has caused no apparent damage.

Light Damage--At this level of damage, no weapon penetration has occurred, but exterior damage is apparent.

Moderate Damage--At this level of damage, weapons have obviously penetrated the bunker.

Severe Damage--At this level of damage, part, but less than one-third of, the bunker roof or sidewalls has collapsed.

Destruction--At this level of damage, more than one-third of the bunker roof and sidewalls have collapsed.

FUNCTIONAL DAMAGE

4-134. Assessing damage to all types of hardened structures requires analysis of aircraft cockpit video (ACV) and a search for blown-off entrance doors, burn marks outside entrances, or smoke from fire or secondary explosions. The results of this analysis must be compared to information about the internal configuration of the bunker to determine the approximate location of the weapon detonation.

4-135. A single weapon is unlikely to collapse or partially destroy a large bunker built with lots of thick concrete, burster slabs, and soil layers. However, a big weapon detonation inside the bunker generally destroys the contents. Knowledge of bunker construction such as dimensions, wall placement, and thickness of roofs, floors, or walls is required to accurately assess the extent of internal physical and functional damage.

4-136. Functional damage to a bunker depends on its mission. If internal compartmentalization allows, a round that penetrates the bunker has a good chance of damaging or destroying sensitive contents such as aircraft or munitions. If the damage is not too great, the contents can be moved. In these situations, depending on the level of physical damage, the contents can be removed and the bunker can be reconstituted to reuse for protective storage of other equipment or supplies.

4-137. A successful weapon penetration and detonation generally damages or destroys both mission and operations in a bunker serving in a production or C2 role. In any of these situations, the extent of functional damage depends on estimates of physical damage to the internal structure, ventilation system, to electronic or communications equipment, and to power supplies, lights, water lines, tools, and equipment, for example. Generally, long recuperation times are associated with this type of internal damage. As with buildings, when reporting recuperation, the team reports both the structure's recuperation and the recuperation of the critical elements.

DAMS AND LOCKS

4-138. Dams and locks have one function: to contain water on the upstream side. Military action can cause--

Chapter 4

NO DAMAGE

4-139. If no damage occurs, no loss of functionality occurs.

DAMAGE

4-140. A breach, break, or puncture in the face of the lock or dam affects functionality immediately in the form of a leak, whose size depends on the amount of damage.

DESTRUCTION

4-141. Loss of the lock or dam causes an immediate flood, which is total functional failure.

DISTILLATION TOWERS

4-142. Distillation tower targets include the tower and all associated equipment.

PHYSICAL DAMAGE

4-143. When reporting physical damage to a specific tower, also report damage level of equipment directly associated with the tower. This equipment usually includes one or more furnaces, heat exchangers, or condensers; and elevated pipe ways. If possible, also report damage level of the control building associated with the distillation tower.

No Damage

4-144. Military action has caused no apparent damage.

Light Damage

4-145. Military action has caused no apparent penetration of tower shell or disruption to piping connections. Portions of the insulation covering the tower shell appear damaged or scorched.

Moderate Damage

4-146. Military action has left the tower shell standing, but has penetrated the tower or deformed or severed piping connections.

Destruction

4-147. Military action has at least partially collapsed or toppled the tower.

FUNCTIONAL DAMAGE

4-148. The effects of damaging a distillation tower on the target's production capabilities depend on the specific functions of the towers such as primary distillation or secondary processing. The team reports functional damage of a distillation tower in terms of the time required to repair or replace it and the specific production capabilities denied in the meantime. The team must also report damage to equipment directly associated with a distillation tower, because the results could compare to significant damage to the tower.

MILITARY EQUIPMENT

4-149. This applies to equipment whether deployed or in depot:

DEFINITIONS

- Armored vehicles include tanks and armored personal carriers.
- Artillery includes field and antiaircraft artillery systems, both towed and self propelled.
- Trucks include all types of nonarmored vehicles, whether used for land transportation, and C2.

- Locomotives and rolling stock include all types of rail transportation.
- Aircraft include all types of fixed- and rotary-wing aircraft.
- Rockets include single-round and multiround rockets and their associated launchers.
- Missiles include surface-to-surface and surface-to-air (fixed and mobile) missiles and their associated launchers.
- Radar antennas include those that stand alone or are attached to a van or trailer. Radars may or may not be associated with a missile site.
- Fire-control components include all vans or trailers (radar, guidance, power and computer) associated with SSM, SAM, and AAA sites.

PHYSICAL DAMAGE

4-150. The team must consider their observations carefully before they report "No damage" to military equipment. They might not observe some of the physical deformations that happened to it. To determine damage level, they must analyze as many sources and types of information as they can observe; for example, a complete lack of either vehicular movement or radio transmissions for an extended period of time. Then, when they are ready to report physical damage to military equipment, they must report the total number of each type of equipment observed, the number of pieces of equipment damaged, and the number destroyed.

No Damage

4-151. Military action has caused no apparent damage.

Damage

4-152. Military action has caused physical deformations to equipment such as holes, exterior scorch marks, or broken or missing exterior equipment or components such as broken tracks or wheel or missing armored plates. However, major components remain intact.

Destruction

4-153. Military action has left the equipment unrepairable, possibly scrappable. This qualifies as catastrophic damage (K-Kill).

FUNCTIONAL DAMAGE

4-154. The level of functional damage of a missile or radar site depends upon the extent of damage, the number of critical elements and their individual levels of damage, and the interconnectivity of the various elements that make up the site. Visible damage might have little or no effect on equipment functionality. Functional damage of equipment includes any damage that partly or completely reduces--

- The ability of the C2 nodes to effectively operate.
- The ability of the logistics nodes to--
    -- Fuel,
    -- Arm,
    -- Fix,
    -- Transport,
    -- Operate, or
    -- Protect.

- The ability of the engineering resources to provide--
  -- Mobility,
  -- Countermobility, and
  -- Survivability support.

TYPES OF EQUIPMENT

4-155. Types of equipment whose reduction in capability can affect the functioning of a site or element include--

Armored Vehicles and Artillery--Functional damage is an elimination of firepower capability (F-kill), prevention of mobility (M-kill), or both, which the crew cannot repair on the battlefield.

Trucks--Functional damage is a reduction in mobility (M-kill) or in ability to use the truck's internal equipment for a number of hours until the crew can repair the equipment or vehicle.

Locomotives and Rolling Stock--Functional damage is prevention of mobility (M-kill) for a number of hours until the crew can repair the locomotive or rolling stock cars. Functional damage can also include damage or destruction of materials within the cars.

Aircraft--Functional damage prevents takeoff (PTO-kill) for a number of hours until the crew can repair it.

Rocket, Missile, or Launcher--Functional damage prevents successful or effective firing of the weapon (F-kill). The crew cannot repair this damage on the battlefield.

Radar Antenna or Its Van or Trailer--Functional damage prevents a radar system from acquiring, firing, or tracking missiles (F-kill) until the system can be repaired.

GROUND FORCE PERSONNEL

4-156. Damaging or destroying an occupied position such as a bunker, trench, or other structure or a vehicle such as a personnel carrier or truck usually causes casualties. FM 6-30 states that for indirect fire, thirty-percent casualties or materiel damage inflicted during a short time span normally renders a unit ineffective. However, a commander will stipulate the desired effects and percentages required for success against specific target categories.

PHYSICAL DAMAGE

4-157. In addition to reporting physical damage levels, the team should estimate the total percentage of the ground force destroyed. For equipment, see the damage definitions provided for military equipment.

No Damage

4-158. Military action has caused no apparent damage.

Damage

4-159. Military action has caused up to 30 percent casualties to visible personnel or to occupied positions or organic equipment.

Destruction

4-160. Military action has caused more than 30 percent casualties to visible personnel or to occupied positions or organic equipment.

FUNCTIONAL DAMAGE

4-161. The attrition of ground forces is influenced by factors in the domains of battle: physical (personnel, weapons systems, and sustainment), cybernetic (C2), morale (will to fight), training, and leadership. Generally, the greater the personnel casualties and damage to their equipment and communications and supply networks, the greater the attrition of ground forces and the lower their combat effectiveness. Desertions or POW losses may also render a unit ineffective. As part of determining enemy combat effectiveness (ability to function), two factors must be addressed in clear and simple terms:

- Reconstitution of forces and recuperation of facilities.
- Residual capabilities to perform defense, assault, and supply missions.

STORAGE TANKS FOR PETROLEUM, OIL, LUBRICANTS

4-162. Although a POL tank may sustain damage, its contents may be retrievable and usable.

PHYSICAL DAMAGE

No Damage

4-163. Military action has caused no apparent damage.

Light to Moderate Damage, Aboveground Tanks

4-164. Military action has punctured top walls, sidewalls, or both; possibly spilled contents; caused no evidence of sustained fire; left structural integrity intact.

Light to Moderate Damage, Partly or Completely Underground Tanks

4-165. Round has penetrated tank, but no secondary explosion or sustained fire has occurred.

Destruction

4-166. Military action has caused at least partial collapse or buckling of side wall; or, a secondary explosion or a sustained fire has occurred, or both.

FUNCTIONAL DAMAGE

4-167. Significant functional damage of a POL storage installation is expressed in terms of storage capacity rendered unusable and time required to repair or replace this denied capacity.

POWER PLANT TURBINES AND GENERATORS

4-168. Power plant turbines and generators may be housed in separate structures or together in a single structure called a "generator hall." Physical damage to the turbine or generator units can be difficult to identify if the generator hall remains relatively intact. Therefore, damage estimates to the units are based upon the location of the weapon detonation and on the physical damage to the building itself. The closer to the floor a weapon detonates, the greater the probability of unit damage. The extent and location of structural damage, as opposed to roof-panel damage, is another indicator of unit damage--the greater the wall damage and structural collapse, the greater the likelihood that the unit(s) is damaged or destroyed under the rubble.

PHYSICAL DAMAGE

4-169. When reporting physical damage, the team reports the number of turbines or generators that are damaged and destroyed out of the total number of units at the facility. When performing combat assessment on a generator hall, they report physical damage to both the building (see previous discussion), and they estimate damage to the turbines or generators located inside.

Chapter 4

No Damage

4-170. Military action has caused no apparent damage.

Damage

4-171. Military action has caused no apparent weapon penetration of unit, but the environmental housing over the unit has sustained damage and is disfigured. The unit may also have been displaced from its foundation.

Destruction

4-172. Military action has breached or penetrated the turbine or generator unit, causing extensive structural deformation, or completely tore the unit apart. This is a catastrophic kill (K-kill).

FUNCTIONAL DAMAGE

4-173. Power plants with free-standing, gas-turbine-generator units can operate independently of each other. These units are housed in light metal structures that provide environmental protection only. Destruction of one unit of these units only partly degrades electrical production. Also, because turbines and generator units exemplify machines that are less vulnerable to damage than their housing, moderately damaging a generator hall can have little or no effect on the units it houses, and thus on their functioning. When reporting recuperation, the team reports recuperation both of the structure and of the turbines and generators.

RAIL LINES AND RAIL YARDS

4-174. Recuperation time for destroyed rail yards may be short in duration, because new rails, repair equipment, and repair personnel might already be onsite or readily available.

PHYSICAL DAMAGE

4-175. These definitions also indicate the rail yard's functional damage.

No Damage

4-176. Military action has caused no apparent damage.

Cut

4-177. Military action has cratered one or more tracks, prohibiting movement around the damaged area, although movement around or past the damaged area (choke point or rail yard) is still possible on undamaged tracks.

Destruction

4-178. Military action has caused multiple cuts to multiple tracks, which keeps rolling stock from moving around or past the damaged area (choke point or rail yard).

FUNCTIONAL DAMAGE

4-179. The location(s) of rail yard "cuts" and the ability of the yard to bypass the damage determine the extent of functional damage to the rail yard.

ROADS

4-180. Where geographically possible, an alternate to damaging a road with crater(s) is to attack the adjacent hillside to cause a landslide to cover the road.

PHYSICAL DAMAGE

No Damage

4-181. Military action has caused no apparent damage.

Cratered

4-182. Military action has cratered the road, but vehicles can maneuver around the damaged section.

4-183. Cut

4-184. Military action has caused so many aligned and close-set craters that vehicles cannot pass.

FUNCTIONAL DAMAGE

4-185. The effectiveness of attacks on roads depends on reducing or stopping traffic flow. Estimating flow reduction and road repair requirements are based on whether and what vehicles the adjacent terrain allows to bypass the damaged road section, on the depth and width of the cratered area, on the availability of repair equipment and personnel (usually readily available), and so on.

RUNWAYS AND TAXIWAYS

4-186. To successfully assess runway or taxiway damage, the team must know the takeoff and landing capabilities of the aircraft located at the airfield. They must also know what type or category of aircraft can or cannot use the airfield. A fighter or bomber base may be considered interdicted if damage prevents normal operation of the aircraft stationed there. However, the airfield may be usable by other aircraft types that can operate on an unimproved runway. The team can refer to the appropriate aircraft documents for specific aircraft minimum clear (takeoff) length (MCL) and minimum clear takeoff width (MCW) dimensions. The team also assesses nearby roads for possible aircraft use.

PHYSICAL DAMAGE

4-187. An assessment of physical damage to runways and taxiways implies its functional capability.

No Damage

4-188. Military action has caused no apparent damage.

Cratered

4-189. Military action has cratered runways or taxiways, but aircraft can maneuver around them.

Cut

4-190. Military action has caused multiple craters in line and close enough together to prohibit aircraft movement around them. However, operations can occur beyond the cut.

Interdicted

4-191. Military action has caused multiple cuts close enough together to prevent any takeoff or landing operations, either between the cuts, or between the last cut and the runway overrun.

FUNCTIONAL DAMAGE

4-192. The effectiveness of an attack on a runway or taxiway depends on whether surface cratering prevents aircraft takeoff or landing.

SATELLITE DISHES

4-193. Before assessing damage to a satellite dish (es), the team must know the dish type (fixed or tracking) and the location of the damage.

PHYSICAL DAMAGE

No Damage

4-194. Military action has caused no apparent damage.

Light Damage

4-195. Military action has blown off a few reflective panels.

Moderate Damage

4-196. Military action has blown off less than 25 percent of the reflective panels and either damaged the dish support structure or the feed horn, or both.

Severe Damage

4-197. Military action has blown off between 25 and 60 percent of the reflective panels, changed the antenna point, and either slightly deformed the dish or damaged its structural components, or both.

Destruction

4-198. Military action has blown off more than 60 percent of the reflective panels, destroyed the feed horn, extensively deformed the dish, or knocked the dish off its base, or any combination of these.

FUNCTIONAL DAMAGE

4-199. Functional degradation to sites depends on damage to the dish or its associated control building(s), or both.

SHIPS

4-200. The types and locations of damage determine the ship's ability to continue offensive and defensive operations as well as its need to return to the shipyard for repairs.

PHYSICAL DAMAGE

4-201. The team must consider certain factors before determining the level of physical damage.

Factors

- Seaworthiness--Is the ship listing, capsized, or sunk?
- Firepower--What are the degrees of damage to the ship's guns, launchers, and magazines?
  - Surface-to-air guns.
  - Surface-to-surface guns.
  - Antisubmarine guns.
- Flight deck.
- Hangars.
- Aircraft elevators.
- Mobility--To what degree is the rudder (steering) damaged? How much does this degrade the ship's sustained speed capability?
- Sensors--To what degree is the ship's search equipment damaged and capability reduced (air, surface, and subsurface)? This assessment considers radar, sonar, and fire-control means.
- Command, Control, and Communications--What percentage, type, and level of damage was inflicted on the pilot house, the bridge, the combat information center, the communications center, antennas, computer systems, and data links? What is the reconstitution time for each?

Levels

No Damage--Military action has caused no apparent damage.

Moderate Damage--Military action has caused physical deformation, holes in the ship or its equipment, reduced the ship's ability to move or maneuver, or any combination of these.

Severe Damage--Military action has destroyed or burned more than one-third of the superstructure or deck area, rendered major subsystems (weapon sensors, radar) inoperable, destroyed the ship's ability to move or maneuver, or any combination of these.

Destruction--Military action has flooded more than one-third of the ship's waterline length. This indicates that the ship is experiencing uncontrolled flooding, and is sinking. In addition, the ship's major subsystem that supports operations is destroyed.

FUNCTIONAL DAMAGE

4-202. When assessing functional damage, the team considers the ship's ability or inability to move and maneuver and the degree of disruption to particular ship subsystems such as its weapon-delivery capability, the functioning of its sensors, and so on.

STEEL TOWERS

4-203. Steel towers transmit electric power and support communications antennas, for example.

PHYSICAL DAMAGE

No Damage--Military action has caused no apparent damage.

Damage--Military action has damaged supports, but tower remains standing.

Destruction--Military action has caused tower to collapse or topple.

FUNCTIONAL DAMAGE

4-204. The level of functional damage associated with a physically damaged steel tower depends on the tower's function and on its connectivity with other target elements.

TRANSFORMERS

4-205. A transformer is a static electrical device that uses mutual electromagnetic induction to convert AC power from one current on one circuit to a different current on another circuit. The team must report the extent of external damage, if any, and the expected effects.

PHYSICAL DAMAGE

4-206. When reporting physical damage, include the total number of transformers, and the number damaged or destroyed.

No Damage--Military action has caused no apparent damage.

Damage--Military action has left the structure of the unit intact, but blackened as a result of a fire or of leakage of oil.

Destruction--Military action has torn the structure apart or greatly distorted it. This is considered catastrophic damage (K kill).

FUNCTIONAL DAMAGE

4-207. The effect of transformer damage on the target's function depends on the facility's power requirements and on the enemy's ability to reroute the power.

TUNNEL ENTRANCES OR PORTALS

4-208. Tunnels at best are dangerous places for people, but some are used only for storage. Tunnels are used for passage, operations, storage, or some combination of these. When reporting damage to tunnel entrances, the team should also include, when possible, the estimated volume and size of the debris or rubble pile that blocks the entrance. This can help in estimating clearing and recuperation times.

Chapter 4

PHYSICAL DAMAGE

4-209. Physical damage to a tunnel entrance generally makes it impassable and can reduce the protection afforded to anything or anyone in the tunnel during follow-on attacks.

No Damage--Military action has caused no apparent damage.

Light Damage--Military action has left the portal (the approach or entrance to the tunnel) intact, but craters and debris partly block access. The doors, if any, operate as before the attack.

Moderate Damage--Military action has left the portal intact, but entrance to the tunnel is completely cut off by craters or debris. The doors, if any, do not operate.

Severe Damage--Military action has partly collapsed the portal, and has completely blocked any entrance to the tunnel.

Destruction--Military action has completely collapsed the portal, rendering access to the tunnel impossible.

FUNCTIONAL DAMAGE

4-210. The extent of functional damage depends on the size of crater(s), the degree of portal collapse, or the amount of debris blocking the entrance. Degree of functional damage also depends on the purpose of the tunnel facility. For example, a storage tunnel is much more vulnerable to entrance damage than a C2 tunnel, which depends less on ingress and egress. The time required to repair tunnel entrances depends on the extent of damage and the availability of personnel and equipment. In addition, the tunnel might have many entrances. How many it has affects the level of functional damage. When reporting the functional damage to tunnel entrances, the team considers accessibility based on how many entrances remain useable.

TUNNEL FACILITY AIR VENTS

4-211. Air vents are vital for some tunnel facilities. They bring in fresh air and remove exhaust fumes and other noxious byproducts. Closing off these vents can sometimes prevent usage of the tunnel or facility altogether. The vents are less critical for facilities used for storage only, but more critical if they contain operating equipment and people.

PHYSICAL DAMAGE

No Damage--Military action has caused no apparent damage.

Damage--Military action has partly blocked the vent opening with craters and debris. The vent structure might not be damaged.

Destruction--Craters or debris completely block the vent opening.

FUNCTIONAL DAMAGE

4-212. Functional damage restricts or cuts off airflow through the vent(s) in the facility. Complete elimination of airflow to or through the facility might require the destruction of multiple air vents. When reporting the functional damage of the facility vents, the team must divide the number of damaged vents by the total number of vents to determine the overall percentage of airflow blockage in the facility.

Section V. TARGET ACQUISITION

Target acquisition is conducted by a combat patrol. The intent of a combat patrol is to make contact with the enemy, in contrast to a reconnaissance patrol where the intent is to avoid enemy contact. LRS teams seldom conduct combat patrols. However, the capability to conduct a target-acquisition mission is inherent within a LRS team. The team can serve as the initial eyes of a long-range targeting asset, by providing terminal guidance using appropriate communication and signal such as a beacon or mirror by marking the target with a laser. Normally, a LRS team is tasked to conduct surveillance or reconnaissance of an NAI. If a target of opportunity is observed, the LRS team can be retasked to engage the target. Under these circumstances, the team guides the munitions or aircraft onto the target, then moves out undetected.

COMBAT PATROL

4-213. The LRS team reorganizes into three elements: acquisition, communications, and security.

- When using a laser designator, the acquisition element has a two-Soldier laser team and a communications element.
- When adjusting artillery, close air, AC-130, or attack helicopters, the acquisition element and communications element can combine.

MISSION PLANNING FACTORS

4-214. Route planning and movement are the same for target acquisition as for a surveillance or reconnaissance mission. A pickup zone close to the planned acquisition point allows for quick removal from the area. The PZ must, at a minimum, support FRIES or SPIES operations, although air landing is preferred. All team members must have a method to illuminate their position to assist in fratricide avoidance. Teams need a positive ground-to-air communications means such as a VHF radio. All attack helicopters and USAF CAS aircraft have VHF capabilities.

- Review all procedures for controlling available fires prior to mission execution.
- During mission coordination, confirm the PAVE Penny codes for laser designators with a USAF or Army aircraft representative.
- Establish and confirm self-authenticators and code words for communicating with the USAF or Navy on a nonsecure net.
- Quickly assess bomb damage before withdrawing.
- Know the rules of engagement.
- Plan for the effect on and reaction of the local populace.
- Coordinate no-fire and restrictive-fire zone for weapons systems operators.

EMPLOYMENT OF LASER DESIGNATORS

4-215. Lasers aid in target identification, location of aim point, site selection, and site illumination.

Aim Point Location—The laser aids in location of aim point by revealing—

— Target reflectivity.

— Ordnance type.

— Method of delivery.

Beam divergence—Offset should not exceed 7 degrees from either side -the aircraft must stay within that cone.

Target Orientation—Laser energy reflects in an arc, but is strongest at the angle where it would reflect if the surface were a mirror. If the laser designator is perpendicular to a surface the reflection can be seen from all angles on the designated side, but can be detected best near the laser designator to target line.

Weather conditions—Visibility of less than three nautical miles restricts the operation, regardless of aircraft type.

Chapter 4

Site Selection—The aircraft must be able to approach within 10-to 60 degrees to the left or right of the team's location. While maximizing standoff, the team needs to ensure they have both optical and electrical LOS to the target. Observer to target distance should not exceed the capabilities of the laser designator being employed.

Site Illumination—The laser aids in illuminating the site.

Function Check—Conduct a laser function check, and recheck the laser codes based on the type of aircraft conducting the attack. Designate the target (paint) for not less than 5 seconds before the aircraft releases its ordnance. For guns or other nonguided ordnance, you can use the laser to identify targets for aircraft without the 5 second (paint).

FIRE SUPPORT

4-216. Fire support, particularly artillery and CAS, are excellent assets to use during target-acquisition missions or in support while breaking out of a site or away from contact. Leaders must fully integrate fire support into their plans. They must understand how fire support can assist or detract from the execution of their assigned missions. They must understand fire support limitations.

INDIRECT-FIRE SUPPORT CAPABILITIES

4-217. Indirect fire support capabilities follow:

- Quick response time.
- Adjustability of fire.
- Variety of munitions, including precision munitions.
- Multiple strike capability.
- All-weather capability.

INDIRECT-FIRE SUPPORT LIMITATIONS

4-218. Indirect fire support limitations follow:

- Range.
- Naval gunfire availability limited to areas with naval assets.

CLOSE AIR SUPPORT (USAF FIXED WING) CAPABILITIES

4-219. Close air support (CAS, or Army fixed wing) capabilities follow:

- Long range.
- Visual target engagement and adjustment of fires.

CLOSE AIR SUPPORT (USAF FIXED WING) LIMITATIONS

4-220. Close air support (CAS, or Army fixed wing) limitations follow:

- Limited time on target.
- Limited compatibility with team radio systems (depending on aircraft type).
- Limited munitions.
- Limited ability to operate in poor weather.

CLOSE COMBAT ATTACK (ARMY ROTARY WING) CAPABILITIES

4-221. Close combat attack (CCA, or Army rotary wing) capabilities follow:

- Medium range.
- Longer time on target than fixed wing assets.

- Visual target engagement and adjustment of fires.
- VHF radio capabilities.

CLOSE COMBAT ATTACK (ARMY ROTARY WING) LIMITATIONS

4-222. Close combat attack limitations follow:

- Limited ability to operate in poor weather.
- Limited munitions.
- Vulnerability to enemy ground fire.

FIRE PLANS

4-223. Teams plan targets on infiltration and exfiltration routes, LZs, DZs, PZs, routes to and from the objective, on the objective, between the objective and surveillance, and between the surveillance and hide sites (contiguous targets work well). Teams plan RFAs and NFAs as needed. Coordination is done using the team target list and coordination checklist. They consider the effects, the mission, the types of targets, and the methods of engagement.

Section VI. URBAN TERRAIN

The LRS teams can be very effective in an urban environment. Their ability to gather information and report timely information about the current situation is vital to the BFSB intelligence-collection plan. The LRS team can help in two ways during an urban operation: First, the commanders can use LRS teams extensively and effectively for surveillance. Second, he can use them on a limited basis for reconnaissance. Before committing a LRS team to a mission in urban terrain, he must consider all aspects of the mission and the environment. Specifically, he must consider the differences between LRSU support to offense, defense, stability and civil support operations.

SURVEILLANCE OPERATIONS

4-224. This is normally the primary mission a LRS team conducts in an urban environment. LRS teams are most often used to report information received along main supply routes to and from the urban environment. Depending on the size of the urban area and location of key buildings, the LRS team might report information on specific buildings, motor pools, and so on. During stability operations, a LRS team might surveil a specific target from inside the environment. Considerations include camouflage, observation, security, and support.

CAMOUFLAGE

4-225. LRS teams are proficient in camouflage techniques specific to urban environments such as window screening and false walls.

OBSERVATION

4-226. In built-up areas, windows provide readily accessible observation ports. However, care should be taken to prevent optics from protruding beyond the window. This is an obvious sign of a surveillance position. The team members must position themselves as far back in the room as possible to keep from being seen. To lower their silhouettes, they can support their positions with a table or sand bags. Another technique is to observe through a hole in the wall. When observing through the window, individuals should stand well back in the shadows. At all times, care must be taken to avoid allowing light to reflect off optics.

Chapter 4

SECURITY

4-227. From the time the team leaves the last secure point until exfiltration, security remains a constant and immediate concern. Teams plan constant and sustained security for every phase of the mission. In the security role, the team can use both active and passive security devices. Teams can stage objects in and around the site that will identify any presence. All security devices must be able to withstand scrutiny.

PROTECTION AND SUSTAINMENT SUPPORT

4-228. In an urbanized AO, the LRS team can achieve protection and sustainment in stay-behind or rollover missions.

RECONNAISSANCE OPERATIONS

4-229. This is the second mission that a LRS team can perform in an urban environment. Due to the increased chance of compromise in an urbanized environment, the commander should consider employing LRS teams in a very limited reconnaissance role. The LRS team plans reconnaissance missions on urbanized terrain in as much detail as possible. Considerations include—METT-TC, the type of mission, planning time, specialized equipment need to accomplish the mission, and specialized equipment weight and size.

PLANS

4-230. Urban areas are categorized in the following manner, by population size:

- Villages (population of 3,000 inhabitants or less).
- Towns (population of over 3,000 to 100,000 inhabitants and not part of a major urban complex).
- City (population over 100,000 to 1 million inhabitants).
- Metropolis (population over 1 million to 10 million inhabitants).
- Megalopolis (population over 10 million inhabitants).

RULES OF ENGAGEMENT

4-231. All LRS team members must receive a detailed briefing on the rules of engagement before deployment in an urban environment.

VISIBILITY

4-232. Weather, smoke, and dust always obscure visibility. Military operations can change urban terrain, shifting shadows and dead spaces around every time a building collapses or a new pile of rubble forms.

SIZE, LOCATION, AND HISTORY

4-233. Within the city, urban terrain differs based on size, location, and history.

Industrial Areas and Residential Sprawl

4-234. Residential areas have some houses or small dwellings with yards, gardens, trees and fences. Streets normally form rectangular or curving patterns. Industrial areas usually have low (one- to three-story) flat-roofed buildings. Most of these are factories or warehouses, and they are generally located on or along major rail and highway routes. Both types of terrain have many open areas.

Core Periphery—A core periphery has narrow streets (12 to 20 meters wide) and continuous fronts of brick- and heavy-walled-concrete buildings. Most buildings are about the same height, ranging between two and three stories in small towns and five to ten stories in large cities.

City Cores and Outlying High-Rise Areas—Typical city cores consist mostly of high-rise buildings that can vary greatly in height. More modern buildings often have more space between them than that do the buildings in older city cores. This open construction style is more prevalent in outlying high-rise areas than in city cores. Streets generally form rectangular patterns.

Commercial Ribbons—These rows of stores, shops, and miscellaneous structures are built on either side of the major (at least 25 meters wide) streets that run through built-up areas. These structures are uniformly two or three stories tall.

INSERTION AND EXTRACTION

4-235. Leaders must consider the distance of the insertion or extraction, and the training of all team members. They consider support assets, their own experience, and internal and external assets. All types of insertion and extraction means available in other environments are still viable in an urban environment; however, considerations for their use may be different. For example, the use of nonstandard tactical vehicles may be a good insertion and extraction platform in urban areas during stability operations. Subterranean corridors such as sewers, subways, underground tunnels, or drainage systems can be used with great effectiveness.

EVALUATION

4-236. When a LRS team evaluates urban terrain, it considers the following factors:

Observation

4-237. Buildings on the edge of a city generally offer better observation than those inside. There, tall buildings with numerous windows often offer the best observation, especially if the buildings have spaces between them.

Avenues of Approach

4-238. The best way to enter a building is from the top. Therefore, the most important avenue of approach is one that quickly leads to the top from fire escapes, drainpipes, or adjacent buildings.

Key Control Points

4-239. Key control points in a building include entrances, hallways, and stairs. Whoever controls these controls the building.

Doors and Fire Barriers

4-240. These are common in commercial buildings. They become obstacles when closed or secured. Furniture and appliances can also become obstacles.

Cover and Concealment

4-241. Buildings with brick walls and a few narrow windows balance cover and concealment. Roofs provide little protection—lower floors offer LRS teams better protection than do areas directly under the roof. Additionally, floor layouts with many small rooms offer more protection than those with larger rooms.

Intercity Distribution of Building Types

4-242. Leaders can generally determine the layout of a city by the distribution of the buildings within the city. In built-up areas, mass-construction buildings (modern apartments and hotels) are the most common structures. They comprise two-thirds of the total area, and they are usually constructed of brick. Steel and concrete-framed multistory buildings comprise the city's core area, its most valuable land. As centers of economic and political power, they have potentially great military significance. Open spaces, such as parks, athletic fields, and golf courses, comprise about 15 percent of the average city's area. Most of this 15 percent is suitable for air assault or Airborne operations. However, approaches to these areas may have obstacles such as tall buildings, trees and wires, and should be carefully considered during

Chapter 4

planning. Additionally, rooftops complement this 15 percent since many can take the weight of aircraft or the impact of men and equipment jumping on to them.

Sources of Information

4-243. Cities offer a wealth of useful information. This information is found in a variety of sources.

- Large-scale city maps.
- Diagrams of underground sewer, utility, transport, and other systems.
- Publicly available information about key public buildings.
- Rosters of key personnel.
- US government studies and data bases that detail—
  — Size and density of the population.
  — Police and security capabilities.
  — Civil defense and air-raid shelters.
  — Fire-fighting capabilities.
  — Utility systems.
  — Medical facilities.
  — Mass-communication facilities.

Equipment

4-244. Some of the items a Soldier might consider carrying into an urban environment include—

- Camera.
- Communications equipment with various antennas.
- Spotting scope with stand.
- Binoculars.
- Dark cloth.
- Tape.
- Glass cutter.
- Complete cleaning kit.
- Multipurpose knife.
- Suppressed pistol.
- Notebook.
- Pencils.
- Tape recorder.
- Sleeping pad.
- Wasp and hornet spray.
- Bungee cord(s).
- Small saw.
- Crowbar.

COMMUNICATIONS

4-245. Probably the most important consideration in planning urban communications is type of antenna placement.

Subsurface Surveillance Site

4-246. A team can split into surveillance and hide sites, with the hide site located outside of the urban area. If so, team members at the hide site should be able to use their HF, UHF TACSAT, and LOS systems normally. However, when the surveillance site operates subsurface, they need an antenna for LOS communications with the hide site. Depending on the situation, the team members at the surveillance site might be able to use the whip antenna that is normally issued with their inter team radio. When the sites are split, the surveillance site can use any one of several methods to establish LOS communications with the hide site.

4-247. Sometimes, the team will have to use remote equipment to communicate with the hide site. Before the mission, they should try to obtain some field-expedient materials.

4-248. If the team is configured in a combined surveillance and hide site, and the entire element is in a subsurface environment, then the team must remote the long-range antenna system to the surface area.

4-249. To make contact with the COB or AOB, the RTO might be able to attach the HF antenna wire to some nearby metallic object that protrudes above the surface such as a light, a fence, or a storm drain gate. The RTO must remember to place the antenna system towards the receiving station. If using UHF TACSAT, the antenna must be placed so that no obstructions stand between it and the satellite.

4-250. Operating from a subsurface situation can be highly risky, because the enemy may be able to detect the antenna.

4-251. Communications are easier to achieve from an elevated position such as a building or other structure. In most cases, depending on distance, the surveillance team can use the whip antennas organic to their radio system.

4-252. If a structure obstructs the LOS view between the two sites (surveillance and hide), then the RTO can make a closed-loop antenna. He fastens Claymore wire to an interior wall in a loop from the socket of the radio's whip antenna to the ground for the radio. He must remember to cut the antenna to at least one full wave-length of the frequency he is using. This is an excellent antenna to use during urban area missions.

4-253. VHF antennas are much shorter than HF antennas. In fact, the antennas used for HF communications can also be used for VHF, except that they must be scaled down for higher frequencies. However, constructing and placing them is much the same as it is for the LOS.

4-254. Other items that can be used for HF communications include—.

- Existing antennas on the structure.
- Existing electrical wiring (team has to test for conductivity).
- Metal plumbing pipes.
- Ceiling grids.
- Metal clothes lines.
- Metal building frames (if power to the building is off, otherwise damage to the radio could occur).

Chapter 4

> **WARNING**
>
> Never construct antennas less than twice their length from power lines and transformers.

HIDE AND SURVEILLANCE SITES

4-255. When selecting a suitable site, Soldiers tend to go for height. In an urban operation, this can be a mistake. The greater the height attained, the more the LRS Soldier has to look out over an area and away from his immediate surroundings. For example, to see the road below a tenth-floor surveillance site, the LRS Soldier must lean out of the window, which reveals his location and exposes him to fire. Though the observer cannot predict where incidents will occur, he can expect that the ranges will be relatively short distances. A surveillance site must cover its surroundings as well as middle and far distances. In urban areas, this is rarely possible: sites are often forced off ground-floor levels by passing pedestrians. However, generally, the team should avoid going above the second floor.

CONSIDERATIONS

4-256. When considering possible hide or surveillance positions, the team should consider old, derelict buildings as they are unlikely to be reoccupied by civilians in the area. Abandoned or unoccupied houses or buildings also offer good sites, but could be booby-trapped or be reoccupied by civilians. The team must search the building after they have occupied it. Buildings that provide protection from weather and small arms are preferred. They should avoid isolated buildings as they tend to be obvious observation positions. After carefully observing the inhabitants' daily routines, the team can occupy private residences. They can occupy the home and establish hides or surveillance sites in the basement or attic, or both, but the homeowner will pose a danger and the team should generally limit their stay to 12 to 24 hours. The team can use shops with empty accommodations on a second floor, but again should limit their stay to 12 to 24 hours.

CONSTRUCTION

4-257. During the reconnaissance phase, the team plans the construction of an urban hide or surveillance site in detail. They must prepare a view aperture, a viewing platform (if needed), and the interior layout. Selection of the viewing aperture takes priority over construction of the viewing platform or any interior work. When construction begins, the team must pull local security to warn of any excess noise or act as early warning. If they have floor plans of the building or house during their planning phase, the team can rehearse the construction and occupation of the site. Before constructing a hide or surveillance site, the team records what the area looks like so that they can return it to normal before departing.

CAMOUFLAGE

4-258. To survive in an urban environment, LRS Soldiers must supplement cover and concealment with camouflage. To properly camouflage their positions, they must study the terrain in the surrounding area. The site must blend in with the terrain. For instance, in an undamaged building, they should not make a loophole for observation. They should use only the materials that they need-excess materials can reveal their position. They should also consider—

Use of Shadows—Buildings in urban areas throw sharp shadows.

Color and Texture—The team needs to break up the silhouettes of their individual equipment. They can use burlap or canvas strips for this. The predominant colors are normally brown, tan, and gray. The team should evaluate the camouflage they need for each location separately.

Section VII. IMAGERY COLLECTION AND TRANSMISSION

The timely and accurate collection and transmission of imagery, video or hand drawn pictures of the objective, plays a key role in the success of follow-on missions. The purpose of imagery collection and transmission is to provide an accurate description of enemy strengths, positions and capabilities so that the commander can make informed decisions. With the introduction of new technology and integration between digital cameras, global positioning system (GPS) devices, laser range-finding equipment and other electronic devices, it is now easier and of more tactical value to take digital imagery of objectives. The use of a digital camera with laser range-finding equipment and GPS helps produce clear, real-time information.

IMAGERY LABELS

4-259. All personnel on the team should be proficient at labeling images according to the unit SOP.

PRINCIPLES

4-260. General principles of labeling imagery include (Figure 4-9 and Figure 4-10) —

- Placement of letters or numbers on the image to correspond with the legend.
- Placing appropriate arrows and other graphics where needed.
- Ensuring the image has a title, including a name, DTG, and grid, for example,
- North-seeking arrow.
- Size of the picture (from East to West, North to South).
- Remarks page for details about what each number or letter means.

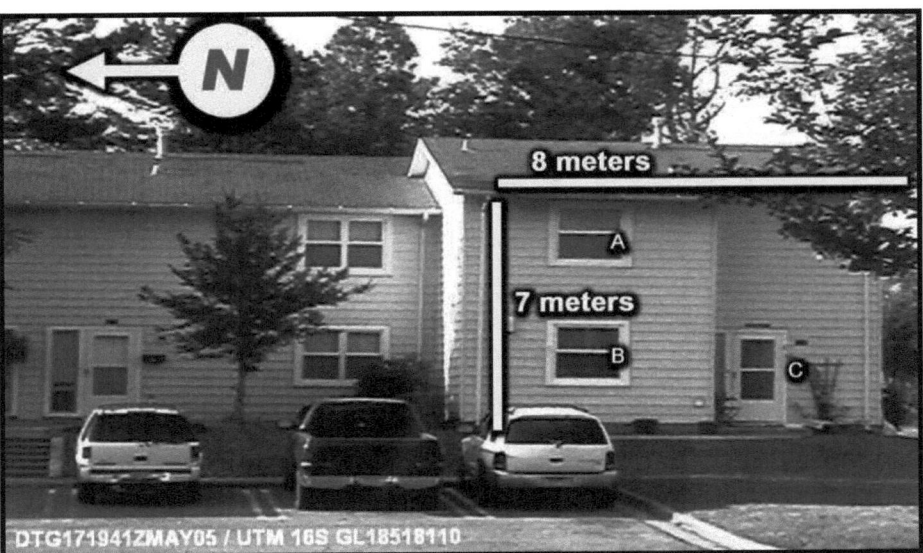

Figure 4-9. Example imagery labels.

Chapter 4

```
1. Date Time Group (DTG): Local, Zulu, Etc.
2. Location: UTM or GEOCOORD
3. Orientation: Azimuth to Target (Magnetic)
4. Distance: Straight line to target / Altitude (mts or ft)
5. Type: Vertical (V), Horizontal (H), Low Oblique (LO), High Oblique (HO), Panoramic (P)

FORMAT: DTG / LOC / ORIENT / DISTANCE / TYPE

Mark Photo with North seeking arrow
```

Figure 4-10. Example imagery legend.

IMAGE-GATHERING EQUIPMENT

4-261. There are many types of cameras and video cameras available. The majority of this equipment becomes dated quickly, so it is important to stay proficient in the equipment available.

CAMERAS

4-262. Several different types of cameras may be used to gather imagery from an objective. Some basic principles should be adhered to when considering what type and or kind to use—

- Durable.
- Waterproof.
- Number of images at a particular resolution.
- Film type, memory stick, or compact flash card.
- Weight and size measurements.

VIDEO CAMERAS

4-263. The introduction of the video camera recorder (a camera capable of taking video and still images) has aided the R&S community in its ability to record information. Considerations for video camera use are—

- Zoom.
- Quality and resolution.
- Battery life.
- Power converter for military batteries (BA-5590).

COMPRESSION SOFTWARE

4-264. The use of compression software can greatly enhance the effectiveness of a unit when it comes to gathering and transmitting imagery and video. Compression software allows compression of files up to 88% smaller then the original file. Being able to compress files aids in decreasing the amount of time required to transmit.

OBJECTIVE SKETCH

4-265. In the case of loss of communication or enemy compromise, an objective sketch may be the only piece of information about that objective available. The ability to draft a proper objective sketch is an extremely important skill. It is important to understand the basics of objective sketch production. Understanding the basics allows a novice or non-artist to draw an understandable objective sketch.

Types

Panoramic Sketches

4-266. Each of these represents an area or object, and is drawn to scale from the observer's perspective. It provides a useful way to record details about a specific area or structure (Figure 4-11).

Topographic Sketches

4-267. Each of these represents a large area drawn to scale as seen from above (bird's eye view). It shows reliable distances and azimuths between major features. A topographic sketch can also be used as an overlay on a range card (Figure 4-12).

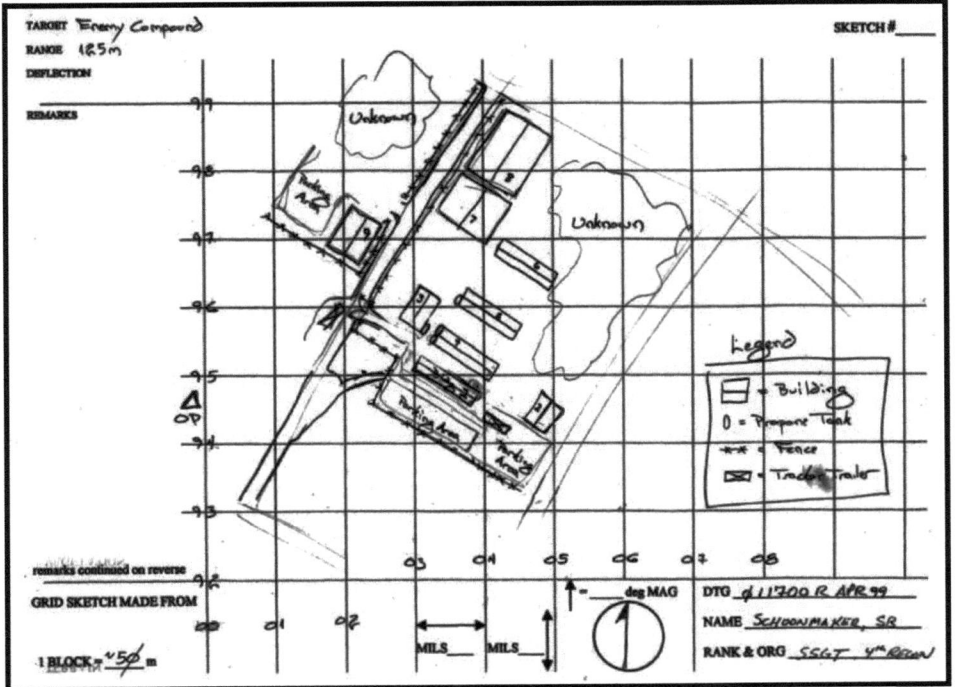

Figure 4-11. Example panoramic sketch.

Chapter 4

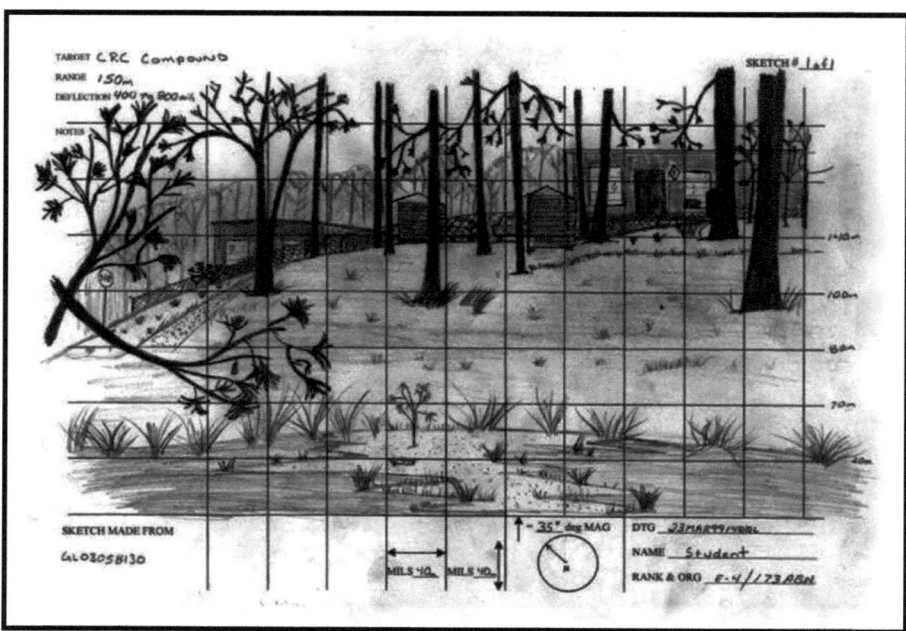

Figure 4-12. Example topographic sketch.

Objective Sketch

4-268.   Each of these, most easily drawn on a printed objective sketchpad, shows all useful information. The pad includes reminders of what specific information should be placed on the objective sketch. The sketch pad includes marginal information and remarks (Figure 4-13).

Team Operations

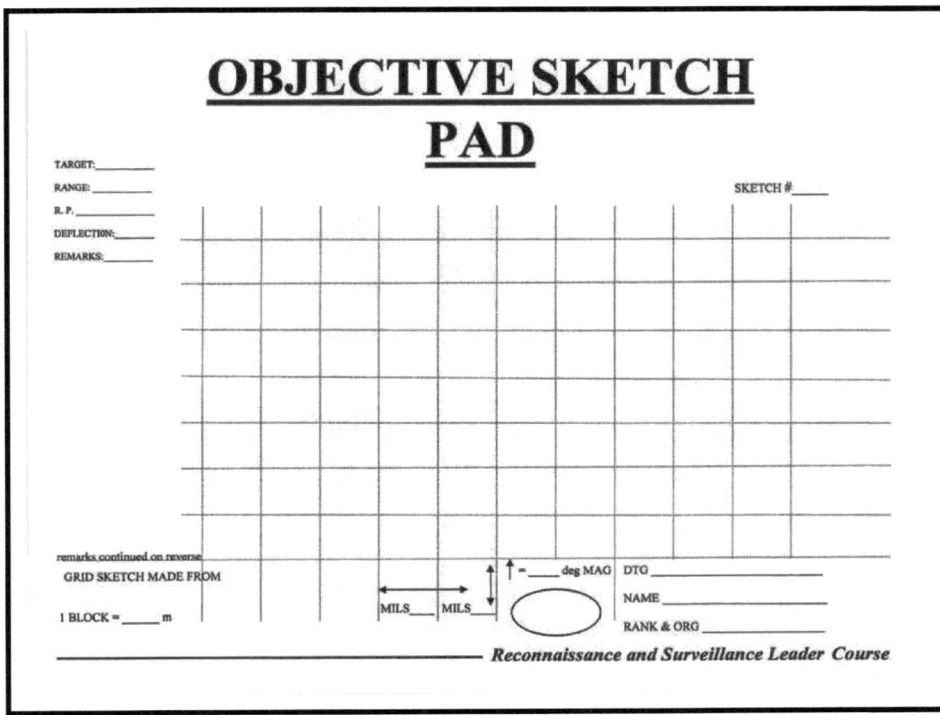

Figure 4-13. Example objective sketchpad.

Marginal Information

- Target number.
- Range from the preparer to the objective.
- Reference point, that is, the point of origin.
- Deflection angle from the preparer to the objective.
- Grid coordinates for the objective.
- Block scale, that is, the size of the blocks printed or drawn on the sketchpad.
- Mils latitude and longitude (for air strikes using fixed wing aircraft).
- Magnetic azimuth from the preparer to the objective.
- Sketch number (if more than one sketch).
- DTG when the sketch was prepared.

Chapter 4

- Name of the preparer.
- Rank and organization of the preparer.

Remarks

- Terrain on the objective.
- Vegetation, that is, whether it can provide concealment.
- Structural composition at the objective, for example, brick, mud, or wood.
- Tactical value of the objective; for example, what purpose it could serve such as prison camp, training camp, or communications site.
- Additional information of tactical relevance such as whether buildings on the objective are elevated and whether their windows have screens, the locations of power lines and construction, types of sidings, building numbers, lighting, gaps, breach points, or any other information that might help the chain of command plan follow-on missions.

Preparation

4-269. Sketching is an easy way to record information about an objective. The following are some useful steps to take when preparing a sketch:

- Work from the whole to the part.
- Use common shapes to show common objects such as roads, buildings, and poles.
- Draw in perspective.
- Use vanishing points.
- Cross-hatch to show depth in the sketch.
- Avoid concentrating on the fine details unless used to clarify the drawing or emphasize something of tactical importance.

Work from the Whole to the Part

4-270. Figure 4-14 shows an example drawing technique for whole to part.

Team Operations

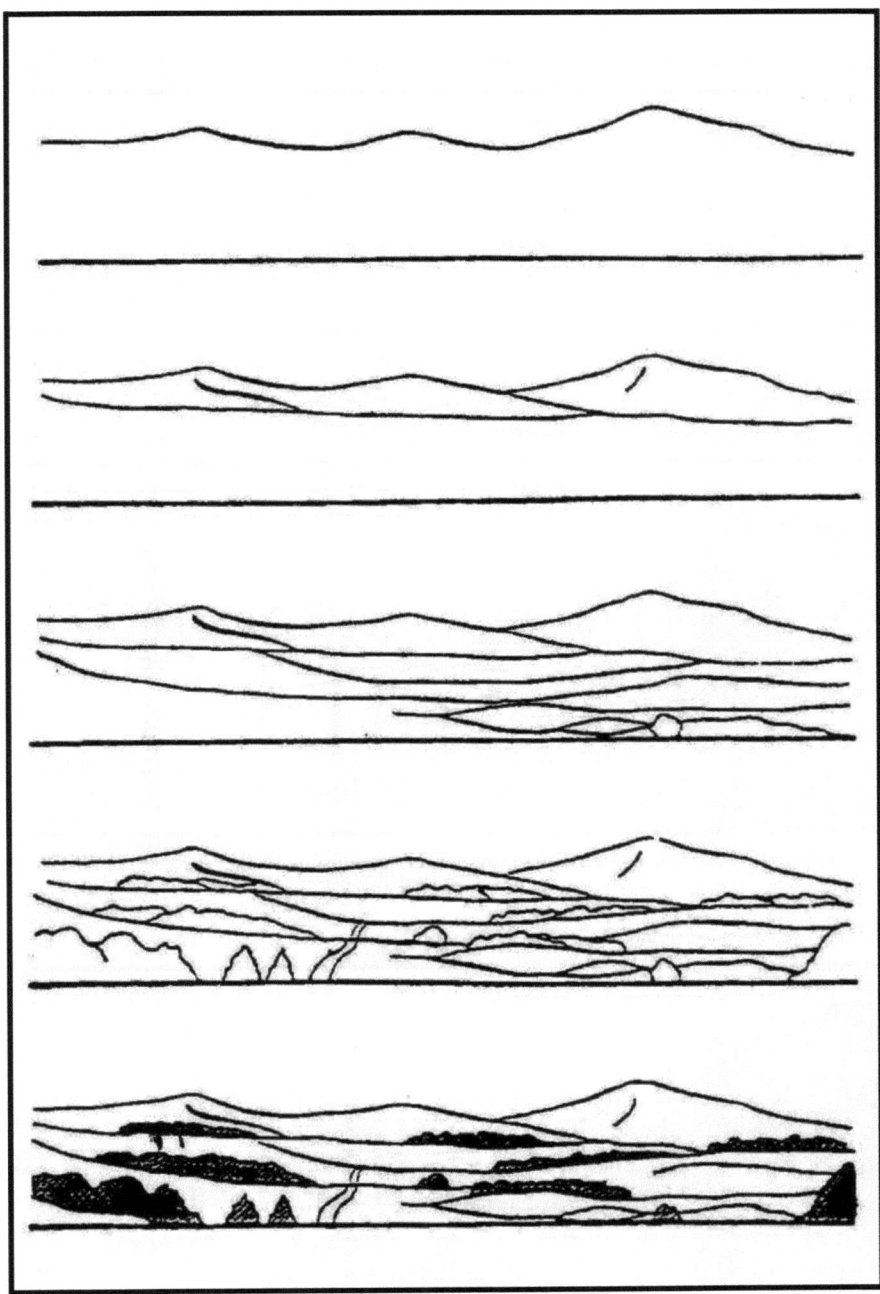

Figure 4-14. Example drawing technique: whole to part.

Use Common Shapes for Common Objects

4-271. Use common shapes to show common objects such as roads, buildings, and poles (Figure 4-15).

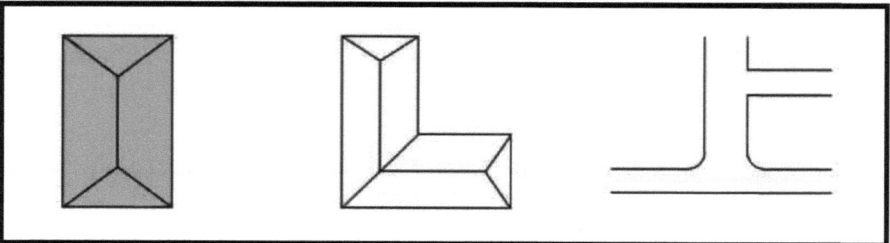

Figure 4-15. Example drawing technique: use of common shapes to show common objects.

Use Perspective Drawing

4-272. Draw in perspective (Figure 4-16).

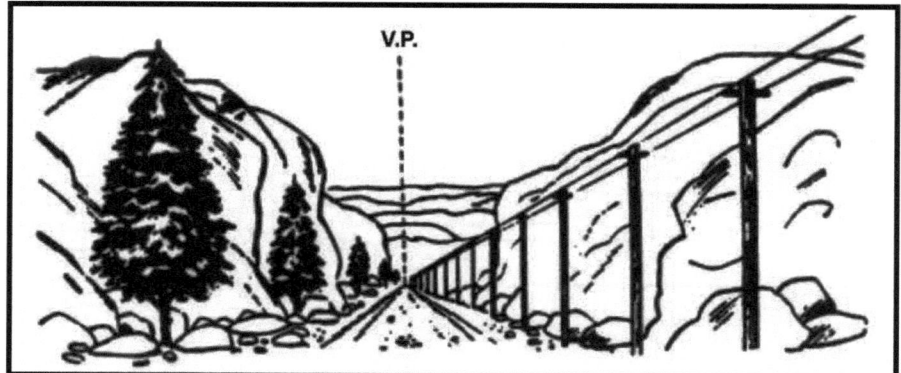

Figure 4-16. Example drawing technique: use of perspective to represent depth.

Use Vanishing Points

4-273. Use vanishing points (Figure 4-17).

Figure 4-17. Example drawing technique: use of vanishing points to indicate distance.

Cross-Hatch to Show Depth

4-274. Use hatching to show depth in the sketch (Figure 4-18).

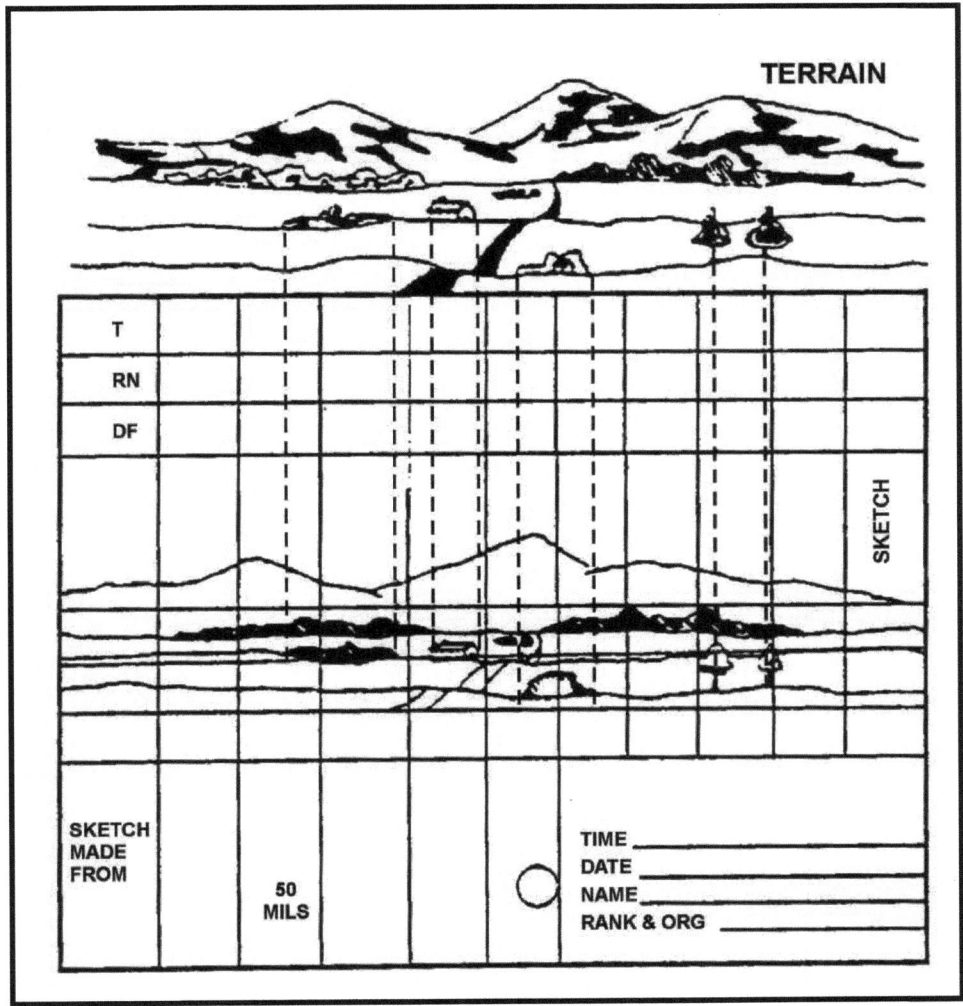

Figure 4-18. Example drawing technique: hatching.

Detail Only the Most Important Elements

4-275. Avoid concentrating on fine detail except to clarify the drawing or to emphasize something of tactical importance.

Section VIII. STABILITY OPERATIONS

This section discusses LRSU activities in stability operations. LRSU are well suited to conduct stability operations, because can they provide both overt and covert combat information. Both of these are critical to success in stability operations. Stability operations occur unilaterally, or with offensive and defensive operations. United States forces can participate in a stability operation while a host nation is at war. In addition, stability can evolve into war, and leaders should be prepared in case this occurs. Stability operations influence

Chapter 4

the political, civil and military environments, and can disrupt illegal activities. Military operations in stability are characterized by indirectness rather than directness.

TYPES

4-276. Stability operations typically fall into ten broad and often overlapping types. For example, a force engaged in a peace operation may at the same time conduct arms control or a show of force to set the conditions for achieving an end state. This paragraph introduces stability operations. (See FM 3-0 and FM 3-07 for more detailed information.) The LRSU can participate in any of the following operations in a stability environment:

- Peace operations (POs).
- Foreign internal defense (FID).
- Humanitarian and civic assistance.
- Foreign humanitarian assistance.
- Security assistance.
- Support to insurgencies.
- Support to counterdrug operations.
- Combating terrorism.
- Noncombatant evacuation operations (NEO).
- Arms control.
- Show of force.

PEACE OPERATIONS

4-277. Peace operations (PO) support strategic and policy objectives and the diplomatic activities that implement them. POs include peacekeeping operations (PKOs), peace enforcement operations (PEOs), and support to diplomatic efforts to establish and maintain peace. Peace operations monitor and ease the implementation of agreements, such as a cease-fire or truce. They can also support diplomatic efforts to reach a long-term political settlement. They usually consist of observing, monitoring, or supervising and aiding the parties to a dispute. The mission of a peace operation is fluid, and is based on a review of METT-TC. LRSU activities supporting PKO include, among others—

- R&S of a demilitarized zone.
- Surveillance of confrontation areas.
- Surveillance of cease-fire areas.
- R&S of refugee camps.
- Damage assessment.
- Monitoring of chemical, biological, radiological, or nuclear activity.
- R&S of smuggling routes.

FOREIGN INTERNAL DEFENSE

4-278. Foreign internal defense (FID) is participation by civilian and military agencies of a government in any action programs taken by another government to free and protect its society from subversion, lawlessness, and insurgency (FM 1-02). The main objective is to promote stability by helping a host nation establish and maintain institutions and facilities responsive to its people's needs. Army forces in FID normally advise and assist host-nation forces conducting operations to increase their capabilities.

4-279. When conducting FID, Army forces provide military supplies as well as military advice, tactical and technical training, and intelligence and logistics support (not involving combat operations). Army forces conduct FID operations in accordance with JP 3-07.1 and FM 3-07. Army forces provide indirect support, direct support (not involving combat operations), or conduct combat operations to support a host nation's efforts.

4-280. LRSU support during FID normally consists of the traditional role of information gathering, but can also consist of training host nation combat information gathering units. LRSU can also provide long-range communications capability to host nations or deployed US forces.

HUMANITARIAN AND CIVIC ASSISTANCE

4-281. Humanitarian and civic assistance (HCA) programs provide assistance to the host nation populace in conjunction with military operations and exercises. The very nature of HCA programs frequently dictates that additional engineer units and support capabilities will augment units participating in HCA operations. In contrast to humanitarian and disaster relief operations, HCA are planned activities authorized by the Secretary of State with specific budget limitations and are appropriated in the Army budget. Assistance must fulfill unit-training requirements that correspondingly create humanitarian benefit to the local populace. HCA programs must be in compliance with Title 10, United States Code, Sections 401, 401(E), (5), and Section 2551. For additional information on selected sections of Title 10, US Code for medical support, see FM 8-42. See AR 40-400 for information on emergency medical treatment for local national civilians during stability operations. Humanitarian and civic actions are limited to the following categories:

- Medical, dental, and veterinary care provided in rural areas of a country.
- Construction of rudimentary surface transportation systems.
- Well drilling and construction of basic sanitation facilities.
- Rudimentary construction and repair of public facilities.
- Specified activities related to mine detection and clearance, including education, training, and technical assistance.

4-282. It is unlikely LRSU would participate in HCA activities. However, LRSU could provide support to units conducting traditional HCA activities by providing search and rescue, and long-range communications support.

SECURITY ASSISTANCE

4-283. Security assistance includes the participation of Army forces in any of a group of programs by which the US provides defense articles, military training, and other defense-related services to foreign nations by grant, loan, credit, or cash sales in furtherance of national policies and objectives (JP 3-07). Army forces support security assistance efforts by training, advising, and assisting multinational and friendly armed forces. LRSU forces are well suited to provide training and advisory services in their normal areas of expertise.

SUPPORT TO INSURGENCIES

4-284. An insurgency is an organized movement aimed at the overthrow of a constituted government through the use of subversion and armed conflict (JP 1-02). At the direction of the President and/or Secretary of Defense, US military forces may assist insurgents or counterinsurgents.

4-285. The US supports selected insurgencies that oppose oppressive regimes who work against US interests. Major considerations include the feasibility of effective support and the compatibility of US and insurgent interests. Because support for insurgencies is often covert, special operations forces are frequently involved. Leaders may call upon general-purpose forces when needed. That is, when the situation requires their particular specialties or when the scope of operations is so vast that conventional forces are required. LRSU activities in support of an insurgency or counterinsurgency consist of but not limited to--

- Early warning for US and host nation forces.
- Reconnaissance for lines of communication (LOC).
- Surveillance of refugee camps.
- R&S of targets for future direct action.
- Training of insurgent forces in their normal areas of expertise.

SUPPORT TO COUNTERDRUG OPERATIONS

4-286. Military efforts support law enforcement agencies (LEAs), other US agencies, state governments, and foreign governments. The goal of counterdrug operations is to stop the flow of illegal drugs at the source, in transit, and during distribution.

- DOD may provide training, assistance, equipment, and facilities as long as doing so does not affect US military readiness.
- The *Posse Comitatus Act* applies only to federalized forces, and *only within the US*. For example, federal military forces may not search US civilians, arrest US civilians, or conduct any related law-enforcement activity involving US civilians.
- *Title 10, United States Code, Sections 371-378*, gives military forces the authority to assist civilian LEAs.
- DOD may pass information collected during normal operations to law-enforcement agencies.
- Military efforts support and complement--not replace--the counterdrug efforts of other entities. (These entities can include US agencies, states, and cooperating foreign governments.) The Army can support any or all phases of a combined and synchronized effort to attack the flow of illegal drugs at the source, in transit, and during distribution. In counterdrug operations, LRS normally supports law-enforcement agencies. LRSU activities supporting counterdrug operations include but are limited to—

—R&S of shipment facilities.

—R&S of shipment routes.

—R&S of marijuana, cocoa, and poppy fields.

—Surveillance of narcotics traffickers.

—Surveillance of air and vehicle traffic.

COMBATING TERRORISM

4-287. Joint Publication 1-02 defines terrorism as "the calculated use of unlawful violence or threat of unlawful violence to inculcate fear; intended to coerce or intimidate governments or societies in pursuit of goals that are generally political, religious, or ideological" (JP 3-07.2).

Categories

4-288. Terrorism is categorized three ways, based on where its control and support originate:

- Non-state supported terrorism.
- State-supported terrorism.
- State-directed terrorism.

Methods

4-289. Antiterrorism refers to defensive measures taken against terrorism. Counterterrorism refers to offensive actions taken against terrorism.

Antiterrorism

4-290. The DOD Dictionary defines this as "defensive measures used to reduce the vulnerability of individuals and property to terrorist acts, to include limited response and containment by local military forces." Typical antiterrorism actions include—

- Crime prevention and physical security actions that prevent theft of weapons, munitions, identification cards, and other materials.
- Physical security actions designed to prevent unauthorized access or approach to facilities.
- Positioning and hardening of facilities.
- Coordination with local law enforcement.
- Policies regarding travel, size of convoys, breaking of routines, host nation interaction, and off-duty restrictions.
- Protection from weapons of mass destruction.

Counterterrorism

4-291. The DOD Dictionary defines this as "offensive measures taken to prevent, deter, and respond to terrorism." Specially organized and trained counterterrorism units usually conduct counterterrorism operations. In some cases, conventional forces provide, at most, limited support. LRSU activities that support combating terrorism include at a minimum—

- R&S to confirm terrorist activity.
- Surveillance of a terrorist safe house.
- Surveillance of suspected or known terrorists.
- Surveillance of individuals on the black, white, and gray lists.
- Force-protection surveillance.

NONCOMBATANT EVACUATION OPERATIONS

4-292. Noncombatant evacuation operations (NEO) relocate threatened civilian noncombatants to secure areas. Normally, these operations remove US citizens from foreign nations where their lives are in danger either from the threat of hostilities or from a natural disaster. In addition to US citizens, relocated civilians can include selected host-nation citizens and third-country nationals. NEO has three basic environments:

Permissive—A permissive environment has no apparent physical threat and no host-nation opposition. Military assistance is normally limited to agency support. This operation (NEO in a permissive environment) is slow and deliberate. It is the least likely environment in which an NEO might occur.

Uncertain—An uncertain environment requires the commander to disseminate the ROE early. The host nation may or may not be in control, but cannot ensure safety. An uncertain environment increases the need for a reaction force.

Hostile—A hostile environment might require a large security element and a large reaction force. The ROEs must be strictly enforced. The host nation or other threat will probably oppose evacuation. The LRSU activities supporting NEO include, at a minimum—

—R&S of NEO sites.

—Early warning for host nation and United States forces.

Chapter 4

- PR and CSAR support.
- Perimeter security.
- Linkup force.
- Pilot and casualty removal from downed aircraft.

ARMS CONTROL

4-293. This stability operation is associated with weapons of mass destruction. For example, on 26 November 1993, Iraq agreed to long-term monitoring of its weapons programs. Under the resolution, international weapons inspectors were authorized to roam Iraq freely and for an indefinite period. Their goal was to prevent Iraq from acquiring weapons of mass destruction. Inspectors monitored numbers, types, and performance characteristics of the weapon systems at issue. The inspection also extended beyond weapons to include C2, logistics support, and intelligence mechanisms. LRSU activities supporting arms control include, at a minimum—

- R&S of ammunition holding areas.
- R&S of motor pools.
- R&S of suspected transshipment sites.
- R&S of weapon cache.
- Chemical, biological, radiological, or nuclear monitoring.
- Inspection of storage facilities.

SHOW OF FORCE

4-294. A show of force is a mission carried out to demonstrate resolve. Units conducting a show of force must deploy rapidly due to the political need for timely action. As the word "show" implies, media coverage is desirable and must be planned. Commanders must prepare in case a show of force evolves into a combat operation.

- A show of force—
  - Bolsters and reassures allies.
  - Deters potential aggressors.
  - Gains or increases regional influence.
  - Defuses a situation that could damage US interests or national objectives.
  - Lends credibility to US commitments and increases regional influence.
- It can take any of the following forms:
  - Combined training exercise.
  - Rehearsal.
  - Forward deployment of military forces.
  - Introduction and buildup of military forces in a region such as in Operation "Golden Pheasant," conducted during a 1988 border incident between Honduras and Nicaragua.
- Some LRSU activities that support shows of force follow:
  - Participation in Airborne operations.
  - Surveillance from fixed observation posts.
  - Route reconnaissance.

CAPABILITIES AND LIMITATIONS

4-295. Planning factors for stability include intelligence, rules of engagement, multinational operations, OPSEC, demography, deception, technology, and COMSEC.

INTELLIGENCE

4-296. The nature of stability operations require detailed intelligence. LRS teams need this intelligence before they infiltrate. It should include the target location and description; enemy equipment and capabilities; any civilian personnel in the area; and a variety of terrain, weather, and other related facts.

RULES OF ENGAGEMENT

4-297. The commander must monitor the ROE to ensure that all teams know when and how to apply force to meet specific situations. He must avoid vague or wordy ROE. Each Soldier must understand the rules as they apply to him. LRS teams must adjust rapidly to changes in the ROE.

MULTINATIONAL OPERATIONS

4-298. LRSUs must be prepared to coordinate and work with the host country's military and paramilitary forces. Every situation is unique and depends on the extent of involvement of US forces and the nature of the operations. Chief considerations when planning multinational operations are C2, intelligence, operational procedures, and sustainment.

OPERATIONS SECURITY

4-299. OPSEC is critical for LRSU in stability operations. Due to the potential for other forces (US or host nation) to operate near LRS teams, LRS commanders must carefully coordinate to reduce the risk of fratricide. This requirement poses an equally dangerous risk to OPSEC for the teams.

DEMOGRAPHY

4-300. LRS commanders must study all aspects of the local population to understand the effect that it might have on teams operating in the area. He can obtain information from a variety of sources, to include area studies, intelligence staff agencies, local government, and the media.

DECEPTION

4-301. To reduce the risk to LRS teams, commanders should consider using deception, particularly during insertion of the teams. Establishing false landing zones and sending dummy radio transmissions are two techniques to deceive the enemy. Deception is limited only by the imagination, but leaders should consider and coordinate all means of insertion or extraction through the R&S squadron S-3.

TECHNOLOGY

4-302. Technology is a proven combat multiplier. Advanced optics, thermal sights, and remote sensors increase the capabilities of the LRS teams. Commanders must weigh the advantages against the inherent disadvantages. These include increased Soldier's load and the impact of emitting the various equipment signatures.

COMMUNICATIONS SECURITY

4-303. The threat of interception and direction finding exists in all levels of conflict. Foreign purchases of threat equipment and relatively inexpensive off-the-shelf technology have enabled many third world countries and indigenous forces to equip themselves with the ability to take advantage of poor COMSEC. LRS commanders and team leaders must take appropriate measures to enforce COMSEC procedures.

Section IX. SPECIAL MISSIONS

Special missions include, among others, CBRN; Pathfinder; and personnel recovery (PR).

CHEMICAL, BIOLOGICAL, RADIOLOGICAL, AND NUCLEAR

4-304. LRSU may be called on to perform limited CBRN reconnaissance duties. The amount or type of equipment available and the current qualifications of the team members help determine the nature of these duties (FM 3-100).

PATHFINDER

4-305. Commanders sometimes require LRSU to perform limited Pathfinder duties. This capability is limited by the team's pathfinding experience, number of radios, and signaling devices (FM 3-21.38).

PERSONNEL RECOVERY

4-306. The LRSU commander, with assistance from the Joint Search and Rescue Center (JSRC), is responsible for conducting PR operations in support of his own operations and should be prepared to do so. He also coordinates with the rescue coordination center (RCC), advising them when his teams might have to evade a threat. He relays information such as isolated personnel report (ISOPREP) cards and an evasion plan of action (EPA) along with overlays of the evasion corridor. After he coordinates with other evasion planning agencies, he might determine that the unit must make its own evasion plans. He starts by identifying the team's evasion corridor and forms an evasion annex with the assistance of the JSRC (JP 3-50.2 and FM 3-50.1).

Chapter 5
Insertion and Extraction Methods

The LRS team must be prepared to insert and extract by various means to accomplish their mission. The team can insert and extract by water (Section I), air (Section II), land (vehicle [Section III] or on foot [Section IV]); or it can stay behind [also covered in Section IV]. Proficiency at inserting and extracting improves the team's likelihood of accomplishing—and surviving—the mission.

A member of the LRSU headquarters should accompany LRS teams on insertions and extractions by air. The presence of a representative from headquarters emphasizes the criticality of the air mission. He can also assist with navigation and other key duties, as dictated by the unit SOP.

Section I. WATERBORNE OPERATIONS

Using inland and coastal waterways can improve the speed, stealth, and flexibility of a LRS team's insertion and extraction.

CONSIDERATIONS

5-1. Waterborne insertion means include using surface craft, swimming on the surface, helocasting, or a combination of these. Whichever they choose, they should execute during limited visibility for maximum stealth. While planning waterborne operations, leaders must consider the following factors:

- Enemy situation.
- Civilian situation.
- Shipping.
- Beach landing site, which must allow the team to infiltrate and support movement to the inland objective.
- Environmental factors; for example, winds, waves, fog, thunderstorms and lightning.
- Equipment.
- Time schedule. Leaders use reverse planning to schedule operational events.
- Drop site. The team debarks a larger vessel at a planned drop site then begins infiltration.
- Launch point. A point where swimmers enter the water and begin infiltration.
- Method of loading. Supervisors inspect to ensure loads and lashings, especially waterproofing, adhere to unit SOP.

COMBAT RUBBER RAIDING RECONNAISSANCE CRAFT

5-2. Most LRS teams use a combat rubber raiding reconnaissance craft, commonly called a "rubber boat," for small-boat operations.

TECHNICAL INFORMATION

5-3. Specifications and other information about the rubber boat follow:

Inflation—This boat is inflated with foot pumps, using four separate valves on the inside of the gunwale. Each valve is used to section off the rubber boat into five separate air-tight compartments.

Chapter 5

Structure—A keel tube runs the length of the boat and two skeg tubes serve as shock absorbers.

Size—Outside, the boat measures about 6 feet by 15 feet. Inside, it measures about 3 by 10 feet.

Weight—The boat weighs 265 pounds, including the deck.

Maximum Payload—The boat can carry a maximum of 2,710 pounds.

Motors—The standard motor is a 40-HP short-shaft outboard motor. Adding a kit to stiffen the transom allows the use of either two 35-HP motors or one 55-HP motor.

CREW RESPONSIBILITIES

5-4.   The crew includes a coxswain, assistant coxswain, timekeeper, two observers (for security), and a navigator. To ensure mission success, each Soldier must track the team's location during movement. The team is positioned as shown in Figure 5-1.

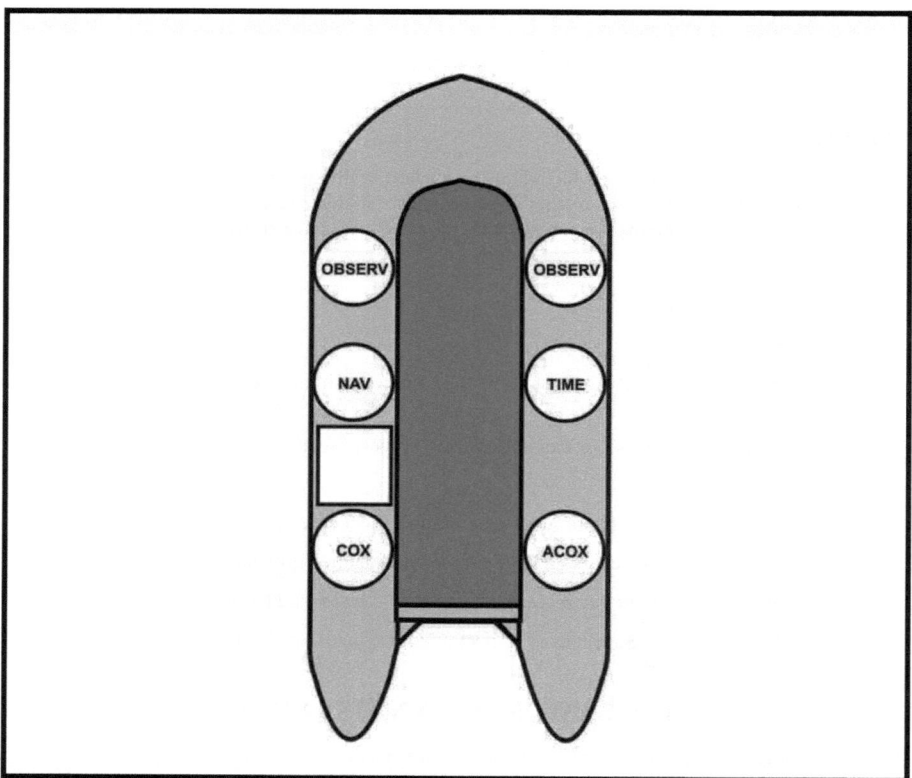

Figure 5-1. Rubber boat.

Coxswain—The coxswain controls the boat and the actions of the crew. He supervises the loading, lashing, and distribution of equipment. He maintains the heading and speed of the boat and gives all commands.

Assistant Coxswain—The assistant coxswain helps the coxswain navigate and control personnel, and, when needed, assumes the duties of the coxswain. He also stows and uses the bowline, and he provides rear security for the crew.

Timekeeper—The timekeeper keeps time during dead reckoning and aids in navigation. He provides flank security during movement.

Navigator—The navigator tracks the team's movement on the chart and reads the GPS as needed. He notifies the coxswain of any changes in heading, and he provides flank security during movement.

Observer—The observer(s) notifies the coxswain of any obstacles and provides frontal security during movement.

PREPARATION OF PERSONNEL AND EQUIPMENT

5-5. Each person dons a work vest, a life preserver, or both. He dons any additional equipment, such as LCE, over his vest. He slings his rifle over his life preserver, where it is readily available. Securely lashing radios, ammunition, and other bulk equipment to the boat prevents loss in case the boat capsizes. To lash equipment to the boat—

- Secure a 5-foot section of 1-inch tubular nylon across the boat at the rear set of "D" rings.
- Securely knot a 12-foot section of 1-inch tubular nylon to the front "D" ring. Place a loop about two feet from the other end of the rope. Using a slip knot, tie off this section to the 5-foot section.
- Attach a snaplink to the top of each rucksack frame and secure the rucksack to the equipment line.
- Place the coxswain's rucksack at the rear of the boat. Route the snaplink through the loop at the rear of the equipment line. If the boat capsizes, the coxswain can release the knot, allowing the rucksacks to float free of the boat while he tries to right it.

LAUNCHING OF BOAT

5-6. When timing the launch, the coxswain observes surf conditions and considers wave intervals. When team members are about thigh deep in the water, the coxswain orders them to board by pairs. As soon as they board, they grab a paddle and help keep the boat perpendicular to the waves as the coxswain starts the motor. After the motor is running, the coxswain orders the rest of the team into the boat. To help avoid capsizing, the boat is kept perpendicular to the waves.

BEACHING OF BOAT

5-7. The coxswain observes the surf to see when to enter. To avoid capsizing, he has the team shift their weight to the rear of the boat before the boat enters the surf.

5-8. As the boat enters the surf zone, all team members work to keep the boat perpendicular to the waves. The coxswain observes the surf and commands the team to vary speed as needed and to avoid plunging into breakers. He periodically looks seaward to observe the waves.

5-9. When the boat reaches shallow water, the coxswain orders the team out of the boat in pairs, for example (short count), "Ones, out; twos, out" (Figure 5-2). On disembarking, each pair immediately grabs the boat handles and pulls the boat to the beach.

5-10. The coxswain directs the team to empty the water from the boat and carry it to higher ground. He has two team members provide security.

5-11. Once the team reaches an area suitable for caching the boat, they conceal it in a surface or subsurface cache, or, if required, they submerge the equipment.

Chapter 5

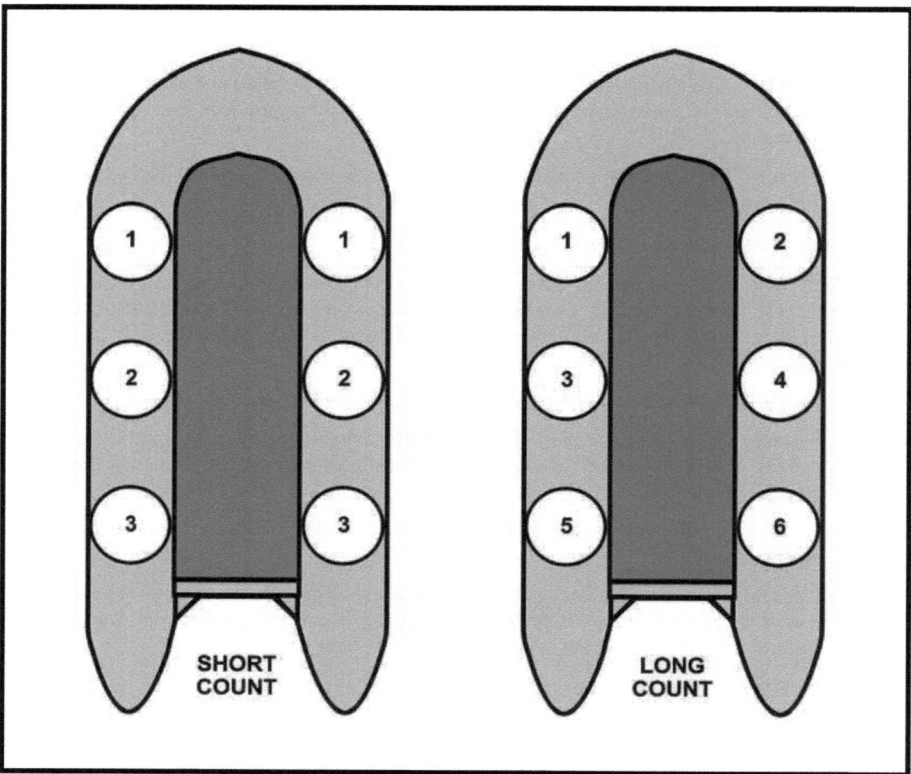

Figure 5-2. Short count, long count.

OFFSHORE NAVIGATION

5-12. Inserting a team from a larger vessel to their small boat might require offshore navigation. To move a long distance in a small boat, the team must be highly skilled in navigating by nautical charts, tides, currents (piloting), and dead reckoning. They must be able to compute for a compensated launch point, using offset navigation to take into account tides and current. Before launching, the team should conduct timed runs at full, half, and a quarter throttle. This helps them determine *the speed of the boat at each setting*, which is a critical part of navigation. To determine the speed (distance/time = speed), they divide the distance by the time, or the team uses a nautical slide rule. The types of navigation follow:

Piloting uses visible references from land or sea, the depth of the water, and other published information. It resembles land navigation by terrain association.

Dead reckoning is the predicted advancement of a vessel's position from a known point (debarking point) to calculate the vessel's approximate position at any time. It uses distance, speed, and time. It does not factor in water currents.

Celestial navigating uses the positions of the sun, moon, other planets, and stars relative to that of Earth. Because it requires a stable platform (for the sextant) and reference publications, celestial navigation is impractical for use on smaller vessels.

Electronic-radio navigating uses radio waves and satellite-based positioning systems such as GPS or Long-Range Navigation—Revision C (LORAN-C).

Note: Express distance in nautical miles.

Maintenance

5-13. The team must wash the boat with fresh water after using it in salty or muddy water.

- Remove all sand and debris.
- Carefully clean between the bottom and the skeg tube.
- Stand the boat upright on its cones against a wall, or suspend it. Using a high-pressure hose, rinse the entire boat. Allow several hours for it to dry.
- Wipe any remaining sand off the boat with a dry rag.
- After the boat dries, check and clean all metallic parts.
- Spray the valves with silicone to prevent freezing.

Storage

5-14. Store the boat in a dry, cool place out of direct sunlight. Avoid storing it near furnaces, steam pipes, boilers, oil or oil-contaminated areas, grease, solvents, or sharp or pointed objects. If possible, store it on a rack. Excessive handling, such as rolling and unrolling, will shorten the life of the boat. When storing it in a carrying bag, be sure to roll the boat from bow to stern.

SCOUT SWIMMERS

5-15. The term "scout swimmers" refers to a pair of surface swimmers assigned a reconnaissance or security mission in advance of the boat or other swimmers. Scout swimmers secure the beach-landing site and reconnoiter it to ensure that it can accommodate the entire team. They must also reconnoiter a suitable assembly area, cache sites, and a position from which to signal the remainder of the team into the beach-landing site. The team must maintain strict noise and light discipline throughout the operation.

Equipment

5-16. Normally, scout swimmers launch from a small boat outside the surf zone. Their equipment includes a life vest, swim fins, dive tool, MK 13 day or night flare, and coral shoes or booties.

Life Vest

5-17. Each scout swimmer wears his life vest under all equipment except his wet suit. It has no quick release, because he must wear it throughout his mission.

Serviceability

- Check the oral inflation tube.
- Inflate the vest and check it for leaks.
- Check the CO_2 inflation mechanism.

Preventive Maintenance

- Wash in fresh water after use.
- Clean and lubricate the CO_2 mechanism.
- Replace the CO_2 cartridge if used.
- Partially inflate the vest.
- Store vest in a cool, dry area.

Swim Fins

5-18. These help propel the swimmer.

Chapter 5

Serviceability

5-19. Check fins for proper fit and broken straps. Check for any rips or tears.

Preventive Maintenance

5-20. Wash the fins with fresh water after each use. Store in a cool dry place.

Dive Tool

5-21. Swimmers keep this tool or knife with them at all times in case they become entangled while swimming.

Serviceability

- Check for rust or corrosion.
- Sharp edges.
- Cracked or broken blade.

Preventive Maintenance

- Wash with fresh water.
- Sharpen.
- Lubricate.
- Store dive tool out of the sheath to dry.

MK13 Day or Night Flare

5-22. This is an emergency signal device.

Serviceability

- Use flare only if seals are intact.
- Check pull-ring lanyard.

Preventive Maintenance

- Wash in fresh water.
- Store according to standing operating procedures.

Coral Shoes or Booties

5-23. The swimmer wears these under his fins. They protect his feet from coral, rocks, or other sharp-edged or poisonous underwater objects.

Serviceability

- Check for rips or holes.
- Check for proper fit.
- Check to ensure the zippers work.

Preventive Maintenance

- Wash in fresh water.
- Dry away from direct sunlight.

Insertion and Extraction Methods

LAUNCH POINT

5-24. Scout swimmers normally move from the debarkation to launch points in inflatable boats with motors. The scout swimmers enter the water at the launch point. The launch point should be at least 400 meters off the beach, out of the range of small arms.

RECONNAISSANCE

5-25. Once the team reaches the launch point, the team leader sends out a scout swim team to reconnoiter the beach-landing site. Before leaving the main body, the swimmers receive last-minute instructions or adjustments to the original plan. The scout swimmers' may either leave their rucksacks with the main body or take them along. The swimmers use a dive compass, or they guide on prominent terrain features or lights on the beach. To allow all-round observation, they swim facing each other, which lets each swimmer observe the area behind the other one.

APPROACH

5-26. As the scout swimmers reach the surf zone, or when they get close to the beach-landing site, they start using the breaststroke to observe the beach. They approach with stealth and caution, keeping a low profile. One of them looks to the rear periodically so he can warn of large waves, which could injure or separate them from their equipment and each other. When they reach shallow water and determine that the situation is safe, they remove their fins.

1. If they can see the wood line easily from the waterline, one scout swimmer remains in the water just inside the waterline and covers the other Soldier's movement across the beach. Once the first scout moves to the edge of the wood line, he covers his partner's movement to the same position.

2. If the topography prohibits easy observation of the wood line from the waterline, the swimmers move by successive bounds.

3. Once both scouts move inland, they use a modified box pattern to reconnoiter and secure the beach. They choose a suitable assembly and cache site that provides the entire team cover and concealment.

4. One scout positions himself at the edge of the wood line. He provides security for the remainder of the team and guides them to the assembly area. The other scout positions himself where he can signal the main body. When he makes visual contact, with the remainder of the team, he moves to the waterline.

5. When the remainder of the team reaches the beach-landing site, the scout at the waterline directs them to the other scout, who guides them to the assembly area. After the last team member passes him, the scout at the waterline disguises any tracks left in the sand and rejoins the team.

6. If possible, the team locates the cache site away from the assembly area. If the enemy discovers and follows the tracks or trails from the beach to the assembly area, they could easily determine the number of personnel involved in the operation by counting swim gear.

HELOCASTING OPERATIONS

5-27. Helocasting can be an effective means of inserting and extracting LRS teams and equipment. The speed, range, and lift capability of rotary-wing aircraft make them excellent waterborne delivery and recovery vehicles. Helocast preparation considerations include--

1. When planning for the number of personnel for each type of aircraft, the leader uses the standard planning figures for loading troops. He can adjust these figures based on aircraft configuration, type of equipment, and casting or recovery procedures. He coordinates these items in advance with the aircrew.

Chapter 5

2. Rehearsals include all jumpers, the crew, the accompanying equipment, and support personnel. During live-casting rehearsals, the leader emphasizes the commands, positions, and timing of body exit and water entry.

3. All equipment attaches to the jumper with 1/4-inch, 80-pound test, cotton webbing. In or on this webbing, he normally carries a mask, fins, web belt with knife, flare(s), and life vest.

4. When using rubber boats, the team must—

    a. Tie down and secure all equipment inside the boat.

    b. Secure the motor in the floor of the boat and pad it with honeycomb cardboard (for UH-60).

    c. Securely attach and isolate fuel cans.

    d. Secure paddles under the gunwales, out of the way of the rest of the gear.

    e. Secure the rucksacks as tightly as possible to the deck of the boat.

    f. Waterproof all equipment in the boat in case of submersion.

    g. Regardless of the type of aircraft used, tie down or secure all equipment. Tape or pad all sharp edges or items.

    h. If using side doors for casting (UH-60 or UH-1H), secure the doors in the open position, and tape all edges.

    i. With a CH-46 or CH-47, ensure the ramp is secured in the open or casting position (10 degrees below horizontal).

    j. To use a wire ladder for recovery, secure it beforehand using a wire "donut" secured to the floor of the aircraft. Use 5/8-inch wire and secure the wire to at least five points using snaplinks.

    k. Ensure all personnel (cast master, pilots, and safety boats) use the same frequency.

    l. Ensure the casting area is clear of all surface and subsurface obstacles.

5-28. When helocasting from a ramp, such as a CH-47, the cast master gives the commands GET READY, STAND UP, CHECK EQUIPMENT, SOUND OFF WITH EQUIPMENT CHECK, and GO. When using UH-60 or UH-1H, he omits STAND UP:

1. The cast master ensures jumpers leave their seat belts on until they hear the command GET READY.

2. If using a rubber boat, the team moves it to the end of the ramp. Just before the command GO, they push out until the boat is about halfway past the edge of the ramp. On hearing the command GO, they push the boat off the ramp.

3. The cast master ensures that when the pilot drops personnel, he flies within 10 feet of the surface of the water at 10 knots or slower.

4. When casting from the ramp, jumpers assume a normal prepare-to-land attitude.

5. When casting from a side door, jumpers cast from a seated door position. On the cast master's command, jumpers push off and face the direction of flight, assuring a normal prepare-to-land attitude.

6. The cast master throws bundles or rucksacks before the jumper exits on the command GO.

7. Upon entering the water, the jumper signals "Okay" to the cast master and safety boat.

8. When using a single rotor aircraft for recovery operations, the cast master lowers a wire ladder to the swimmers, who line up at 50-meter intervals in the recovery area.

Insertion and Extraction Methods

9. As the aircraft flies over, each swimmer hooks the lowest rung on the ladder with his leading arm and climbs to a designated height, where he hooks up (with snaplink and rope seat) to the ladder.

10. All CH-46 or CH-47 aircraft land in the water. If using a rubber boat with a motor, the team drives the boat up to the ramp. When the rubber boat is not using a motor, the cast master uses the aircraft's winch to lower a rope with a 10-pound-padded weight attached to it. He lowers the rope behind the boat and drags the rope over it. The swimmers secure the rope, and the winch pulls the boat in.

11. When swimmers are in the water without a boat, they either go up a ladder or, if the aircraft is on the water, they swim up to the ramp.

12. For SPIES recovery, swimmers put on their harnesses before the helicopter arrives. The helicopter hovers over the group of swimmers as they attach their harnesses to the "D" ring.

5-29. Due to the hazards involved, the leader emphasizes safety in all aspects of planning and executing helicopter casting and recovery operations:

- Immediately before a helocast and recovery operation, the leader—
    — Physically reconnoiters the casting area.
    — Verifies water depth and the absence of obstacles and debris.
- He ensures that the water is at least 15 feet deep.
- He ensures motorized safety boats are in the water with motors.
- He establishes radio voice communications between the safety boats and the drop aircraft.
- He ensures the cast master has voice communications with the pilot.
- He ensures one dive supervisor and two divers, with complete scuba gear, are in a safety boat.
- He ensures that a qualified medic/dive medical technician is in one of the safety boats.
- He ensures drop altitude stays within 10 feet of the surface of the water.
- He ensures drop speed remains at or slower than 10 knots indicated airspeed.
- If an injury occurs, the leader ceases until he determines the cause and extent of the injury.

Section II. HELICOPTER OPERATIONS

Helicopters provide a variety of methods for inserting and extracting teams. (FM 3-05.210, TC 21-24 and USASOC Reg 350-6 provide more information.) This section only covers the operational requirements and procedures for SPIES, FRIES, Army aviation, air assault, pick-up and landing zones, and the UH-60 loading sequence. Training requirements are covered in FM 3-05.210 and USASOC Reg 350-6.

SPECIAL PATROL INSERTION/EXTRACTION SYSTEM

5-30. The SPIES should be used only when the team needs immediate extraction or cannot move to a clear (open) position suitable for helicopter landing. The SPIES works best for extracting LRS teams over short distances. Teams are almost never inserted by SPIES because doing so would expose them to observation and fire throughout the insertion. Before a SPIES operation, the leader thoroughly briefs participants. Before inserting, the team receives extensive training in SPIES operations. Personnel supporting the SPIES operation receive a complete preoperational briefing. This is most crucial when the operation involves assets other than the extraction helicopter such as gunships, aerial observers, or artillery.

Chapter 5

FAMILIARIZATION

5-31. When time and situation permit, personnel unfamiliar with SPIES should watch or help with the rigging of the helicopter. Initial training with SPIES is without combat equipment. This builds confidence in the equipment and procedures, and it aids in more comprehensive training of new SPIES masters.

SAFETY

5-32. This safety briefing should cover at a minimum--

- Area hazards.
- General Aircraft safety.
- SPIES equipment and its characteristics.
- Preoperational inspection of equipment.
- Proper donning of the harness.
- Method of insertion and extraction.
- Hand-and-arm signals and emergency signals.
- Medical coverage.
- Communications requirements.
- Operational requirements for limited visibility.

COMMUNICATIONS

5-33. Helicopter noise necessitates radio communications be the primary means of communications between the Soldiers on the ground and the helicopter. Light and arm-and-hand signals are the alternate means of communication. The SPIES master uses the inter-cockpit communication system on the helicopter. Soldiers practice the appropriate arm-and-hand and light signals in case radio communications fail.

5-34. Radio operators should use headsets and voice suppressors (if available) rather than handsets. This frees the ground RTO's hands so that, when the helicopter hovers, he can hook up faster and more safely.

EXTRACTION

5-35. After the extracting team is located on the ground, the SPIES master helps direct the helicopter the proper distance above the team. On order of the pilot, the SPIES master drops the rope.

1. The team leader positions himself to move and approach the rope as the SPIES master drops it. Once the rope is clear of any obstacles, the team leader directs the team to move to their assigned positions along the hookup points.

2. Each member attaches the primary (harness) snaplink to the "D" ring on his side of the line. Using the safety line and snaplink, he hooks into the alternate or secondary hookup point on the opposite side of the rope. He faces forward along the line so that, when the aircraft ascends, he is looking in the direction of travel. He holds up the SPIES rope and routes it over the shoulder closest to the rope. With the other hand, he gives a thumbs-up signal to the team leader and SPIES master signaling he is ready to go.

3. After all team members signal they are ready, the team leader physically inspects (if time and situation permit) each team members hook up. The team leader then hooks himself on to the lowest point with the RTO. He then gives the thumbs-up signal to the SPIES master. He continues this thumbs-up signal, which at night is an arranged light signal, until the helicopter reaches a safe altitude (about 3 meters above the tallest obstacle at the extraction site).

4. During extraction, the team RTO maintains communications with the extraction helicopter. Because he is near the bottom of the rope with the team leader, he can assist in giving verbal confirmation of all light and arm-and-hand signals, and relay information between the team leader, the SPIES master and the aircraft crew.

> **CAUTION**
>
> Not all chemical lights are visible at night using image intensifying night vision devises. Coordination with the pilots during the air mission brief to finalize the types and colors of chemical lights to use is required.

EMERGENCY PROCEDURES

5-36. During the flight, from the time the team extracts until they safely detach from the SPIES rope, each team member watches for any problems that might arise from above or below. The Soldier above checks the Soldier below. At the first sign of danger, or if an emergency occurs, the team leader or a team member places his free hand on the top of his head. When the SPIES master sees this, he instructs the pilot to make an emergency landing in the nearest and safest area.

DISMOUNTING PROCEDURES

5-37. When the extraction helicopter has reached a tactically safe dismount area, the pilot transitions to a hover and then begins a vertical descent. The SPIES master continuously provides information to the pilot on the distance from the ground to the lower end of the SPIES rope. As team members reach the ground, they immediately move away from the aircraft. For a UH-60, the team walks to a 90-degree angle to the front of the aircraft. For a UH-1, the team walks to the direction of the front of the aircraft. In both cases, the pilot or the crew can see the team is out from under the aircraft. The team ensures the SPIES rope does not interfere with the aircraft and that the aircraft does not land on the rope. All team members rapidly unhook themselves and their teammates who need assistance. Once unhooked, they move away from the area and set up security, or help clear the rope if the helicopter is going to land.

OPERATIONAL TRAINING

5-38. In preparing for an operation, if the leader thinks the situation, mission, or terrain indicates the need for a SPIES extraction, then he should include a SPIES harness in each Soldier's equipment list. If the mission or insertion precludes team members from wearing the harnesses, they should carry them inside their packs. As soon as they request helicopter extraction, they can retrieve and don their harnesses.

1. The extraction helicopter(s) proceeds to the area and the pilot establishes radio or visual contact with the team. The backup helicopter, equipped with the SPIES, remains aloft and away from the area, maintaining visual contact with the LZ and monitoring radio communications.

2. The SPIES master deploys the rope, and then notifies the pilot the rope is out. The pilot normally cannot see the team nor determine the most suitable position for the aircraft. The SPIES master gives the pilot vertical and lateral corrections until the aircraft reaches the desired position. He commands, LEFT, RIGHT, FORWARD, or REAR, along with the estimated distance. For example, LEFT, 10 FEET. The SPIES master counts down as the pilot responds, for example, "*Ten, nine, eight, seven, six, hover. Hold, ropes out.*" The SPIES master informs the pilot of any unexpected drift that could pull the team into an obstruction. The crew chief maintains his attention to the safety of the aircraft and watches for any possible interference with the tail rotor.

3. To avoid losing weapons during a SPIES extraction, team members sling them over their shoulders and attach them to their bodies with safety lines. They secure other weapons and equipment against the wind as well. After observing this, the team leader gives them the thumbs-up signal.

Chapter 5

4. During the extraction, the team RTO maintains communications with the extraction helicopter. He verbally confirms the thumbs-up signal and relays any other relevant information during the flight. He should position himself near or at the bottom hookup point. This ensures he can give accurate information about the extraction, the clearing of obstacles, and the descent.

5. The extraction aircraft must lift off vertically until the SPIES rope clears all obstacles. If needed, team members can fire their individual weapons from the hip with their barrels directed downward at a 45-degree angle and outward from the team.

6. Once the aircraft clears any vertical obstacles, the RTO, who is the lowest Soldier on the SPIES rope, signals the pilot all team members have cleared the obstacle. This is especially important during limited visibility. Even if the pilot uses night vision goggles, his depth perception is poor when looking 120 feet below the aircraft.

7. On descent, both the RTO and the SPIES master inform the pilot of his altitude, drift, and forward speed. They also tell him whether their ropes are oscillating enough to potentially injure team members on impact. The RTO counts down in 10-foot increments ("Fifty, forty, thirty, twenty, ten, nine, eight…one; one man down, two…") until the whole team is down. During limited visibility, the SPIES master might not be able to see this.

8. Once the team is on the ground, the SPIES master monitors drift. Sudden lateral shifts can drag team members before they can disconnect from their ropes.

WATER-EXTRACTION PROCEDURE

5-39. The SPIES also works well for extracting LRS teams from the water. Three inflatable life vests or other flotation devices are attached to the SPIES; one to each end of the attachment points, and one to the middle of the attachment point area, just above the middle two sets of "D" rings. Each team member wears a SPIES harness under his life vest. In amphibious operations, he may also wear swim fins, mask, and snorkel. This simplifies hookup to the SPIES rope in the chop and spray caused by the helicopter.

1. After the pilot stabilizes the aircraft above the team members, he gives the order and the SPIES master drops the SPIES rope (with flotation attached).

2. When the team members finish hooking up to the SPIES rope, the team leader signals the SPIES master to start liftoff.

3. The aircraft must lift off vertically until all team members and the bottom end of the rope clear the water. During initial liftoff, the aircraft may drag team members through the water. They must be ready to roll onto their backs until the aircraft lifts them clear of the water.

4. The aircraft should fly at the same speed and altitude as it would over land. Dismounting procedures also remain the same, except for shipboard landings: on a ship, all team members take their orders from the personnel in charge of the deck.

PERSONNEL DUTIES AND RESPONSIBILITIES

5-40. SPIES training and operations require the designation of key personnel to perform assigned tasks. The positions are unit commander, SPIES master, ground safety officer (GSO) or NCO, air mission commander, and pilot in command.

QUALIFICATIONS OF SPIES MASTER

5-41. Selection of personnel for qualification as SPIES master should be based on the individual's demonstrated leadership capabilities, maturity, and knowledge of SPIES operations. Individuals selected must participate in at least three SPIES operations (observe twice and execute SPIES master duties once under the supervision by a qualified SPIES master). For example, the SPIES master candidate configures the hookups in the helicopter, helps prepare for an operation, and conducts a successful operation under the supervision of a qualified SPIES master. He must be able to give an effective pilot's brief, use the aircraft communication equipment, and understand aviation terminology. Additional qualifications requirements include--

- Holds the rank of sergeant or above (may be waived).
- Completed initial SPIES training.
- Knows all aspects of a SPIES operation.
- Has received instructions on and demonstrated proficiency in:
  — Rigging of the helicopter.
  — Inspection and preparation of SPIES.
  — Donning of the SPIES harness.
  —Coordination responsibilities.
  —Soldier or aircrew briefings.
  —Organization of the personnel to be extracted.
  —Instruction to pilots in maintaining the aircraft position over the target.
  —Throwing and retrieving SPIES.
  —Hand-and-arm signals.
  —Emergency procedures.

GENERAL DUTIES OF THE SPIES MASTER

5-42. The SPIES master is responsible for the safe conduct of the SPIES operation. Preflight, he—

Equipment

5-43. Inventories and inspects all SPIES equipment.

Briefs

5-44. Briefs pilots and others concerned about the details of the operation, concentrating on extraction and dismounting procedures.

Inter-Cockpit Communications System Helmet

5-45. Ensures that he has an inter-cockpit communications helmet and a gunner's belt or, lacking a belt, a sling rope instead. He connects, then checks the operation of the inter-cockpit communication system. On all SPIES operations, he, the crew chiefs, and the pilots must establish interagency communications using this system.

Rope

5-46. Attaches the SPIES rope to the helicopter as previously described.

Loose Items

5-47. Checks for loose items that could fall on a team member during flight.

Axe

5-48. Ensures that the axe is available, sharp, and securely stored so that it presents no danger to the Soldiers on the SPIES rope. He also ensures that he has an alternate means of cutting the rope in case of emergency.

Chapter 5

DUTIES OF THE SPIES MASTER DURING EXTRACTION

5-49. On arrival at the team's estimated position, the SPIES master helps the pilot determine the exact locations of the team members, and—

1. As the aircraft approaches the team's location, he helps the pilot (using the clock system) position the aircraft directly above the team.
2. He requests permission from the pilot to drop the SPIES rope when the aircraft is hovering above the team.
3. He drops the rope, taking care to avoid striking team members on the ground.
4. He notifies the pilot when the rope is down, and reports any altitude corrections necessary to ensure that the team members can reach all of the SPIES attachment points.
5. He watches for the team leader to give the thumbs-up signal.
6. When he sees it, he tells the pilot that the team is ready for extraction, and he requests a vertical liftoff.
7. He advises the pilot of the team's approximate position, the locations of any potential obstacles, and the avoidance of horizontal movement.
8. If a team member becomes entangled with an obstacle during the extraction, he notifies the pilot to stop the vertical lift immediately. If the situation is critical, he prepares to cut the SPIES rope (the anchor point or cargo straps) after team members are secured to the obstacle or on the ground.
9. When he is sure the team has cleared all obstructions, he advises the pilot. The pilot flies to a safe altitude (about 350 feet above ground level for training purposes, or, in combat, however high the situation dictates), or transitions to forward flight.
10. At frequent intervals during the flight, he advises the pilot on the safety status of all team members. He constantly watches the team and checks the security of the SPIES attachments.

DUTIES OF THE SPIES MASTER DURING DISMOUNTING

5-50. On arrival at the dismount area, the SPIES master tells the pilot the approximate height of the lower roper from the ground and—

1. Once the pilot starts the vertical descent, the SPIES master continually informs the pilot of the approximate distance between the lowest roper and the ground.
2. He informs the pilot of any horizontal drift or oscillation in the rope, and of any obstructions.
3. He tells the pilot when the rope is about 25 feet above the ground and again when it is 10 feet above the ground.
4. He ensures that the rate of descent is slow enough to enable the team members to land safely.
5. He reports initial touchdown of the lowest roper, when the last team member lands safely, and when all team members are disconnected.
6. On order of the pilot, he either retrieves or disconnects and drops the SPIES rope. With the UH-60/1H helicopter, he can only retrieve the rope if he has previously attached a recovery rope (a 12-foot sling rope or two 6-foot sling ropes joined together) about 5 or 6 feet below the cargo hook or cargo strap hookup point. The recovery rope must be attached using a self-tightening knot such as the Prussik knot. The standing end of the sling rope may be fastened to the deck tie-down or to a snaplink. Although the line should be kept out of the way, it must be long enough to control any oscillation in the SPIES during flight.

INSPECTION

5-51. A certified SPIES master or rigger inspects SPIES at a minimum every 6 months or whenever the serviceability of the equipment is in doubt. Out-dated, spliced, abraded, or cut ropes are removed from service. The SPIES master or rigger inspects the ropes as follows:

1. Inspects harness and suspension sling webbing for signs of contamination from oil, grease, acid, rust, cuts, twists, fading, excessive wear, or fusing (indicated by unusual hardening or softening of webbing fibers), fraying, burns, abrasions, and loose or broken stitching (in excess of three stitches). The inspector removes damaged harnesses or suspension slings and returns damaged equipment to supply for appropriate disposition. In some cases, riggers may be able to repair deficiencies.

2. Inspects all hardware for signs of corrosion, pitting, ease of operation, security of attachment, bends, dents, nicks, burrs, and sharp edges. Replacing any hardware (except the chest strap adapter) that requires unstitching the webbing, makes the harness unserviceable.

3. Replaces the "V" ring by cutting the strap above the stitching. A qualified rigger can fold and stitch a new end section for leg straps. If straps are damaged, the harness or suspension sling is unserviceable and must be returned to supply for appropriate disposition.

4. Ensures rope is free of splices.

5. Inspects the surface of the rope for splices, cuts, excessive abrasions, and snags. Cuts are considered excessive when four or more strands in any 5-inch length are cut. The two-to-one braided rope has 12 pair (24 strands) around the circumference. Abrasion is extensive when torn yarns are equivalent to that of four strands of any 5-inch length. A rope that has been subjected to heavy loads might display glazed areas where it has rubbed against hard surfaces. Painted or fused fibers can both produce a glazed appearance. A rope can also get fuzzy on the surface from long use. The effect on the strength of the rope by glazing or surface fuzz is negligible.

6. Inspects the rope for signs of contamination by acid, alkaline compounds, salt water, fire extinguishing solutions, and petroleum-based solvents. Although ropes gradually and uniformly change color with use, this does not decrease its strength. However, exposure to strong chemicals, usually indicated by spotty discoloration on the rope, can affect the strength of the rope.

7. Ensures the eye loop at the attachment point is not broken, frayed, or loose.

SERVICE LIFE

5-52. The SPIES master and riggers check ropes, harnesses, and suspension slings for expiration of service life or total life. Expiration of service is 7 years from the date the manufacturer's package is opened, and total life is 15 years from the date of manufacture.

REPAIR AND CLEANING OF ROPE

5-53. To repair the rope, the SPIES master must--

Note: You *must* repair loose or broken stitching.

1. Wash contaminated ropes with a mild detergent (such as liquid dish soap) and cold water, followed by a rinse in clean, fresh water. Dry the rope at room temperature (not to exceed 140 degrees Fahrenheit).

2. Remove stubborn oil, grease, hydraulic fluid, and other petroleum stains with the cleaning agent xylene (Grade A or B, TT-X 916). Use the cleaning agent as directed.

Chapter 5

> **WARNING**
>
> AVOID TRYING TO REPAIR NONREPAIRABLE DAMAGE, TO INCLUDE ACID CONTAMINATION, CUTS, OR FRAYS ON HARNESS OR SLING WEBBING.

STORAGE

5-54. The SPIES master stows the SPIES as follows:

1. Protect nylon materials from direct sunlight as much as possible to avoid ultraviolet deterioration.
2. Stow the SPIES rope in an aviator's kit bag for protection when not in use.
3. Use bins or similar facilities to store SPIES equipment. Use shelves that are at least 4 inches from the walls and 12 inches from the floor. Ensure that storage areas are well ventilated and free of oil, acid, cleaning compounds, and other contaminants. Avoid stowing equipment above or near hot water pipes or other heat sources.

RIGGING OF A UH-1H AND UH-60

5-55. To rig the UH-lH or UH-60 helicopter (with cargo hook) for a SPIES Operation, use the following equipment:

- One SPIES rope with deployment bag.
- Two 11-foot or 9-foot, three- or four-loop, Type 26 cargo slings.
- Two Type IV connector links.
- Nine locking snaplinks.
- One 12-foot sling rope.
- One roll of heavy-duty ("100 mile-an-hour") tape.
- One block of wood measuring 4 by 4 by 18 inches.
- One fire ax (for use during emergency cutaway procedures).

Note: For UH-1H or UH-60 helicopters without cargo hooks, use the same equipment *plus* two additional 9-foot or 11-foot, three- or four-loops, Type 26 cargo slings, two additional Type IV connector links, and a total of 16 locking snaplinks.

5-56. The primary attachment point for the SPIES rope is the cargo hook. At the end of the SPIES rope, the polyurethane-encapsulated eye is attached to the cargo hook. The two 9- or 11-foot-long, cargo suspension slings joined together by a Type IV link to form one continuous sling. The team stretches out this sling on the helicopter deck. They take one end under the helicopter and through the eye of the SPIES rope, and they connect it on the other end of the sling using a Type IV link assembly. The straps are taped with 100-mph tape at 12-inch intervals (Figure 5-3). On the UH-1H only, the sling is passed between the helicopter skids and the fuselage. For both models, they can use locally procured padding to protect the sling from damage around the edge of the cargo hatch.

Insertion and Extraction Methods

Figure 5-3. SPIES rope rigging on UH-60

5-57. Once the team secures the SPIES rope and cargo straps, they secure the straps running across the deck of the helicopter. To do this takes eight snaplinks. The team spaces the snaplinks evenly across the deck, alternating from one side of the strap to the other, and from top and bottom. The first snaplink goes from the rear of the strap around the bottom two straps, and the next snaplink goes from the front of the cargo strap around the top two sections of the strap. This continues until the team establishes at least four points (Figure 5-4).

Figure 5-4. Rigging of snap links.

5-58. The 4-inch by 4-inch block of wood is taped down along the right edge of the doorway so the cargo strap crosses the block perpendicularly at the middle. The wood block serves as a chopping block pad in case of an emergency cutaway (Figure 5-5).

Chapter 5

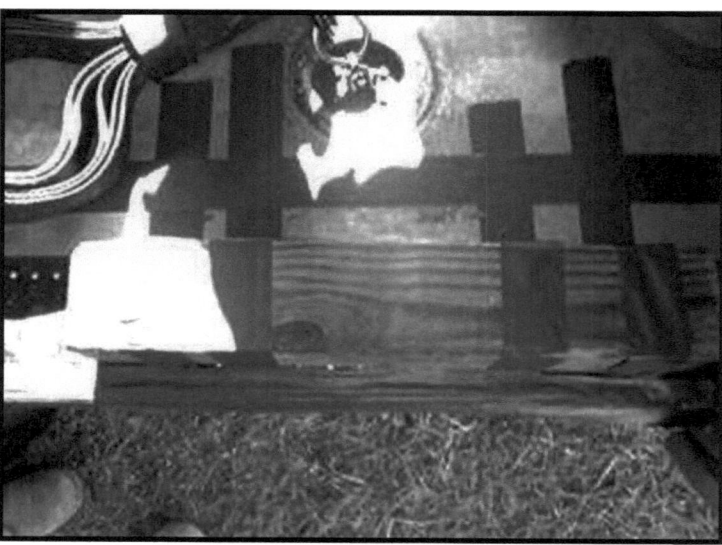

Figure 5-5. Rigging of wood block.

5-59.   Once the cargo strap and wood block are secured, the excess cargo strap is gathered on the opposite side of the wood block and taped to the floor of the aircraft (Figure 5-6).

Figure 5-6. Excess cargo straps secured.

5-60.   The sling rope is then tied to the SPIES rope by a Prusik knot about 2 to 3 feet below the cargo hook (Figure 5-7). A bowline knot with a half hitch is then tied to the running end and the ninth snap link is inserted and connected to a cargo ring in the middle of the aircraft floor. This line serves as a recovery line for the rope so that the aircrew can retrieve the rope into the aircraft. The recovery rope should be long enough so the weight on the SPIES rope is hanging from the cargo hook and not the recovery rope.

Insertion and Extraction Methods

Figure 5-7. Recovery line with Prusik knot.

5-61. If the SPIES rope is not to be used immediately, it is neatly coiled and placed on the opposite side of the aircraft from the wood block. This ensures it will not become tangled or interfere with normal aircraft operations.

5-62. If the helicopter has no cargo hook, or if the hook is not working properly, the team can still use the SPIES by doubling up on the cargo slings and Type IV links. This places two cargo straps side by side, for a total of four slings and four Type IV links.

5-63. The team must use caution when using the UH-1H, because different configurations of the helicopter may make rigging difficult. Some of them have steps, which get in the way during both installation and operation. Others have rocket pods or mounted machine guns. The team might have to hook up two different UH-1s in two different ways.

RIGGING OF A CH-46 OR CH-47

5-64. The CH-46 and CH-47 require—

- One SPIES rope with deployment bag.
- Two 9-foot or two 11-foot, three- or four-loop slings.
- Four Type IV connectors.
- Eight oval snap links.
- Heavy-duty tape (100-mph tape).
- A 12-foot length of tubular nylon or one 12-foot sling rope.

5-65. As they do with the UH models, the team attaches the SPIES rope using the slings and snaplinks (Figure 5-8). They pass the slings through the eye of the rope and attach it to the outboard cargo tie-down rings on the aircraft floor. They use two tie-down rings for each sling. They can use locally procured padding around the edge of the cargo hatch to protect the slings from damage.

5-66. The team arranges the cargo straps to form two U-shapes. They place one strap forward of the cargo hole in the center of the aircraft floor, and the other one aft, toward the rear of the helicopter. However, varying positions of the tie-down rings could require them to adapt their rigging techniques accordingly. Once they finish the rigging, the cargo straps should hold the SPIES rope comfortably centered and slightly below the cargo hatch. Attaching snaplinks close to all four tie-down points serves as a backup in case of a faulty tie-down ring, and reduces the amount of movement in the cargo suspension straps. The team should use eight snaplinks, two at each point, with swing gates reversed, for added security.

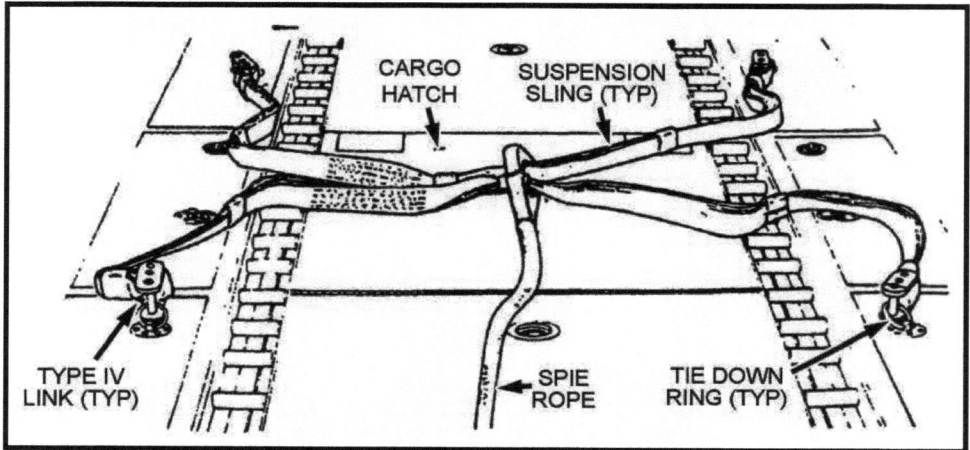

Figure 5-8. SPIES rigging procedures for CH-46 or CH-47.

FAST-ROPE INSERTION/EXTRACTION SYSTEM

5-67. The FRIES, also know as "fast rope," is a polyester rope, consisting of three olive drab 1 3/4-inch strands, and comes in 20-, 40-, 60-, 90-, and 120-foot lengths. The top of the main rope has an 8-inch eye spliced in, which allows the team to attach the rope to specially equipped helicopters. Before conducting a fast-rope operation, the team must thoroughly inspect the rope. The following paragraphs cover the requirements and procedures for FRIES operations; however, FM 3-05.210 and USASOC Reg 350-6 cover training requirements.

INSPECTION

5-68. The team lays out the rope so they can inspect the full length for fraying and the eyelet on the end for excessive wear. Snags from normal use will not significantly weaken the rope. A rope with several strands frayed in one particular spot or any single strand cut halfway through cannot be used. If the fast rope gets wet, team members "S"-fold it or hang it in a dry, warm area out of direct sunlight. It must dry thoroughly before the next use. If the team uses the fast rope in salt water or it becomes imbedded with dirt or mud, the rope must be washed in fresh water within 72 hours and then dried as described above. Inspect the rope for contamination by acid, alkaline compounds, salt water, fire extinguishing solutions, or petroleum-based solvents. Although ropes gradually change color uniformly with use, this does not necessarily indicate a decrease in strength, unless the change is due to contact with strong chemicals. Chemicals usually cause spotting, not uniform discoloration. A DA Form 5752-R (Rope Log (Usage and History)) for each rope must be maintained. See TM 10-1670-262-12&P for detailed maintenance and inspection information on FRIES equipment. The aviation unit is responsible for installing, removing, storing, and maintaining the FRIES mounting bars.

Rigging of the UH-60

5-69. The aviation unit is responsible for rigging the aircraft (Figure 5-9 and Figure 5-10). The FRIES master and selected personnel may rig or assist in the rigging under the supervision of the aircrew. Personnel—

- Ensure the aviation unit has removed the center row, which has nine seats.
- Ensure the aviation unit has provided in-flight floor restraints for fast-rope personnel. These restraints can include seat belts, sling ropes, or CGU straps.
- Remove both of the storage pins, and allow the bars to rotate down.
- Extend the fast-rope bars out to their desired length, fully extended for insertions, and insert the storage pin in the correct hole.
- Inspect the bar for cracks and for security of nuts and bolts.
- Rig the fast rope to the fast-rope attachment point, as follows:
  — Remove safety pin from the fast-rope release system and apply upward pressure to cabin wall-mounted release handle, releasing the gate.
  — Insert woven loop the fast rope into the attachment point.
  — Insert the gate through the woven loop of the fast rope and into the receptacle.
  — Apply a downward pressure to cabin wall-mounted release handle while pushing the gate out until the gate is fully seated in the receptacle (locking position).
  — Back-coil the fast rope and secure it to the cabin floor; or, insert the fast-rope retention strap through the coil, and suspend the fast rope from the ceiling of the fuel tank. Finally, use the safety pin to secure the quick-release mechanism (Figure 5-10).

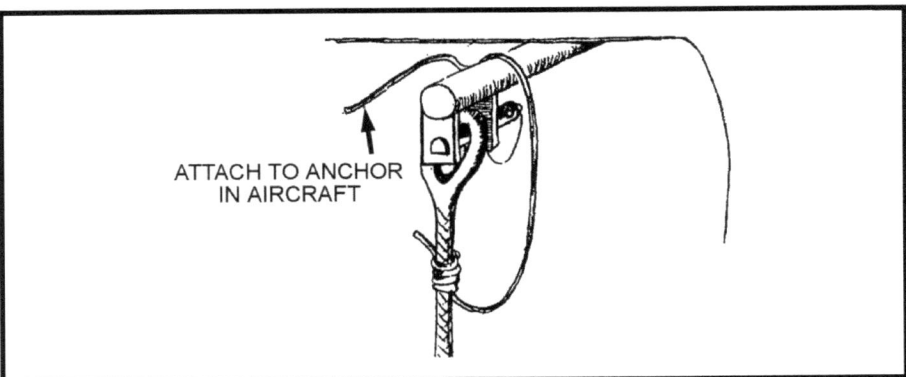

Figure 5-9. Fast-rope rigging procedures for UH-60.

Chapter 5

Figure 5-10. UH-60 rigged for fast roping

OTHER AIRCRAFT

5-70. The CH-47, CH-46, RH-53, and HH-53 aircraft use the same type of fast-rope bar, only double, for use with the ramps (Figure 5-11).

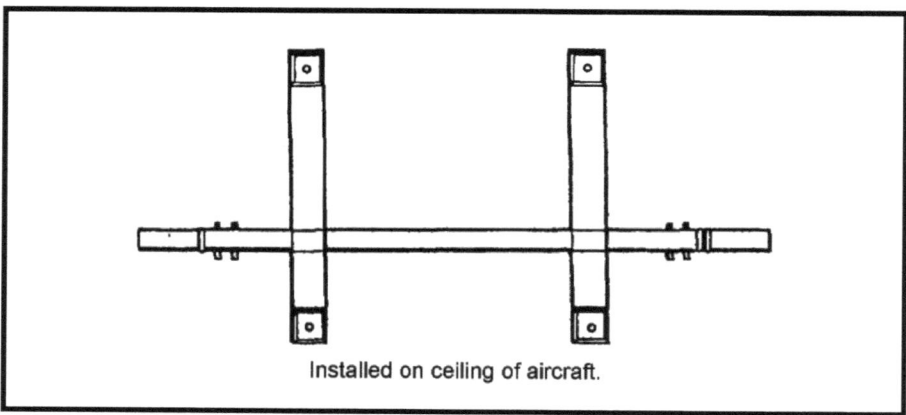

Figure 5-11. Fast-rope rigging procedures for other aircraft.

SAFETY

5-71. All personnel involved in FRIES operations are responsible for identifying hazardous situations and preventing injuries of personnel. Anyone who observes an unsafe condition or act is authorized to halt the operation and inform the FRIES master or the pilot in command. See USSOCOM Reg 350-6 for the most current safety requirements.

Emergencies after Roping Starts

5-72. In case of an emergency, personnel follow emergency procedures:

Unsafe Drift or Premature Liftoff

- FRIES master, assistant FRIES masters, safety, or roper stops the stick.
- Ropers stop descent and lock in.
- FRIES master or crewman informs the pilot in command and guides him in, moving the aircraft back on target.
- Unit continues operations.

Rope Hung or Snagged

- Safety ensures ropers are off the rope and clear.
- Aircraft descends or lands, as needed.
- Ground personnel free the rope.
- Unit resumes the operation.

Premature or Unintentional Deployment of the Fast Rope

- Notifies the pilot in command.
- Follows the aircrew's instructions.

Lost Communications

5-73. During all training and operations, the pilot in command or the crewmembers and the FRIES master must use the intercom. If the intercom fails, they use hand-and-arm signals until they can clear the rope and restore the intercom:

Stop Stick--A clenched fist touching the chest.

Ropes--Open palm toward the door in a horizontal motion.

Aircraft Movement--An open palm moving and facing in the direction required.

Stop Aircraft Movement--A clenched fist.

> **DANGER**
>
> HOLD ONTO THE SAFETY LINE DURING THE CRITICAL TIME BETWEEN THE ONE-MINUTE WARNING AND BEFORE "ROPERS AWAY." WHEN THE DOORS ARE OPEN, ANY SUDDEN AIRCRAFT MOVEMENT COULD THROW YOU OUT OF THE AIRCRAFT.

DUTIES OF FAST-ROPE MASTER

- Brief members of his team and aircrew.
- Inspect team members to ensure that their equipment is configured correctly. For example, ensure that each member has his work gloves and has tied down all equipment on his person.
- Backbrief, and then inspect aircraft rigging.
- Install the FRIES rope in the aircraft and conduct safety checks.

Chapter 5

- Relay 10-minute, 6-minute, and l-minute warnings to team members. Use time warnings as a tool to help synchronize aircrew and ropers' actions. Modify time warnings according to user needs, but always issue 1-minute warning.
- Break chemical lights, if required, at the 6-minute warning. During night operations, mark the rope with six chemical lights:
    —Two at the mount.
    —Two at the end.
    —Two 15 feet from the end.

Note: Not all chemical lights are visible at night, even through image-intensifying night vision devices. The FRIES master must coordinate with the pilots during the air mission brief to finalize the types and colors of chemical lights to use.

- Ensure the rope is properly configured for deployment (back-coiled to prevent tangles).
- Ensure the team members are in order of exit no later than the 1-minute warning.
- Confirm target on final approach.
- Deploy the rope and ensures it reaches the ground.
- During night operations, wear NVG, see and verify that you see two horizontal chemical lights.
- Deploy personnel, advising the pilots by announcing--
    —"*Rope out*" when deploying the rope over the target.
    —"*Ropers away*" when the first roper exits on the fast rope.
    —"*Rope clear*" to inform the pilots the aircraft is clear for flight.
    —"*Hold*" to inform the pilots to hold the aircraft position.
    —"*Move, [left, right, forward, or back]*" as needed.
- Account for personnel and signals aircrew.

INDIVIDUAL ROPER

- Understand all aspects of the insertion and emergency procedures.
- Configure his individual equipment correctly to prevent snagging and injuries.
- At the command STAND BY (given at 1-minute warning), check self one last time and prepare to exit the position.
- At the command GO, maintain an orderly formation and exit rapidly.
- Grasp the rope firmly before exiting--never jump for the rope.
- On exit, rotate your body 90 to 180 degrees to ensure your equipment clears the aircraft.
- Exit at 1-second intervals. Begin to slow descent about halfway down to avoid landing on the other ropers.
- Descend the rope, controlling your speed.
- Brake two-thirds of the way down to avoid landing on ropers that preceded you.
- Prepare to land just before reaching the ground by spreading your legs about shoulder-width apart, and with your knees slightly bent.
- At landing, quickly move clear of the rope to avoid colliding with descending ropers.

ARMY AVIATION AND AIR ASSAULT

5-74. Army aviation can increase LRSU mobility as well as flexibility. Once inserted behind enemy lines, LRS teams gather combat intelligence that can lead to rapid and decisive action by friendly forces. A successful air assault derives from carefully analyzing the factors of METT-TC and from detailed and precise reverse planning. The latter actually consists of five basic plans, all of which leaders develop for each air assault. To make the best use of available time, R&S squadron insertion and extraction section in coordination with the LRSC headquarters normally coordinate and develop these plans. If time is limited, planners can compress the steps of planning, or they can conduct them concurrently. They can also supplement detailed, written plans and orders with SOPs. Normally, a battalion is the lowest level that plans, coordinates, and controls air assault operations. Even when companies and lower conduct operations, most planning occurs at battalion or higher.

GROUND TACTICAL PLAN

5-75. The commander's ground tactical plan forms the foundation of a successful air assault operation. All other plans must support it. It specifies actions in the objective area to accomplish the mission, and it also addresses subsequent operations.

LANDING PLAN

5-76. The landing plan supports the ground tactical plan. The landing plan sequences elements into the AO. It ensures units arrive at the designated locations on time and are prepared to execute the ground tactical plan.

AIR MOVEMENT PLAN

5-77. The air movement plan is based on the ground tactical and landing plans. It schedules the movement of Soldiers, equipment, and supplies from PZs to LZs by air.

LOADING PLAN

5-78. The loading plan is based on the air movement plan. It ensures that Soldiers, equipment, and supplies are loaded on the correct aircraft. Planning aircraft loads helps ensure unit integrity. Cross-loading may be necessary to ensure survivability of C2 assets and the mix of weapons arriving at the LZ ready to fight. The detachment or team leader should ensure the aircraft is loaded so that dismounting Soldiers can react promptly and contribute to mission accomplishment.

STAGING PLAN

5-79. The staging plan is based on the loading plan. It prescribes when and in what order ground units (Soldiers, equipment, and supplies) will move to the PZ (order of movement).

PICKUP AND LANDING ZONES

5-80. Pickup and landing zone size requirements depend on the type and number of helicopters and the minimum acceptable distances between aircraft. Small unit leaders should be skilled in selecting and marking of PZs and LZs.

MARKING TECHNIQUES

5-81. During the day, a ground guide marks the PZ or LZ for the lead aircraft by holding his individual weapon over his head, by displaying a folded VS-17 panel chest-high, or by other identifiable means. At night, an inverted "Y" marks the landing point of the lead aircraft. Chemical light sticks or beanbag lights help maintain light discipline (Figure 5-12). Each additional aircraft that lands in the same PZ or LZ requires an additional light. For an observation, utility, or attack aircraft, the exact landing point is marked with a single light. For cargo aircraft (CH-47, CH-53, CH-54), each additional landing point is marked with two lights. The two lights are placed 10 meters apart and aligned in the aircraft's direction of flight.

Chapter 5

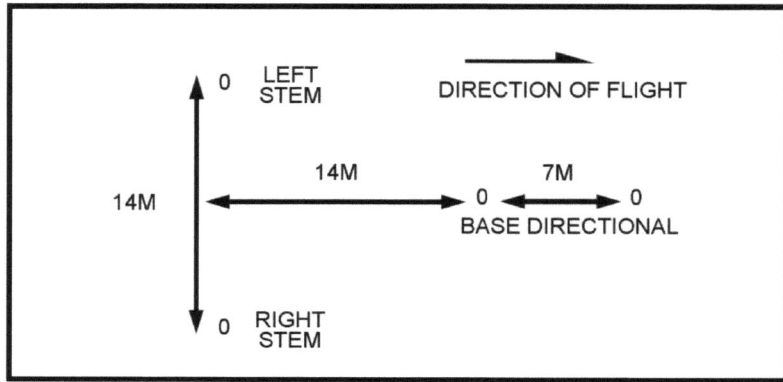

Figure 5-12. Marking procedures for landing and pickup zones.

> **CAUTION**
> Not all chemical lights are visible at night, even through image-intensifying night vision devices. The FRIES master must coordinate with the pilots during the air mission brief to finalize the types and colors of chemical lights to use.

LANDING POINT

5-82. Each aircraft requires a circular landing point separate from those designated for other aircraft, and free of obstacles. Table 5-1 shows the minimum recommended landing point diameters (in meters).

Table 5-1. Minimum recommended landing point diameters.

Aircraft	Minimum Landing Diameter
Observation helicopters	25 meters
UH-1, AH-1	35 meters
UH-60, AH-64	50 meters
Cargo helicopters	80 meters

SURFACE CONDITIONS

5-83. The surface of the PZ or LZ should allow clear visibility of the touchdown point. It should be free of landing hazards such as blowing sand, dust, or snow. It should contain no obstacles that could damage landing aircraft such as trees, stumps, or large rocks. The surface must be firm enough to support the traffic. It should have adequate drainage to allow rainfall to run off. Unacceptable levels of CBRN contamination can preclude the use of an area. If part of an area falls short for any reason, that part is not used.

GROUND SLOPE

5-84. Generally, if the ground slopes 0 to 6 percent, then the pilot should land upslope. If the ground slopes 7 to 15 percent, then he should land side slope. Over 15 percent, he should not touch down at all, but he may, if conditions allow, hover to drop off or pick up personnel or equipment.

Obstacles

5-85. For planning purposes, use an obstacle clearance ratio of 10 to 1 on the approach and departure ends of the PZ and LZ. That is, a helicopter approaching or departing directly above a 10-foot tall tree needs 100 feet of horizontal clearance. Mark obstacles within the PZ and LZ only if the enemy cannot see the markings. At night, mark them with red lights, but turn them on only when the PZ or LZ is in use. In the daytime, use red panels instead of lights.

Approach and Departure

5-86. Analyze the terrain surrounding a possible PZ or LZ for air traffic patterns. In a tactical situation, avoid repeatedly approaching the PZ or LZ over the same ground. Choose approaches that are free of obstacles. Pilots should land into the wind, but away from the sun. Ideally, they approach and depart along the axis of the LZ, over the lowest obstacle, and into the wind.

Load Size

5-87. When a helicopter is loaded to near maximum lift capacity, it needs more distance to lift off and land. It cannot ascend or descend vertically. The nearer the load to maximum, the larger the PZ and LZ must be to accommodate a flight.

Operations

5-88. Before the aircraft arrives, the PZ control party secures the PZ. Both the PZ control party and the Soldiers and equipment are positioned in the LRS team PZ or ORP. When occupying the team PZ or ORP, the team leader should (Figure 5-13) —

- Maintain all-round security of the PZ or ORP.
- Maintain communications (ground-to-air).
- Brief the marking team for the exact aircraft landing point, and check their equipment.
- Establish priority of loading for each Soldier.
- If time permits a detailed plan, use a coordination checklist.

I. GENERAL
1. Mission:
2. Units Participating:
3. Threat Forces:
4. Weather:
   Wind Direction   Visibility   Wind Speed
   Sunrise   Temperature
   Moonrise   Sunset
   Prevalent Weather Condition   Moonset   Illumination

II. FLIGHT DATA
1. Troop Load:
2. Equipment Load:
3. Pickup Zone: Time:
4. Liftoff:
5. Alternate Pickup Zone: Time:
6. Landing Zone: Time:
7. Alternate Landing Zone: Time:
8. Deception Measures:
9. Penetration Points:
10. Flight Route to Objective:
11. Flight Route from Objective:
12. Alternate Routes to Objective:
13. Alternate Routes from Objective:
14. Checkpoints and Descriptions:
15. Aircraft Linkup Point:
16. Air Control Points:
17. Downed Aviator Pickup Points:
18. Abort Criteria:
19. Enemy Air Defense Locations:
20. Suppression of Enemy Air Defense Measures:
    A.
    B.
    C.
    D.
21. Gun-Target Lines:
22. Prominent Features on Routes or LZ:
    A.
    B.
    C.
    D.
23. Aircraft Formation:
24. Aircraft Speed:
25. Aircraft Altitude:
26. Aircraft Crank Time:
27. Pathfinders:

Figure 5-13. Example coordination checklist.

28. Extraction Time:
29. Extraction Pickup Zone:
30. Alternate Extraction Pickup Zone:

III. TACTICAL PLAN
1. Ground Tactical Plan:
2. Fire Support Plan:
3. Air Cavalry:
4. Attack Helicopter:
5. Lift Aircraft:
6. Tactical Air:
7. Ordnance:
8. Handoff Point:
9. Aircraft Security Force:

IV. COMMUNICATIONS   FREQUENCY   CALL SIGN
1. Commander:
2. Pickup Zone Control:
3. Pathfinders:
4. SOI in Effect: Time Change:
5. Challenge: Password:

V. MARKINGS
1. Panels:
2. Strobes:
3. Bean Bags:
4. Pyrotechnics:
5. Smoke:
6. Light Gun:
7. Flashlights with Filters:

VI. CODE WORDS
1. Clean:
2. Secure:
3. Hot:
4. Cold:
5. Abort:
6. ALZ:
7. APZ:
8. Request Resupply:
9. Fire Preparation:
10. Request Extraction:

VII. SYNCHRONIZATION OF WATCHES   Time Zone:

VIII. MISCELLANEOUS   Air Movement Table:

Figure 5-13. Example coordination checklist (continued).

Chapter 5

LANDING ZONE OPERATIONS

5-89. The following priority of action applies when landing on an LZ:

1. The LRS team leader gets the landing direction from the pilot, and then alerts all team members before landing. This helps orient them to the LZ, particularly at night.

2. When the aircraft lands, personnel immediately unbuckle their seat belts and exit the aircraft with all equipment.

3. As soon as the crew chief or pilot directs them, the LRS team unloads the aircraft.

4. The team moves 15 to 20 meters away from the side of the aircraft and assumes the prone position facing away from the aircraft, weapons at the ready position, until the aircraft departs the LZ.

5. The team moves to a predetermined location using techniques that fit the terrain. Once the team reaches the concealed assembly point, the team leader quickly counts personnel and equipment, and then proceeds.

6. The team moves quickly to an assembly area out of sight and hearing of the LZ. They remain only long enough to adjust their senses to the surrounding environment and to verify the location of the LZ.

7. If planed and coordinated during the air mission brief, the insertion aircraft may be loitering nearby in case the team is compromised and needs hasty extraction. This is critical if the team is engaged by enemy forces on the LZ.

8. If the team makes contact on or near the LZ, they immediately execute the appropriate battle drill.

9. The LRS team leader calls for CAS, CCA or fire support, if available.

10. Once the team disengages from the enemy force, the team leader moves the unit to a covered and concealed position, accounts for personnel and equipment, and decides whether to continue with the mission.

11. If the team leader decides to call for emergency extraction.

    a. The team leader gives a direction and distance to the emergency extraction site from the insertion site.

    b. As the aircraft approaches, the team leader initiates a directional signal using, for example, pen gun flares, or a strobe light with a directional funnel attached.

    c. This ground-to-air signal lets the pilot determine a clock direction and distance from the aircraft to the team's location. The pilot identifies the signal initiated by the team.

    d. After confirming the signal, the pilot forms his approach, assisted by the team leader.

OBSTACLES

5-90. These include any obstructions, such as trees, stumps, or rocks that could interfere with aircraft operation on the ground. During daylight, the aircrew is responsible for avoiding obstacles on the PZ or LZ. For night and limited visibility operations, all obstacles are marked with red lights. The following criteria are used to mark obstacles:

1. Mark the near and far sides of the obstacle on the aircraft approach route.

2. If the obstacle is on the aircraft departure route, mark the near side of the obstacle.

3. If the obstacle protrudes into the PZ or LZ, but is outside of the flight route of the aircraft, mark the near side of the obstacle.

4. Mark large obstacles located on the approach route by circling the obstacle with red lights.

Insertion and Extraction Methods

5. (Signalman) use arm-and-hand signals to guide aircraft in for landing. Stand to the right front of the aircraft, where the pilot can see him best. At night, use lighted batons or flashlights in each hand. When using flashlights, avoid blinding the pilot. Keep the batons and flashlights lit at all times when signaling. The speed of the arm movement indicates the desired speed of aircraft compliance with the signal.

UH-60 LOADING SEQUENCE

5-91. To maintain communications with the pilot, the team leader--

1. Uses the aircraft troop commander's handset or requests a separate headset. Initiates movement once the aircraft has landed.

2. When the far- and near-side teams move to the aircraft, in file, leads the near-side group (Figure 5-14).

3. Ensures that all personnel wear and carry rucksacks on the aircraft.

4. Notifies the crew chief when all team members board and prepare for liftoff.

5. Ensures that all personnel buckle up as soon as they reach their assigned seats.

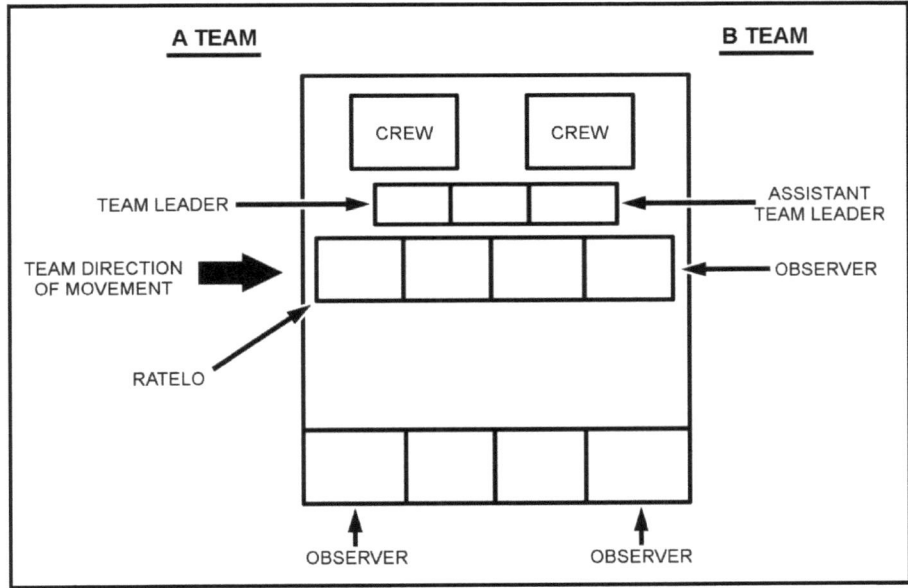

Figure 5-14. UH-60 loading sequence.

Section III. VEHICLE OPERATIONS

The LRS team can move by vehicle from the planning area to a point of departure in a secure area. Traditionally the team normally dismounts at the FLOT, makes final preparations, and conducts a forward passage of lines. LRS teams now have the capability to conduct mounted operations providing relatively rapid and secure operational assets within the AO. A LRS team no longer need rely on outside support in order to insert, extract, infiltrate, and exfiltrate in the AO.

Chapter 5

MOBILITY PLATFORMS

5-92. LRSU use a variety of vehicles to support themselves when conducting operations. The use of all-terrain vehicles (ATV), variations of the HMMWV, and nonstandard tactical vehicles (NSTVs) allow the commander to rapidly employ teams with reduced resupply requirements while conducting operations.

5-93. The ATV's primary mission is short-range mounted reconnaissance. The ATV provides the capability to conduct surveillance and reconnaissance missions over a 48-hour or 250-mile range (carrying extra fuel) without resupply in austere environments over difficult terrain.

5-94. The different variations of the HMMWV such as the army standard M1025A2 or the M1114 are good platforms to conduct long-range R&S in a desert environment.

5-95. NSTV come in a variety of configurations such as four-door pickup trucks and sport utility vehicles. The NSTV are also used to conduct R&S missions, but allow the LRS teams flexibility when operating in areas that limit the use of standard military vehicles.

PLANNING CONSIDERATIONS

5-96. Planning and preparation for a mounted mission starts long before the LRS team is alerted. Preparations include training and rehearsals such as mounted battle drills, laager/hide sites and vehicle maintenance. The distance from the FOB to the operational area, or even the staging (launch) site may require additional transportation. Various infiltration combinations of aircraft, rail line, or surface ships may be required to get the mounted team positioned to insert into an operational area. These infiltration combinations may also be used to increase the operational range of the mounted team by decreasing the required distance for overland insertion. When an operation requires either aircraft and surface ships or other combinations, a rendezvous must take place to transfer the team. The method selected should be one that will land or position the element with the least chance of detection as close as possible to its AO and as simply and rapidly as possible. Factors to consider include—

- Security.
- Size of the element.
- Operational requirements relating to the overt or covert nature of the mission.
- Capabilities of personnel and equipment loads.
- Availability of transport and delivery capabilities.
- Weather, terrain, hydrographic, and astronomical data, and conditions in the delivery area.
- Enemy and friendly situation in the delivery area and AO.
- The team may be delivered into the staging area, delivery area or the AO via—
  — Surface ships.
  — Amphibious landing craft.
  — Fixed- or rotary-wing aircraft.
  — Rail lines.
  — Line haul transport.
  — Any combination of the above.

PRE-MISSION CONSIDERATIONS

5-97. Elements should consider the following factors when planning for a successful infiltration.

Mission

5-98. The mission determines what and how much ammunition and supplies are necessary, including special equipment.

Enemy and Friendly Situation

5-99. Order of battle (OB) affects the routes, communications procedures and capabilities, external exfiltration capabilities, and sources of resupply.

Troops Available and Training Level of Unit Personnel

5-100. LRS teams are proficient in air infiltration and dismounted operations. However, mounted operations require additional training such as cross-country and night driving with and without night vision aids and vehicle maintenance, recovery operations, and use and care of mounted weapon systems.

Terrain and Weather

5-101. Terrain and weather affect route planning, personal equipment, and special equipment needs. Light conditions determine the time available for movement.

Time and Distance

5-102. These factors primarily affect the amount of required fuel for the vehicles and subsistence for team members, since distance and duration are similar.

Civilian Populace

5-103. Mission planning must consider the local civilians in the AO and what to do in case of mission compromise.

Equipment and Supplies

5-104. The pre-mission considerations help determine the teams' logistical needs. The team must plan for the minimum levels of all needed supplies. Mission essential equipment and supplies take priority in the allocation of space. During planning, the team may find that pre-positioned equipment is available in the AO. This equipment can range from fuel and water to a complete HMMWV with weapons, communications equipment, and repair parts. The availability of pre-positioned supplies greatly reduces the number of vehicles and amount of equipment the unit must deploy with, and reduces the deployment timeline. Additionally, when planning for deployment, the unit must allocate time to inspect and prepare the equipment when it arrives in country.

Collective and Individual Training

5-105. LRS teams are capable of operating in all types of terrain and using various insertion and extraction techniques. However, all teams require training to become and stay proficient.

Collective Training

5-106. Training required for the mounted LRS teams include cross-country and night driving (with and without night vision aids), vehicle navigation, vehicle infiltration, garage site, MSS and hide site establishment, vehicle maintenance, recovery operations, mounted battle drills, and dismounted crew battle drills. Priority for team collective training for the vehicles must always include maintenance.

Individual Training

5-107. The following paragraphs address suggested individual training team members:

> *Team Leader and Assistant Team Leader*--Mounted mission planning, detachment mounted training concepts, mounted employment, battle drills, load planning and vehicle maintenance management. The assistant team leader should also be hazardous material certified.

> *Senior Scout Observer*--Mounted mission planning, mounted employment, battle drills, vehicle maintenance, load planning and hazardous material certified.

Chapter 5

Radiotelephone Operator (RTO)--Mounted employment, battle drills, electrical wiring techniques, and vehicle maintenance.

Assistant Radiotelephone Operator (ARTO)--Mounted employment, battle drills, electrical wiring techniques, and vehicle maintenance.

Scout Observer--Mounted employment, battle drills, and vehicle maintenance.

Cross-Training

5-108. LRS team members require thorough cross-training. Each vehicle crew must be able to operate independently for extended periods of time.

Vehicle Preparation

5-109. LRS personnel prepare as necessary for airland, paradrop, waterborne, and overland insertions. They plan for and spend sufficient time preparing their vehicles for the assigned mission, from infiltration to exfiltration. They must be prepared to conduct all maintenance and repair operations in the field.

5-110. Team members cross load each vehicle so that if required it can act independently during the mission. Total weight of the vehicle, cargo and personnel is a prime consideration during operations. An overloaded vehicle handles poorly, consumes fuel at a higher rate, and will experience more maintenance problems. Items having the greatest effect on weight are fuel, water (50 pounds per 5 gallon container), ammunition by type (including shipping containers), and personal equipment.

Equipment and Personnel Preparation

5-111. An important aspect to pre-mission preparation is vehicle maintenance and keeping all equipment in a ready status. Members must inspect and exercise their vehicles even while in garrison. The assistant team leader is responsible for status of the team's vehicle. Preventive maintenance checks and service are normally conducted, at a minimum, weekly while in garrison. This includes road testing the team's vehicles. This test should include on- and off-road operation in all gears. Check for wheel alignment and listen for any unusual noises.

5-112. Keep the basic equipment common to each mission on the vehicle at all times (Figure 5-15). This equipment includes tools; petroleum, oils, and lubricants (POL); spare parts; recovery items; tire repair kits; and other miscellaneous items. This method will not only save loading time and storage space, it reduces the chance that these items will be forgotten. Prepare each vehicle using a unit standardized vehicle load plan. This list is compiled from unit SOPs, experience, and mission requirements. The unit vehicle load plan standardizes the location of equipment common to all in each vehicle. This ensures that anyone assigned to the unit can go to any vehicle and locate or pack team equipment. Control and assist the preparations after alert using pre-mission checklists.

5-113. Leaders conduct inspections to ensure the vehicles are loaded properly. Upon receipt of a notice to deploy, inspect the unit's vehicles as soon as possible to ensure mechanical reliability. Conduct this inspection at least 30 days before vehicle shipment (or as early as possible) to allow motor pool personnel time to correct deficiencies. Motor pool personnel normally help inexperienced team personnel perform this inspection. It is key that the team personnel be present at the vehicle maintenance inspection. Test-drive each vehicle to ensure mechanical reliability. Make sure the inspection takes the vehicle up to operating temperatures. Check climbing ability, winch operation with load, transmission and transfer case performance through all gears on challenging terrain, engine performance, and wheel alignment. Also, listen for any unusual noises or rattles. After the inspection and test, rate each vehicle by performance. The stronger vehicles should perform the more challenging aspects of the mission. Avoid overloading or hauling trailers with the weaker vehicles.

5-114. The next inspection should take place 3 to 5 days before load out or during planning. Inspect the items normally kept on the vehicle and all mission-related equipment for accountability and serviceability. The last inspection should be the normal final inspection or spot check done during the last few hours before the infiltration or shipment of the equipment.

5-115. Plan for sufficient fuel supplies. Fuel trucks, fuel points or resupply may not be available in the mission area. Frequently, it is difficult or impossible to get any kind of resupply. For a HMMWV, a good figure is 9 miles per gallon (mpg) for initial estimation of fuel requirements. Plan for and take adequate water. Minimum water planning figures are 4 to 6 quarts per man per day for mounted operations in a desert environment. Take additional water for dismounted missions within the mounted role. Omit the water carried on individual load-bearing equipment (LBE) for this requirement. Team members use a vehicle water bottle for the crew. They never use the water supplies on their LBE unless separated from the vehicles during dismounted operations or when placed in a survival or evasion situation. As a rule, consume water from the vehicle's stores first before using personal stores.

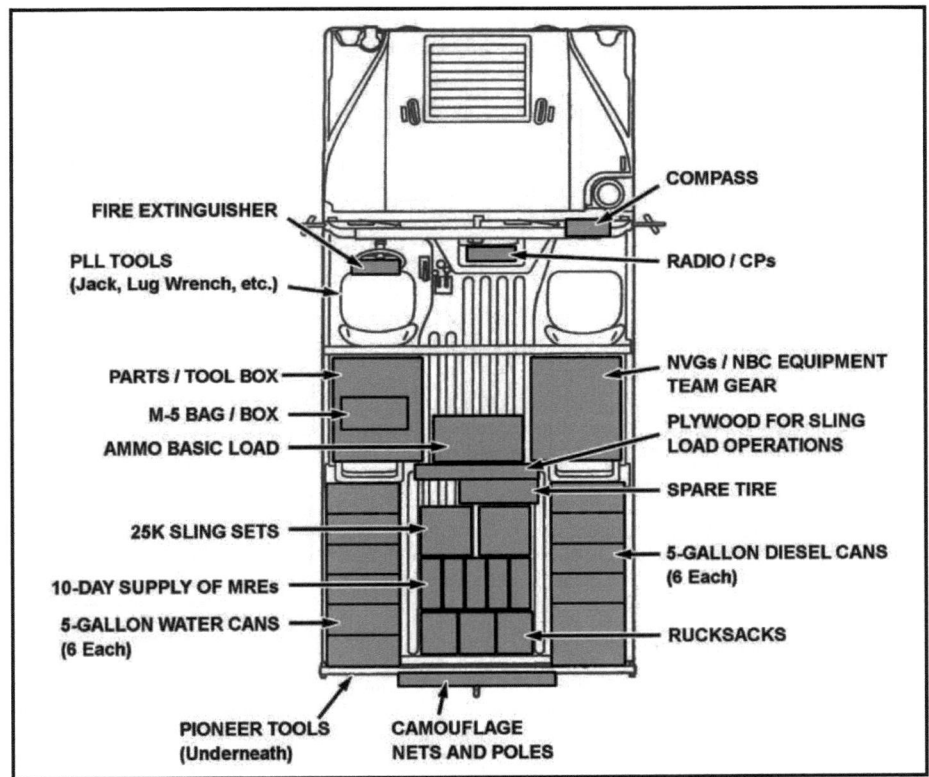

Figure 5-15. Example vehicle load configuration.

5-116. Plan for and take adequate food supplies. Remember that food consumption in hot, dry climates is generally less than in other climates. Individuals should pack most of their food items in a food bag (ditty bag) instead of their rucksack to limit the extent of unpacking their rucksack when getting meals. A ditty bag ensures they will have a minimal kit of food, survival and evasion items on hand. Construct the ditty bag from a durable bag large enough to hold three days of food, minimal sleeping gear, personal evasion and survival gear, first aid kit, and personal toilet articles. Pack at least three meals in the rucksack, so individual team members will have a food supply if required to abandon the vehicle rapidly. If several cases of food are packed on the vehicle, the crew must only open one at the time. This aids in estimating how long the food will last, and prevents the constant shuffling of equipment. Place ammunition where it can be accessed quickly. Secure large ammunition cans or containers to prevent injury in accidents due to shifting loads. Construct and position a vehicle destruction kit for quick accessibility. Each member should have three basic loads of small arms ammunition: one on the LBE (primary), one in rucksack (secondary),

Chapter 5

and one in an ammunition can positioned in the vehicle (contingency). The ammunition can in the vehicle should contain all contingency ammunition for the crew. Position basic signaling ammunitions near the navigator's position. These would include colored smoke and colored star clusters to aid in identification.

5-117. Plan for maintenance and repair contingencies based on the mission, the terrain and weather in the AO, mission duration, and maintenance experience. A detachment normally carries one general mechanic's toolbox, with metric supplement. In addition, each vehicle carries its own operator vehicle maintenance set and a small supply of motor oil, transmission fluid, and brake fluid for basic maintenance needs.

5-118. For long duration missions, a trailer towed by a second and third vehicle can carry additional maintenance supplies. Each vehicle should also carry one *complete* replacement set of fluids, including motor oil, transmission fluid, brake fluid, and antifreeze. Each vehicle should also carry basic spare parts such as fan belts, upper and lower radiator hoses, and main fuel tank drain plugs. Construct a battle-damage repair kit with such items as a tire plug kit, automotive liquid metal, assorted hose clamps, and radiator repair kit (FM 4-30.31).

5-119. On long-duration missions requiring trailer usage, construct an additional spare parts box to carry such items as starter, alternator, half shafts, glow plugs, and battery. The team will normally carry enough POL and replacement parts to repair any maintenance problem in the field within their capability. Once everything is packed and ready for deployment, strap down and secure all equipment and supplies. All equipment must be tied down securely for cross-country driving.

Operational Employment

5-120. The success of the mission and survival of the LRS team lies in its ability to infiltrate, move, conduct operations, and exfiltrate--all without being detected. In mounted operations, survival depends on moving mostly during times of limited visibility and using proper camouflage techniques at all other times.

Infiltration and Exfiltration

5-121. The threat to each method of infiltration and exfiltration is different. The following paragraphs illustrate typical threats to a mounted team when infiltrating by air or by ground.

Air infiltration and Exfiltration

5-122. Mounted LRS teams infiltrating and exfiltrating by air must avoid an extensive and integrated enemy air defense system. Such a system provides complete coverage at all altitudes with a high redundancy of coverage in heavily defended areas. Soviet doctrine, currently used by many nations, has tried to improve low-altitude detection.

Ground Infiltration and Exfiltration

5-123. Mounted LRS teams infiltrating and exfiltrating by land must avoid security forces. These forces employ sensors, minefields, barriers, patrols, checkpoints, and other populace control measures to detect clandestine movement across closed borders. Once the mounted team crosses the border, it still faces rear area security threats.

Planning Considerations

5-124. The following paragraphs address the planning considerations for air and ground infiltration:

Air Infiltration

5-125. The mounted LRS team can use several platforms to infiltrate its mission area.

Insertion and Extraction Methods

C-17A

5-126. The C-17A Globemaster III can be used as an infiltration platform. Planning considerations include—

- Travel time to operational area is greatly reduced.
- Team can carry a mounted weapons system.
- Vehicle is mission-ready, except weapons system is unloaded.
- Aircraft can carry an entire team with trailers loaded and prepared for movement.
- Team can carry spare parts and additional mission equipment such as trailers and pallets.
- Runway must be C-17A-capable, with a packed dirt runway at least 916 meters long.

C-130/MC-130

5-127. The C-130 Hercules aircraft has a great deal of advantages as an infiltration platform. Important planning considerations include—

- The team can fit two vehicles per aircraft.
- Weapons systems can be mounted.
- Vehicle is mission-ready, except weapons system is unloaded.
- Everyone can ride on the aircraft.
- On the C-130, vehicle fuel tanks must be half-empty (without waiver).
- On the MC-130, vehicles allowed on board with a full tank, with prior coordination.
- These aircraft require a C-130-capable dirt strip (916 meters).

CH-47 Helicopter, Internal Load

5-128. The basic HMMWV will fit inside a CH-47 helicopter with two inches of clearance around the vehicle (Figure 5-16). This clearance makes for a very tight fit and must be carefully rehearsed with the aircrew. Planning considerations include—

- Crew must rig the vehicle.
- No objects may extend from the top or sides of the vehicle.
- The weapon system is stored.
- The team cannot use trailers.
- Available rehearsal time with driver and aircrew.
- A requirement for flat LZ or PZ. Any surface undulation will cause the internal frame of the CH-47 to bend. This bend will lock the GMV in the helicopter or prevent it from being loaded or unloaded.

Chapter 5

Figure 5-16. Procedures for loading HMMWV into CH-47 for infiltration.

MH-47 Helicopter, Sling Load

5-129. Using procedures developed with 5th SFG(A) and 160th Special Operations Aviation Regiment, the MH-47 can land, hook up the vehicle, and load the vehicle crew on the same aircraft. The procedures for working with an MH-47 are different from conventional sling load operations and require coordination and rehearsals. Planning considerations include the following:

- Need for additional sling sets.
- Rigging the vehicle.
- Rehearsal with aircrew.

Ground infiltration

5-130. The HMMWV leaves a unique vehicle signature, and its tracks are difficult to conceal. Take extreme care during route selection.

Range

5-131. Mission planning range for a mounted LRS team with a full combat load is 10 days or 1,000 miles without resupply. A combat loaded HMMWV can expect to carry at least six additional 5-gallon fuel cans and six 5-gallon water cans on the back of the vehicle. This planning range can be extended to 10+ days or 1,000+ miles if using trailers to carry more supplies (fuel, food, and water), or if using an advanced operational base or caches for en route resupply.

Rigging of Vehicle

5-132. A common mistake is to take everything when using a vehicle. Take care to properly load and configure the vehicles for a long distance movement.

Trailer(s)

5-133. These can be taken for use en route or cached.

Fundamentals of Movement

5-134. When planning and conducting movement, consider the below listed fundamentals of movement to reduce the chance of enemy observation and contact.

Cover and Concealment

5-135. Use terrain features and vegetation that offer protection from enemy observation. When using cover and concealment to its full advantage, personnel will usually need to compromise between security and speed of movement.

Skylining

5-136. Avoid skylining. Select routes that avoid high ground that may silhouette the vehicles.

Chokepoints

5-137. Avoid chokepoints. Chokepoints or areas where the terrain naturally channels routes are often sites for ambushes or areas that the enemy may have under observation. If a chokepoint proves impossible to avoid, then reconnoiter it thoroughly before moving through it.

Populated Areas

5-138. Avoid known or suspected populated areas. For example, in desert environments, this means all water holes, because the populace--and the enemy--know all the water holes. A mounted LRS team cannot move covertly if people know they are in the area.

Movement Discipline

5-139. Practice movement discipline. Movement discipline means adhering to your light, noise, litter, and interval rules. It also means keeping your speed slow enough so that you do not leave a large dust signature (usually 10 to 12 miles per hour [mph] on most surfaces at night, slower during the day).

Security

5-140. Maintain 360-degree security at all times to avoid being surprised. The team leader or the unit SOP assigns a sector of fire and observation to each vehicle during movement and at halts.

Routes and Contingencies

5-141. Make sure all team members know the route and contingency plans.

Methods of Travel

5-142. There are two methods of travel in the operational area. Either on existing tracks, trails, or roads, or traveling cross-country. Each has advantages and disadvantages:

> *Tracks, Trails, or Roads*
>
> > *Advantages* include speed of movement, hard-packed trails that do not easily yield readable prints and signs of passage, quietness of movement, less stress on vehicles and tires, and sometimes easier navigation.
> >
> > *Disadvantages* are a greater chance of being seen or compromised, natural lanes of observation and fire for the enemy, and more probable mechanical or manual ambushes. The HMMWV leaves a distinctive tire trail unlike any other truck.
>
> *Cross-Country*
>
> > *Advantages* to traveling cross-country include less chance of enemy observation or contact, usually more cover and concealment, and less chance of an ambush.
> >
> > *Disadvantages* are slower rates of movement, more noticeable vehicle tracks and signs of passage, greater tire failure and vehicle stress, and more difficult navigation. The team must rehearse cross-country movement in terrain as close as possible to that of the target area before deployment.

Chapter 5

MOVEMENT TECHNIQUES AND MOVEMENT FORMATIONS

5-143. Movement techniques combined with movement formations allow units to conduct tactical movement in any METT-TC situation. These combinations can be used when all elements are mounted, or when there is a combination of mounted and dismounted forces, regardless of what type of vehicle is used.

Movement Techniques

5-144. The three standard movement techniques are traveling, traveling overwatch, and bounding overwatch. Different movement techniques are used based on the likelihood of enemy contact.

> *Traveling*--The traveling movement technique is used when enemy contact is not expected or likely.
>
> *Traveling Overwatch*--The traveling overwatch movement technique is used when enemy contact is possible.
>
> *Bounding Overwatch*--The bounding overwatch movement technique is used when enemy contact is likely or expected.

Formations

5-145. The mounted LRS team can employ a number of different movement formations depending on the number of vehicles and the situation.

> *Column and Staggered Column*--Use this formation when speed is essential as it moves on a designated route. The column offers good protection to the flacks, but little to the front and rear. The lead vehicle or section normally controls column movement by following the planned route and speed. The staggered column is used in open terrain. Use the visibility rule for interval.
>
> Illumination conditions, terrain and vegetation, and night vision equipment affect this rule. The driver keeps the vehicle to his front in sight.
>
> *Line Formation*--Use this formation is best used when maximum reconnaissance forward is needed.
>
> *Wedge and VEE Formations*--Use these formations when immediate mutual support and depth is desired. In the wedge formation, the vehicle(s) in the middle of the formation are forward. In the VEE formation, the vehicle(s) on the flanks are forward. These formations can also be used with extremely wide intervals, determined by visibility, to conduct reconnaissance operations (Figure 5-17).
>
> *Diamond Formation*--Use this formation when crossing extremely large open areas. Each section forms a side of the box when moving forward. Visibility determines the interval between vehicles in each section. The interval between sections should not be greater than 900 to 1,000 meters. This formation is hard to control; therefore, the sections plan for and designate rally points before they separate.

Figure 5-17. HMMWVs in wedge formation.

Actions at Halts

5-146. Any time the team conducts a planned halt (short or long), it will also conduct a coordinated shutdown of all vehicles. The team leader initiates the shutdown using hand and arm signals. He exits his vehicle and stands where everyone can see him. He waves his arm in a circle over his head, and then drops it toward the ground. This signals all vehicles to shut down their engines at the same time. When the halt is over, he uses the same procedure to signal all drivers to start their engines at the same time. If neither the leader nor the assistant leader can visually signal all of the vehicles at the same time, then either may use the radio to indicate engine shutdown or engine on. Although radio use should be avoided to lessen the team's radio signature, it can be conducted safely if done properly. Once the vehicles have been shut down, and before any other functions take place, the team conducts a security listening halt. The commander sets the duration for halts during planning or in team SOPs. Short-duration halts are used to communicate with higher headquarters, make necessary repairs, or establish a satellite position fix on a GPS receiver. For halts of less than 15 minutes, the team remains in formation. Personnel man all vehicle weapons and establish 360-degree security. For halts of more than 15 minutes, the team tries to move off its direction of travel and reform. During the halt, the team performs necessary tasks. Each Soldier receives a briefing on the present location. An updated contingency plan is issued if needed.

Coil

5-147. Use this formation when moving in a column formation or along a road or trail. The team moves into a partial perimeter along the route of march. Members of each vehicle observe their assigned section of the perimeter. The terrain determines vehicle interval, but it is seldom less than 50 meters.

Chapter 5

Laager Sites

5-148. A laager site is a secure vehicle encampment. Mounted teams should use this site to maintain vehicles, rest crews, plan missions, and hide during daylight.

Types

5-149. The two types of laager sites are *short duration* (occupied for only one period of daylight) or *long duration* (occupied for longer than one period of daylight). During route planning, select tentative primary and alternate laager sites on the primary and alternate routes. The team should arrive in the general area of the laager sites about two hours before begin morning nautical twilight (BMNT). This allows time for a proper reconnaissance and emplacement and camouflage of the vehicles before first light. Upon reaching a tentative laager site, or before first light, the ATV element or a dismounted element can reconnoiter it. Once the site is selected, the assistant team leader enters the site on foot and directs incoming vehicles into position. As each vehicle moves into position, its members receive their areas of responsibility. After the team is in place, it conducts a listening period to determine if there is any activity in the area.

Tasks

5-150. After the listening period, tasks (in order of priority) are—

- Security.
- Launch a dismounted patrol to erase signs of vehicles into the laager site out to a predetermined distance set by unit SOP.
- Camouflage one vehicle, or one vehicle per section, at a time. The others provide security.
- Confirm sectors of fire and prepare range cards.
- Emplace early warning devices or claymores.
- Establish observation posts (OPs) or listening posts (LPs), if necessary.
- Establish field telephone communications to each vehicle.
- Reduce security, refuel, perform maintenance, and attend to personal hygiene.

Description

5-151. The laager site need not resemble a circle. The terrain and vegetation play a role in locating each vehicle. All vehicles may be placed in the perimeter if necessary. When conducting detachment operations, the detachment sergeant's vehicle (number 2), is normally located in the center of the laager site. This formation resembles a triangle and allows a greater arc of fire if attacked. When selecting and preparing a LRS team's laager site, the priority is concealment, remaining undetected, and if compromised, breaking contact rapidly (Figure 5-18).

Occupation

5-152. The LRS team may have to occupy the laager site for more than one period of daylight. Such an occupation is most common when it needs to wait for more advantageous weather or light conditions before moving, has deployed a dismounted element on a mission and must remain in the area, or is in a situation where repairs to equipment must be made before resuming the mission. When occupied for more than one period of daylight, additional tasks include—

- Enhancing early warning measures.
- Improving continuously defensive positions (to include defensive minefields as necessary).
- Conducting reconnaissance and establishing surveillance of the area.
- Upon vacating the laager site, the team sterilizes the site as much as possible to deny the enemy intelligence on its operations.
- Continuing to enhance concealment of the site, even if doing so reduces its potential evacuation routes.

Insertion and Extraction Methods

Figure 5-18. Single camouflaged HMMWV.

Camouflage

5-153. Mounted LRS teams operating behind enemy lines need to stay undetected to complete the mission. In an unsupported role in a desert environment, a key to remaining undetected is to use proper camouflage measures. The team's ability to hide in the desert is limited only by the imagination and resourcefulness of its members (Figure 5-19).

Figure 5-19. Multiple camouflaged HMMWVs.

Chapter 5

Camouflage Theory

5-154. The biggest threat to the team is detection. Detection can be by—

Direct Observation--Where the observer sees the subject with his eyes, either aided or unaided.

Indirect Observation--Where the observer sees an image of the subject and not the subject itself. Indirect observation uses photography, radar, infrared, thermal imaging, and televideo.

Factors of Recognition

5-155. Regardless of the method of observation, certain factors help the eye and brain identify an object. The six factors of recognition are—

Position—This factor relates to the position of the object in relation to its surroundings. In addition, position is space relative to one object and another.

Shape—Experience teaches people to associate an object with its shape or outline. At a distance, the outline of objects can be recognized long before the details of its makeup can be determined. Trucks, guns, tanks, and other common military items all have distinctive outlines that help to identify them.

Shadow—Shadow may be even more revealing than the object itself. This fact is true when viewed from the air. Sometimes it may be more important to break up or disrupt the shadow than the object itself.

Texture—Texture refers to the ability of an object to reflect, absorb, and diffuse light. It may be defined as the relative smoothness or roughness of a surface. A rough surface reflects little light and will usually appear dark to the eye or in a photo. A smooth surface such as an airstrip, although it might be painted the same color as its surroundings, would show up as a lighter tone on a photo. One of the most revealing breaches of camouflage discipline is shine. Shine attracts attention by reflecting light such as sunlight or moonlight.

Contrast—Color is an aid to an observer when there is a contrast between the object and its background. The greater the contrast in color, the more visible the object is. Usually darker shades of a given color will be less likely to attract an observer's attention than the lighter shades.

Movement—The last factor of recognition is movement. Although this factor seldom reveals the identity of an object, it is the most important one of revealing location. Movement is detected easily and usually through the observer's peripheral vision.

Concealment of Objects

5-156. Hiding is the concealment of an object by some form of physical screen.

Hiding—Using thick vegetation or terrain features that screen vehicles from ground observation. In some cases, the screen itself can be invisible to detection and, at times, it is the overt screen that protects the activity or equipment from observation.

Blending—Arranging or applying camouflage materials on, over, or around an object so that it appears to be part of the background. Blending distinctly man-made objects into a natural terrain pattern is necessary to maintain a normal and natural appearance.

Disguising—Simulating an object or activity so that it looks like something else. Clever disguises will mislead the enemy as to identity, strength, and intention.

Camouflage in the Desert

5-157. Camouflage challenges encountered in the desert require special attention to overcome. The lack of natural overhead cover, the increased range of vision, and the bright tones of terrain all require emphasis on sitting, dispersion, and camouflage discipline to achieve concealment. Cast shadows are notably conspicuous. Deserts generally have large areas of sand, little tall vegetation, brilliant sunlight, and extreme temperatures. Rocky areas, steep wadis, and washes characterize desert environments. The density of vegetation coverage is often as high as 80 percent. Most of the vegetation is low, averaging about 30 inches high in flat areas, while in the wadis and at higher elevations, it can average close to 10 feet. When viewed from the air, the desert floor appears spotted or pockmarked in many areas. Vegetation commonly found in the desert includes colors ranging from pale yellow to dark gray and dark brown. Although green and brown are the principal colors of most desert vegetation, it is important to study the target area vegetation and terrain to formulate a proper vehicle camouflage plan. No one camouflage system or pattern will work for every desert or even different parts of the same desert. Only with detailed planning can a mounted detachment plan for and prepare the materials necessary to properly conceal their vehicles.

Further Camouflage Considerations

5-158. In preparing for desert operations, position selection, reflection reduction, and concealment are conditions the team must consider--

Position Selection

5-159. Site or position selection is of critical importance in any environment but particularly so in the desert. Site positions that fit into the existing ground pattern with minimum alteration to the terrain are ideal. The sites selected should suppress ground observation. Some areas such as valley floors might have sparse vegetation, but adjacent wadis could offer thicker vegetation with opportunities for defilade and enhanced potential for concealment from aerial threats. Day laagers should not be areas that would be obvious to enemy patrols. The team leader usually positions the vehicles to provide 360-degree security and good concealment, and to allow rapid egress from the position.

Reflection Reduction

5-160. Reducing surfaces that reflect light is a measure that starts in garrison before deployment. It involves removing mirrors and covering headlights and taillights. The windshield can be left on so that it provides protection from blowing sand, dust, and rocks thrown up by the vehicle in front. The other option is to remove the glass and have team members use eye protection. The windshield frame should not be removed because it provides rollover protection. Team members cover all reflective surfaces with a close-weaved, non-see-through cloth such as canvas or target cloth. They leave a sight portal open for driving. If cloth or other material is unavailable, they mix water and dirt to get mud, and apply it to the reflective surfaces.

Concealment

5-161. Usually the best way to conceal vehicles is with nets. Ideally, use the Lightweight Camouflage Screening System (LWCSS) in the desert. These nets provide concealment from visual, near infrared, radar, and target-acquisition devices. This net is not intended as a complete camouflage system as it depends on imitation of the ground surface, both in color and texture, to be effective. In some deserts, the woodland pattern would blend in better. Alternatives to using the LWCSS are--

- Use open-weaved cloth with color patches to match the terrain in the operational area. This type of net might be the best choice in an area consisting mostly of sand dunes.
- Garnish a large fishing net with burlap to suit the color of the operational area.
- Add vegetation to this net to enhance concealment.

5-162. In open areas, drape the net over the vehicle and slope the sides gradually to the ground. Break up the outline of the vehicle by placing props or poles underneath, and then intertwine vegetation into the net. Eliminate shadows caused by the vehicle or net. In broken country, use the drape to tie the net to some irregularity in the terrain such as next to a mesquite or brush mound. Break up the outline and eliminate shadows. After placing the net, cut and place brush into the net to add realism, texture, and similarity to the terrain and to help break up the outline.

Chapter 5

Maintenance and Recovery

5-163. Preventive maintenance is critical to being able to execute mounted operations. Long supply lines and minimum stocks on hand will increase the time needed to get vital replacement items and repair parts. Proper maintenance must be performed on equipment throughout the whole spectrum of service, that is, before, during, and after operations.

Organization

5-164. R&S squadron units conduct operator level maintenance as with any other unit. Organization level maintenance is provided by the BSC. The R&S squadron receives organizational support from a maintenance team provided by the BSC. The LRSC also receives support from this BSC maintenance team. As a result, it is highly unlikely a LRS team conducting a mounted mission will be accompanied by a mechanic from the BSC. Therefore, the mounted LRS team should prepare itself to handle all operator and unit maintenance during a mission. In addition, some depot-level knowledge may be necessary. Team members regularly attend maintenance courses for the mobility platforms the unit uses.

Preventive Maintenance Checks and Services

5-165. The vehicles assigned to a mounted LRS team must be properly maintained and serviced. Its members must perform routine PMCS on their vehicles before, during, and after all operations. The vehicles also require regular operation. The team must perform post-operations maintenance procedures immediately after the conclusion of each mission.

Desert Environmental Effects

5-166. Several factors affect mounted operations in a desert environment:

Rough Terrain

5-167. Severe terrain consisting of rough, uneven ground, steep mountains, and loose sand and rocks will cause vibrations and result in the loosening of nuts, bolts, fuel, and hydraulic lines. It could also disrupt electrical components. Rough terrain can severely affect tires, wheels, transmissions, and suspension systems. Therefore, frequent inspections are necessary to ensure vehicles function properly and to prevent long downtime due to repairs.

Sand and Dust

5-168. The abrasive effects of sand and dust adversely affect equipment. Any moving part faces the probability of being damaged or impaired by sand or dust. Brakes, recoil systems, bearings, hydraulics, and relays are all susceptible to incapacitation by sand or dust. Also, sand and dust mixed with lubricants turns into an abrasive paste that can easily wear and score moving parts. Cover equipment when not in use. Frequent preventive maintenance will help to alleviate these problems to a manageable degree.

Heat and Low Humidity

5-169. Surface temperatures can reach 140 degrees and reflect heat under and into vehicles. Surface temperatures heat parts and accessories making them untouchable without protection. Such intense heat coupled with low humidity can overheat the vehicles and batteries, and can degrade the seals and tires. Frequent inspections, protection with covers, and regular maintenance can aid in reducing the effects of these environmental factors.

Vegetation

5-170. In some deserts, thorny and spiny plants pose a serious problem for tires, and can puncture radiator hoses. Use of proper individual driving techniques is the first preventive measure for stopping flats.

Section IV. OTHER OPERATIONS

The team can also be inserted by other means such as by Airborne operations, stay-behind operations, and foot operations.

AIRBORNE OPERATIONS

5-171. Air insertion is the fastest way to infiltrate. LRS teams and equipment may insert by parachute, by static line, or by free-fall techniques.

PLANNING CONSIDERATIONS

5-172. Units must plan—

- To coordinate for the suppression of enemy air defenses along the infiltration corridor.
- To determine whether enemy air defense artillery lies within artillery or naval gunfire range.
- To coordinate with the transporting unit.
- To consider and prepare for in-flight emergencies.
- To use an adverse weather aerial-delivery system during limited visibility or adverse weather.
- To dispose of parachutes, once assembled.
- Lost or dead Soldier.

LANDING PLAN

5-173. Leaders plan the operation using reverse planning. The ground tactical plan drives the other plans. The landing plan includes—

- Place of delivery.
- Time of delivery.
- Assembly area.
- Method of delivery (type of parachutes).
- Sequence of delivery. Team may be transported on an aircraft with personnel dropping on a different DZ.
- Load in order of the sequence of drops.
- Door bundles.

AIR MOVEMENT PLAN

5-174. The air movement plan includes the manifest, load plan, flight routes, in-flight checkpoints, flight times, load time, station time, takeoff time, and time on target.

MARSHALING PLAN

5-175. The jumpmaster gives his briefings. The team conducts sustained Airborne training. Leaders plan all joint tactical operations and support. The LRS team, equipment, and supplies are moved to departure airfield. Leader must know the answers to the following questions:

- Aircraft location.
- Transportation to the airfield.
- Linkup point for transportation.
- No later than team arrival time at a specified location.

STAY-BEHIND OPERATIONS

5-176. The stay-behind team lets the enemy bypass so they can perform a specific mission behind enemy lines.

PLANNING CONSIDERATIONS

5-177. When friendly forces expect an enemy offensive and friendly defensive operations, or when friendly forces are conducting limited offensive or reconnaissance operations, a stay-behind operation might offer the best way for a LRS team to infiltrate. In both cases, the forward friendly unit escorts the LRS team to the AO and provides security during site preparation.

SITE PREPARATION

5-178. Because the enemy is expected to overrun and occupy the LRS team's AO, they must prepare a good subsurface site. The team can stock enough supplies to operate for an extended period in a subsurface hide site. Engineer support is highly desirable in the construction of such a site (Appendix J).

FOOT MOVEMENT OPERATIONS

5-179. When traveling on foot, the LRS team departs as usual from a secure area. The team can move on foot alone, or can combine foot and vehicle movement. They normally move during limited visibility. They always depart from a secure area. To prevent enemy detection, they travel over rugged terrain normally not occupied by enemy forces.

PLANNING CONSIDERATIONS

5-180. Route planning requires extensive intelligence on enemy unit locations. The team needs fire support during movement.

INTELLIGENCE

5-181. Ground surveillance radar (GSR) can help them avoid enemy units, and radar-detection systems alert them when the enemy uses it. Tactical communication-intercept systems can warn them of actual enemy along the infiltration route.

SUPPLIES

5-182. The team can only carry enough supplies to move short distances for short periods of time, normally not more than a few days. Because the team's supplies may be depleted once they arrive at the AO, the parent unit must place a priority on resupply.

Chapter 6
Communications

This chapter discusses the networks (Section I), operations (Section II), radios, computers and base radio station (Section III), reports (Section V), electronic warfare (Section VI), antennas (Section VII) and operational environments (Section VIII), LRSU use to send and receive near real-time information. It also discusses communications in electronic warfare (Section IV) and unusual environments (Section VII).

Section I. NETWORKS

The LRSC must use several communications networks simultaneously. For example, the COB communicates internally, to the AOB, to higher, and to deployed teams. The AOB maintains nets to the deployed teams and the COB, and must be ready to communicate with the R&S squadron S-2, BFSB S-2, G-2, or J-2, if needed. The LRSC maintains an internal communications net with deployed teams. The deployed team must maintain a net to higher echelons and a team internal net.

ARCHITECTURE AND FREQUENCY MANAGEMENT

6-1. The LRSU have sophisticated and powerful communications equipment. They must also have access to multiple frequencies in multiple spectrums. Both are needed for the LRSU to send and receive near-real time information over many types of digital and analog systems.

ARCHITECTURE MANAGEMENT

6-2. The LRSU will need frequencies in the HF, VHF and UHF spectrums. Current communications systems operate in all three spectrums. The LRSU need multiple high frequencies for HF radio systems ever-changing optimum frequency of transmission (FOT) as well as multiple channel assignments for automatic link-establishment (ALE) radios.

FREQUENCY MANAGEMENT

6-3. Such complex communications require extensive frequency management. The BFSB S-6 is responsible for requesting frequencies with the JTF, corps or division G-6 to ensure that the unit is allocated a sufficient amount and type of frequencies to accomplish the mission The R&S squadron S-6 and the LRSC signal platoon leader submit all frequency requests thru the BFSB S-6.

OPERATIONS BASES

6-4. Three primary networks and two backup networks are normally established for communications between operating bases:

PRIMARY

6-5. This includes--
- Internal wire net with tactical switching system (landline telephone).
- Tactical local area networks (LAN) for communication by computer or voice over internet protocol (VoIP) phones.
- Combat net radios (single channel ground and Airborne radio system (SINCGARS)) and AN/PRC-148.

Chapter 6

BACKUP

6-6. This includes--

- HF radio.
- UHF tactical satellite radios.
- Secure cellular/satellite phones.

TEAMS

6-7. For internal communications, the LRS teams use secure LOS combat net radio systems. Secure, handheld, lightweight radios like the multipurpose and multiband inter/intra team radio (MBITR) incorporate frequency hopping (FH) and embedded communications security (COMSEC) that are compatible with the SINCGARS. These radios also allow communications with other Army and joint elements, including aircraft, and thus are ideally suited to LRS operations.

Section II. RADIOS, COMPUTERS, AND THE BASE RADIO STATION

R&S units that see everything and cannot report what they see are a wasted resource. The ability to communicate is the lifeblood of LRSU, and radios are the heart that make this possible. LRSU must be experts in the use of multiple radios systems and in the three primary military radio frequency spectrums: high frequency (HF), very high frequency (VHF), and ultra high frequency (UHF). LRS Soldiers must be highly proficient in programming, troubleshooting, and maintaining many types of radios.

ELEMENTS OF SUCCESS

6-8. Successful communications depend on—

- The type of emission.
- The amount of transmitter power output.
- The characteristics of the transmitter antenna.
- The amount of propagation path loss.
- The characteristics of the receiving antenna.
- The amount of noise received.
- The relative sensitivity and selectivity of the receiver.
- An approved list of usable frequencies within a selected frequency range.

HF, VHF, AND UHF RADIOS

6-9. These three radio wave spectrums combine to provide the primary and alternate means for LRSU to effectively communicate on the battlefield.

HF RADIOS

6-10. High frequency radios are harder to maintain than the commonly used LOS radios. However, they provide an unbeatable combination of reliability, economy, transportability, and versatility. Under ideal conditions, a HF radio using only 20 watts of transmitter power can successfully communicate over thousands of miles. Knowledgeable operators, backed by well-designed antennas and by propagation predictions from a propagation-engineering service, are key to successful HF radio system performance. Modern HF radios, such as the AN/PRC-138 and AN/PRC-150, incorporate the technologies of ALE, link quality analysis (LQA), embedded COMSEC, and digital modems are ideal for LRSU operations. These radios simplify HF communications and increase reliability and interoperability (Table 6-1).

VHF Radios

6-11. These are generally simple to use and provide reliable and clear, short-range tactical communications. The SINCGARS series of radios provide tactical units excellent communications that is easy to secure from enemy eavesdropping.

Table 6-1. Radios that work with AN/PRC-150 in various security modes.

	HF	VHF	PT	CT	KY-57 External	LOS	NLOS
AN/PRC-148 MBITR		X	X	X		X	
AN/PRC-152		X	X	X		X	
AN/PRC-119 SINCGARS		X	X	X		X	
AN/PRC-117A/D/F		X	X	X		X	
AN/PRC-113	X		X	X	X	X	X
AN/PRC-138	X	X	X	X			
AN/PRC-112A/C	X		X			X	
MX-300B6/B12		X	X	X		X	
TR720A/B/C		X	X			X	
Saber 5/G6		X	X	X		X	
PSC-5C/D		X	X	X		X	X
LST-5C			X	X	X		X

UHF Radios

6-12. These provide reliable tactical (LOS), operational, and strategic communications. However, due to the high demand and to potential interoperability problems with other units, it is not always practical for LRSU to use this spectrum.

PRIMARY, ALTERNATE, AND CONTINGENCY RADIOS

6-13. The COB and AOB maintain long-range communications with employed teams using HF and UHF TACSAT radios. For single-channel HF radio systems, each team should have a separate frequency. However, due to ever-changing ionosphere conditions and competition for frequencies, two teams might have to share a single frequency. If so, the COB should set up primary, alternate, and guard frequencies; use the primary and alternate frequencies for scheduled communications traffic; and use the guard frequency only for priority traffic--

- To report ISR tasks.
- To request extraction and fire support.
- To request medical evacuation.

6-14. The LRSC communications platoon leader must carefully design HF networks that use ALE. To ensure network reliability, he must analyze in detail the number of deployed teams, the availability of frequencies, the distances between stations, and the configurations of radio sets. Since ALE and 3G-capable radios automatically choose the best frequency for a particular radio path, he should program separate day and night channel groups to speed link establishment.

Chapter 6

FUNDAMENTALS

6-15. The team RTO transmits important information over the HF radio system. He continually adjusts it to keep up with changing conditions and missions. Successful HF communications depends on his knowledge; the type of emission (voice or data); the transmitter power output; selection of the best possible antenna and antenna site; proper antenna construction; propagated frequencies; terrain and weather; and atmospheric conditions. The variable over which he has the most control is antennas. To help eliminate skip zones, the RTO can achieve the NVIS effect with any HF-friendly antenna. This lets him establish communications with the COB or AOB. Extensive training of team members on HF radio systems and antenna construction is essential to mission success (TC 9-64, FM 6-02.74).

BEYOND-LINE-OF-SIGHT EQUIPMENT

6-16. In addition to communicating with many other types of digital and analog equipment, the LRSU also requires equipment that can communicate beyond line of sight (BLOS). Tactical VHF radios like the SINCGARS, are LOS only. The LRSU must be experts in the use of HF and TACSAT systems. Only HF allows long-range communications without the use of terrestrial or satellite relays. The LRSU can send either secure voice or data over HF.

SPECIALIZED RADIO MODEM

6-17. The ALE controller (modem) automatically controls a HF receiver and transmitter. This allows the radio to establish the best possible link with one or more HF radio stations. Each ALE controller (radiotelephone) can be embedded (internal) or external to modern HF radio equipment. It works on the principles of LQA and--

- Has in memory a predetermined set of frequencies, each properly propagated for conditions.
- Continuously scans its memory channels, typically about two channels per second.
- Has call signs programmed in, including own (SELF) and network's.
    -- Network (NET) call signs.
    -- Group (GROUP) call signs.
    -- Individual (IND) call signs.
- Transmits LQA, each of which sounds the programmed frequencies to find the one with the best link quality factors on a regular or automated schedule or when initiated by the operator.
- In a listening mode, logs each station's call sign and ranks the station's associated frequencies and channels based on the quality of the link.
- When someone at the station wants to place a call, tries to link to the outstation using the data collected during ALE and sounding activities. In the absence of this data, seeks the station and tries to link a logical circuit between two users on a network with all channels working. When the receiving station hears its address, the ALE controllers stop scanning channels and remain at that frequency. Each station notifies users that it has found the other station and is checking to confirm communications compatibility. This is called a "handshake." Once the handshake is complete, each station notifies its users that it is ready for traffic. Figure 6-1 shows communications between two stations during the "handshake" and LQA.

Handshake Process	Call Station	Message	Receive Station
	Ranger 4	To Ranger 6 "This is Ranger 4"	Ranger 6
	Receive Station	Message	Call Station
	Ranger 4	"To Ranger 4" "This is Ranger 6"	Ranger 6
	Call Station	Message	Receive Station
	Ranger 4	To Ranger 6 "This is Ranger 4"	Ranger 6
	System Linked		

Figure 6-1. Automatic link sequence.

- At the end of a link session, the ALE controllers send the link command TERMINATION, and returns to scanning mode to await further traffic. Built-in safeguards ensure that ALE controllers return to scanning mode if contact is lost.

AN/PRC-150(C) ADVANCED HF OR VHF TACTICAL RADIO SYSTEM

6-18. The AN/PRC-150(C) is a HF transceiver that covers the frequency range from 1.6 to 60 MHz in SSB and FM modes. Embedded COMSEC allows secure communications between ground and aircraft as well as with the Army's SINCGARS radios. The AN/PRC-150(C) also has an internal, high-speed, Military Standard 188-110B serial-tone modem, which sends and receives data at speeds up to 9,600 BPS; an embedded military standard 188-141A; ALE; digital voice 600 (DV 600); and frequency hopping (electronic protection). The AN/PRC-150(C) belongs to a family of interoperable software-designed radios. This family also includes the AN/PRC-117F(C), which is the manpack test platform for Step 2B of the Joint Tactical Radio System program. The AN/PRC-150(C) gives units BLOS communications without the need to rely on satellites from a crowded battlefield. The systems' manpack and vehicular configurations ensure reliable communications and allow rapid transmission of data and imagery. The AN/PRC-150 replaces the AN/PRC-138.

AN/PRC-148

6-19. The MBITR AN/PRC-148 is a lightweight, durable, compact radio. Its secure multiband voice and data communications are ideally suited for use by LRS teams. It interoperates with a wide variety of existing military and civilian systems, while providing the LRS team leader internal C2. The MBITR's built-in emergency beacon and a GPS interface with PLGR can serve as an emergency radio during escape and evasion (E&E) operations. It transmits in the 30- to 512- MHz frequency range and allows communications in the following bands:

- VHF FM and AM.
- UHF AM (air to ground).
- UHF FM (LOS).

AN/PRC-152

6-20. The AN/PRC-152 is a multiband, lightweight, handheld radio. An optional, built-in GPS receiver allows time tracking and position reporting. Embedded COMSEC supports Vinson, advanced narrowband digital voice terminal (ANDVT), AES, Fascinator, and KG-84. The AN/PRC-152 operates between 30 and 512 MHz, and is compatible with many military and civilian radio systems. VHF/UHF line of sight supports AM and FM modulation as well as high-performance waveform and TACSAT communications.

Chapter 6

AN/PRC-117F

6-21. The AN/PRC-117F is a multiband, multimission, 30 to 512 MHz radio. All -117Fs (manpack, vehicular, marine, and base station) have COMSEC, UHF TACSAT, ECCM, and DAMA capabilities. The AN/PRC-117F works with many communications systems, including SINCGARS, AN/PRC-148, AN/PRC-112, AN/PSC-5, and most civilian handheld radios. Like the AN/PRC-150, the AN/PRC-117F interfaces with many data devices, to include ruggedized laptop computers. Although the AN/PRC-117F is microprocessor-based, it is controlled by software rather than hardware. It can retransmit voice or data across traditional frequency bands and waveforms with two antenna ports and data rates up to 64 Kbps.

AN/PSC-5C/D

6-22. The AN/PSC-5 is a multiband, multimission communications terminal. It provides excellent interoperability with military, marine, and civilian radio systems. It operates in the VHF and UHF frequency spectrum (30 to 512 MHz), and supports line of sight (LOS), TACSAT (5K, 25K and DAMA), SINCGARS and Havequick I and II. It has an embedded COMSEC engine, which allows the sending of secure voice and data. It can achieve data rates of 76.8 Kbps.

INTEROPERABILITY

6-23. Table 6-2 shows the interoperability capabilities and characteristics of the radios commonly used by LRSU.

Table 6-2. Radio interoperability capabilities and characteristics.

Radio	Freq Range	Max Power Output	Data Controller	TACSAT	Vehicle Kit
AN/PRC-150	1.6-60 MHz	20 Watts	Yes	N/A	AN/VRC-104
AN/PRC-148	30-512 MHz	5 Watts	N/A	5K	AN/VRC-111
AN/PRC-152	30-512 MHz	5 Watts	Yes	5K, 25K	AN/VRC-110
AN/PSC-117F	30-512 MHz	20 Watts	Yes	5K, 25K, DAMA	AN/VRC-103
AN/PSC-5C/D	30-512 MHz	20 Watts	Yes	5K, 25K, DAMA	Multiple kits available
AN/PRC-119F	30-88 MHz	4 Watts	N/A	N/A	VRC-89/90/91/92

RETRANSMISSION

6-24. Retransmission can greatly extend the range of a radio LOS network. Traditionally, SINCGARS retransmission networks are used with two different frequencies or net IDs, called F1 to F2 retransmission. This requires planning and establishment of triggers where radios will have to switch frequencies based on their location on the battlefield. With the ASIP radio, users can use same net retransmission using the same frequency or net ID. This is called F1 to F1 retransmission. Most current radios support retransmission operations with the use of a retransmission cable. If the range between two networks it too great for ground wave radios, two LOS networks can be connected using TACSAT radios. Both the AN/PSC-5 (all models) and the AN/PRC-117F will connect two LOS (VHF/UHF) networks by way of satellite communications.

VIDEO TRANSMISSION

6-25. Each deployed LRS team uses a lightweight video-reconnaissance system to send and receive real-time images.

RUGGEDIZED COTS LAPTOP

6-26. The LRS teams use this ruggedized, standard laptop to send and receive text messages and images over the radio. A serial port or external data controller card connects the laptop by cable to the data port of the radio. A data controller card is needed to send data via radio waves. The card manages the data reducing errors and transmission times. Some radios use "tactical chat" or wireless messaging-terminal software as the Graphic User Interface for sending and receiving data. Radios without an internal data controller card require an external card. Either an inline cable controller or Personal Computer Memory Card International Association (PCMCIA) card will work with the laptop as long as it supports the cards interface. In the field, special charging kits allows a team to operate and charge a laptop with a BA-5590 or BB-390 / BB-2590 battery.

COMMUNICATIONS BASE RADIO STATION PLATFORM

6-27. Each LRSC is authorized eight BRS platforms. The BRS is a multifunctional communications platform currently in development. In addition to HF communications, each BRS provides;

- TACSAT communications with the AN/VRC-103(V1 or V2).
- VHF communications via AN/VRC-92, AN/VRC-110, or AN/VRC-111.
- Network capability with interface to existing secure and nonsecure networks.
- Modular and reconfigurable to meet changing mission requirements.
- Dismountable and can be stored or transported in transit cases.

Section III. OPERATIONS

BRS comprise the most critical part of the LRSU communications network. It is the primary link between the commander and his deployed teams. Each BRS monitors all deployed team frequencies and channels.

TACTICAL EMPLOYMENT

6-28. All LRSU BRS are based on the Army's Transformation High-Frequency Radio System (THFRS). This system is in turn based on the AN/PRC-150(C) manpack radio (Figure 6-2). The THFRS can be configured in the same basic manner as the older AN/TSC-128 in an S-250 communications shelter. The THFRS works with various power amplifiers, couplers, antennas, software, and ancillaries to build various vehicular and BRS configurations.

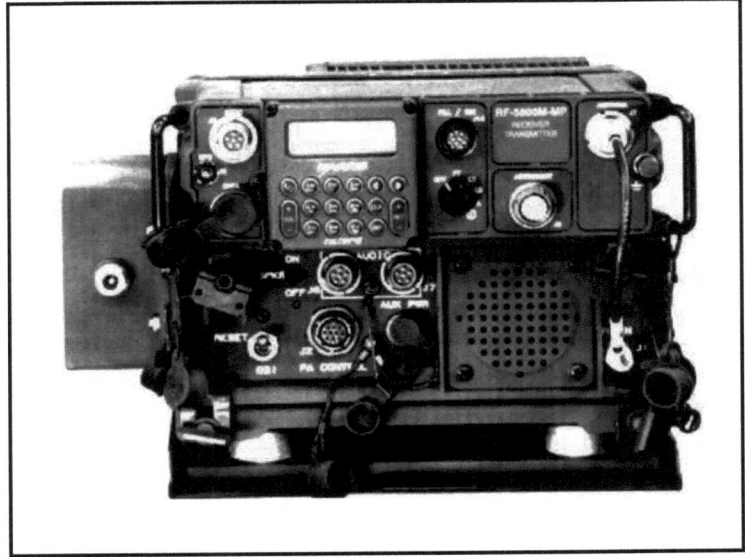

Figure 6-2. AN/PRC-150(C) in vehicular AN/VRC-104 (V)3 configuration.

Chapter 6

COMPANY OPERATIONS BASE

6-29. When space allows, the BRS should be physically located with or in close proximity to the COB TOC. This allows a direct link between the operations cell and the BRS operators. If circumstances prohibit this employment, the BRS is linked to the TOC by VHF, UHF TACSAT, tactical LAN, or field wire. A new BRS configuration is in development.

ALTERNATE OPERATIONS BASE

6-30. This base relays communications between the LRS teams and the COB BRS. It links to the COB through joint or Army tactical switching systems.

- The COB and AOB use HF and UHF TACSAT radios as secondary means of communication.
- Message traffic between the two stations should travel by the fastest, most secure means available.
- Due to variables such as terrain and interference, the AOB can sometimes receive messages the COB cannot.
- The AOB BRS is normally positioned farther from the deployed teams than the COB BRS and can operate mounted or dismounted. It is prepared to assume the mission if the COB, if the COB displaces or is destroyed.
- The communications platoon leader recommends to the LRSC commander the approximate distance and location of the AOB based on, among other factors--
  -- Probability of communications with the deployed teams and the COB BRS.
  -- Available local support or unit capability to support.

HF OR UHF TACSAT RADIO

6-31. These are the surveillance team's primary means of communication with the BRS. Data-burst equipment and compression software shorten transmission times. Encryption prevents the enemy from deciphering radio transmissions. Lightweight digital photo or video systems allow transmission of near real-time imagery.

- The team leader selects a communication site using METT-TC. The site should allow antenna construction and terrain masking.
- Teams transmit and receive routine messages during scheduled communication windows. For messages requiring transmission outside the schedule, the team first establishes a link with the COB or AOB in ALE mode or with the guard frequency, if in single sideband (SSB mode), and then transmits the message.
- Internal team communications is via secure VHF radios and visual signals. Leaders ensure everyone takes the proper OPSEC and COMSEC precautions.

SITE SELECTION

6-32. The reliability of radio communications depends largely on proper radio site selection. The communications platoon leader and the BRS team leader must ensure that both primary and alternate sites satisfy technical, tactical, security, and other performance criteria.

ALL OPERATIONS BASES

6-33. The site needs good cover and concealment, and a location free of interference (man-made or natural). Moving the site may be necessary if interference becomes a problem. Common sources of interference include; high-tension power lines, over population of antennas, electronic countermeasures, thick vegetation and terrain.

COMPANY OPERATIONS BASE

6-34. The COB is the primary link between the deployed teams and BFSB S-2 fusion element. COB BRS is normally located well within the security umbrella of the BFSB. It should be close enough to the BFSB and R&S squadron S-2 sections to permit a land wire network for reporting purposes.

ALTERNATE OPERATIONS BASE

6-35. The AOB may collocate if communications are established and maintained between the deployed teams and the COB. For increased survivability and redundancy, the AOB may be located elsewhere in the AO. If communication cannot be established or maintained between the teams and the COB, the AOB is moved in order to establish communication with the deployed teams and the COB. When the AOB is used as the primary reporting link, it must maintain a constant communication path with the COB, while the COB generally moves with the BFSB or R&S squadron.

TACTICAL SATELLITE

6-36. UHF TACSAT radio is a reliable communications system with unlimited range. It comes in both manpack and vehicle configurations. The best systems for LRS missions are multiband, multimission, multisystem-compatible UHF TACSAT systems with--

- Embedded demand-assigned, multiple access (DAMA) capabilities.
- Satellite communications modems.
- Diverse communications and transmission security capabilities.

6-37. Understandably, satellite channels and UHF TACSAT systems are in high demand and are also in short supply. Because the priority for UHF TACSAT channels goes to division HQ and above, joint and special operations units, LRSUs usually must share satellite channels. For this reason, the HF radio remains the primary means of communication. When LRSU do get satellite access, they must carefully manage it for airtime and message precedence.

Section IV. REPORTS

Teams communicate with the BRS at specified times or per-designated communications windows, with each team having a separate window. The number of scheduled times used by the LRSU depends on METT-TC. Scheduling windows too often places a team at risk, while scheduling windows too seldom can reduce the relevance of time-sensitive intelligence.

MESSAGES AND REPORT FORMATS

6-38. To accomplish their mission, LRS teams must send timely and accurate messages, properly formatted, to the COB, AOB or MSS BRS. Each team does this during assigned "windows," based on METT-TC. Using too frequent windows raises susceptibility to enemy interception and direction-finding capabilities; however, using too few windows reduces the relevance--and usefulness--of time-sensitive intelligence. For the purpose of this manual, a message refers to the information sent from one station to another. Most messages follow a report format.

MESSAGES

6-39. Each BRS logs in detail all messages it sends and receives. The unit SOP specifies how done. The BRS team chief in the COB ensures that all messages for committed teams originate with the operations section and that they are properly formatted.

Interoperability

6-40. Report formats provided below are based on the standardized formats in FM 6-99.2. LRSC should base unit SOPs on these report formats in order to gain rapid interoperability between LRSU.

Chapter 6

Incoming Messages from Team

6-41. When the BRS receives a message from a team, it is logged and forwarded to the operations section for decryption. Intelligence reports are generally sent directly to the BRS located at the COB, then to the BFSB S-2 fusion element and the R&S squadron S-2 after being logged and examined by the LRSC TOC. The LRSC TOC neither delays nor changes any intelligence report. Sometimes, the AOB BRS receives a message that the COB BRS does not receive. When this happens, the AOB logs the message and sends it, exactly as received, by the fastest, most secure means to the COB (Figure 6-3).

Outgoing Messages to Team

6-42. The LRSC operations section formats and encrypts any message going out to a team. The BRS then transmits it during that team's next scheduled communication time after the BRS team chief ensures the message is properly formatted.

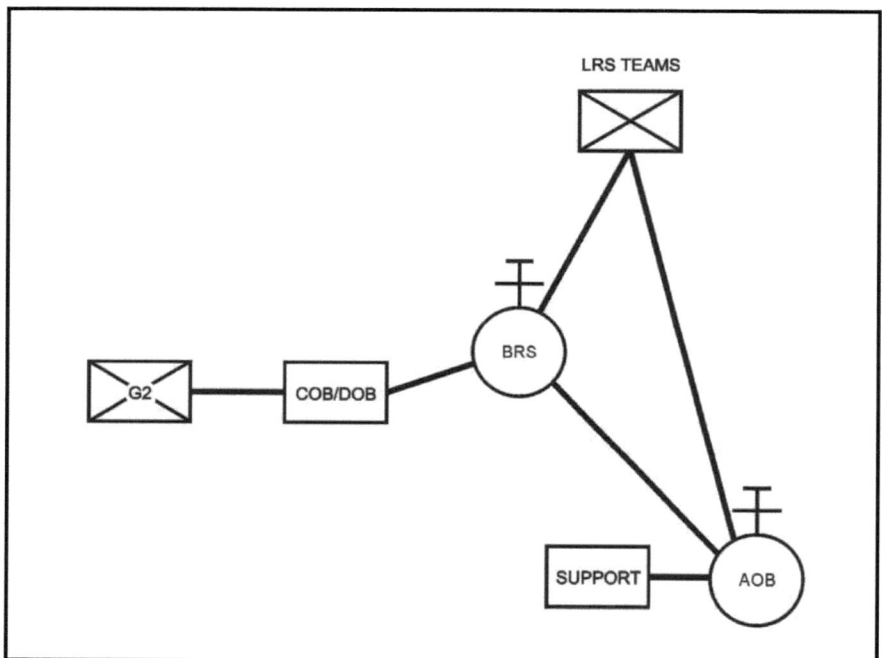

Figure 6-3. Communications data wire diagram.

Code Words or Letters

6-43. Code words or letters are used by transmitting stations to send vital information quickly, and in a secure manner. These letters and code words are given to the team during mission planning. They keep transmissions short. They inform the receiving station of the situation on the ground without long descriptions. Code words are also used to send vital information in a secure manner. Without knowledge of the meaning of the code words/ code letters, the meaning of message will not be known to any intercepting station or person.

Duress Codeword

6-44. A duress code is a simple word placed in a message to indicate the sending station is not under duress (Table 6-3). A duress code requires planning and rehearsal to ensure an appropriate response. This code is normally changed after each mission to avoid compromise.

- Only the team, the COB, the AOB and--if used, the MSS--know the duress code.
- The sending station inserts the code into a precise location in the message so the receiving station will know they did so deliberately, not under coercion. *Each* team and BRS has a *different* duress code.

Situation Normal

- Sender includes duress code in the correct location.
- Recipient responds to content of message.

Situation Compromised

- Sender omits the duress code.
- Recipients ignore content of message and responds to the emergency by initiating compromise procedures.

Table 6-3. Procedure for use of duress codes.

Situation	Sender	Recipient
Normal	Include duress code	Respond to content
Compromised	Omit duress code	Initiate compromise procedures

REPORT FORMAT

6-45. Information is placed into a report format (Figure 6-4) to aid encryption, decryption and information recognition. Using a report format makes even partially received messages useful, because the information is more recognizable. The message is divided into three parts.

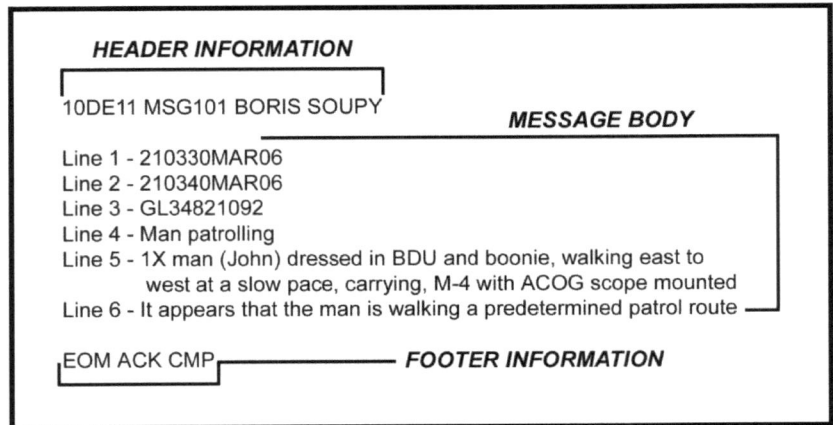

Figure 6-4. Report format.

Chapter 6

Header Information

6-46. Messages are numbered in sequence of transmission, the first number being the team number. If messages include pictures, they are named using the message number along with an alphabetical designation to match the picture with the corresponding message, for example, 101A, 101B (Figure 6-5).

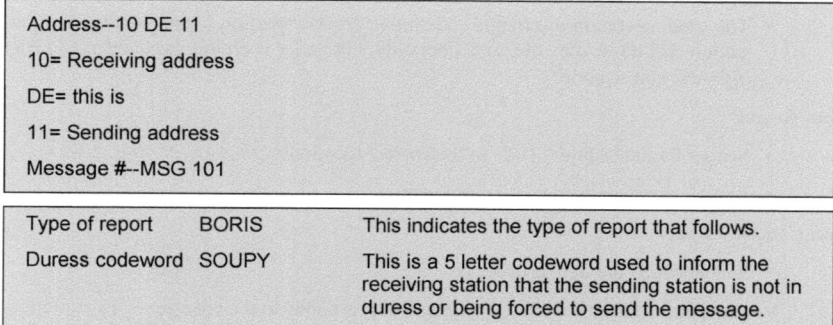

Figure 6-5. Example message header.

Message Body

6-47. The message body varies depending on the report format. Recipients must be able to recognize, understand, and react quickly to the information contained in the message. This means the unit SOP must provide for short, standard message language. This serves three purposes: observer guidance, speed and communications security. The format gives the observer a tool to report specific information. Knowing the format speeds the writing and reading of the message. Keeping messages short decreases transmission time and helps avoid enemy radio direction finding (RDF) units.

Footer Information

6-48. Particular information goes into a report footer:

End of message	EOM	This tells the receiving station that the message is complete.
Acknowledgement Requested	ACK	This requests that the receiving station acknowledge receiving the message.
Signature	CMP	Initials of the RTO responsible for transmitting the message.

REPORT FORMAT TYPES

6-49. LRSU use five basic report formats (Table 6-4):

Communications

Table 6-4. Report formats.

Proword	Actual Report Title	Purpose
Angus	Initial Entry Report	To alert the BRS and operations of the status of the insertion, the team's initial situation, and possible deviations from infiltration plan.
Boris	Intelligence Report	To report PIR/intelligence requirements/SIR and intelligence tasks when observed.
Cyril	Situation Report	To report the team's situation (excludes PIR/intelligence requirements/SIR and intelligence tasks).
Under	Cache Report	To report an emplaced cache.
Crack	Battle Damage Assessment Report	To report battle damage on a specified target.

Angus (Initial Entry) Report

6-50. The RTO normally sends the Angus (Initial Entry) Report as soon after the insertion as the tactical situation allows. This is usually completed within four hours of insertion. If the Angus is not transmitted within this window the LRSC operations section may assume the mission is compromised and initiate emergency procedures. This message alerts the operations section of the status of the insertion, the team's initial situation, and possible deviations from the infiltration plan due to previously unknown conditions on the ground. Table 6-5 shows the typical format of an Angus report and the message information contained in it.

Table 6-5. Typical format for an Angus (Initial Entry) Report.

Line No.	Content	Example
1	DTG	152307NOV06
2	Team status (use code words)	Green
3	Current location (6-digit grid with grid zone identifier)	GL098569
4	Possible deviations from briefed plan (inform higher of pending changes to team plan)	Due to restrictive terrain, the team will deviate more to the North on primary infiltration route
5	Remarks	None

Boris (Intelligence) Report

6-51. The RTO normally sends the Boris (Intelligence) Report to the BRS as soon as the LRS team has PIR to report. Other ISR tasks are normally sent during prescribed communications windows. Table 6-6 shows the typical format for a Boris report and the message information contained in it.

Chapter 6

Table 6-6. Typical format for a Boris (Intelligence) Report.

Line No.	Content	Example
1	DTG	131844SEP05
2	DTG of observed activity	131506SEP05
3	Location of observed activity	West side of hill GL96578354
4	Observed activity	Preparing radio to transmit and receive Manning reinforced fighting/defensive position
5	Description of personnel, vehicles, weapons, and equipment	4 pax in military uniform outside reinforced fighting position. The pax are called A, B, C, and D. Three of them, A, B, and C, are wearing PCs. Pax D is wearing a boonie hat. Pax A is manning a radio that is carried inside a rucksack and placed on top of the fighting position. (The radio has a long, whip-type antenna.) Pax B (fair skinned) is talking on the radio. Pax A and Pax B are both standing on the West side of the position. Pax C is standing on the North side of the position. Pax D is on the North side of the position, but is walking toward the East. Pax C and Pax D are carrying unidentifiable assault rifles. Pax D is wearing load-carrying equipment. The fighting position is a poured concrete structure, built into a berm, whose [wooden?] roof has with a small overhang. About 3 feet of the structure is visible aboveground. The structure is about 7 feet long (North to South). In the middle of the East and West walls are viewports. A triple-strand concertina wire obstacle runs North to South about 5 meters to the East of the fighting position. Triple strand wire is set up on the West side of an 8-foot tall chain link fence, which also runs North to South. V-shaped barbed wire runs along the top of the fence.
6	Team assessment	Believe A and B have weapons, though not observed. Pax B seems to be the leader of the group, because he is talking on the radio. The enemy seems to be preparing to man the fighting position for an unknown period of time. The enemy also appears to be in a nonaggressive posture, because their weapons are slung. Assume additional ammunition and possibly explosives are cached in the position. The layout of the obstacles and the location of the position suggest the position is used for observation and early warning.

Cyril (Situation) Report

6-52. The RTO must send the Cyril (Situation) Report during, and only during, scheduled communication windows. The Cyril reports the team's situation, status (medical, team equipment, food, water, batteries), past, current, and planned activity. Table 6-7 shows the typical format and content of a Cyril report. The team must send a Cyril report during every communications window.

Table 6-7. Typical format for a Cyril (Situation) Report.

Line No.	Content	Example
1	DTG	131844SEP05
2	Current location (8-digit grid with identifier)	JL14593487
3	Medical status of team (code words)	Green
4	Status of team equipment	Green
5	Status of food, water, and batteries (food and water per person)	3xMREs, 4xQts water, 14x 5590s, and 3xAAs
6	Team activity since last communications window	Pulled surveillance on objective. Moved survey site due to poor visibility on objective.
7	Team activity until next communications window	Broke down equipment and prepared for exfiltration
8	Team leader remarks	Weather deteriorated, dropping distance of standoff and visibility of objective

Under (Cache) Report

6-53. The LRS team RTO or the COB normally sends the Under (Cache) Report to report caches of personnel records, intelligence documents, personnel burials, and so on. After the team infiltrates, the LRSC operations section reports caches such as ammunition, demolitions, barter items, weapons, food, and water. Table 6-8 shows the typical format of an Under Report and the message information contained in it.

Table 6-8. Typical format for an Under (Cache) Report.

Line No.	Content	Example
1	DTG (date-time group)	131844SEP05
2	Type of cache (surface, subsurface, or submerged)	Subsurface
3	Contents	7xMREs 1,000 rds 5.56
4	Location (10-digit grid with identifier)	FT 7404620956
5	Initial and Final Reference points (IRP and FRP)	IRP: Intersection at 34590216 (300m E) FRP: Stream intersection at 34650236 (20 m N)
6	Depth	3 feet
7	Additional information	Cache is buried at the base of 50-foot oak tree that has scratch marks at knee level on the West-facing side

Crack (Battle Damage Assessment) Report

6-54. The Crack (Battle Damage Assessment) Report is used to provide a timely and accurate estimate of damage resulting from the application of military force, either lethal or nonlethal, against a predetermined objective. Table 6-9 shows an example for a Crack report and the message information contained in it.

Table 6-9. Typical format for a Crack (Battle Damage Assessment) Report.

Line No.	Content	Example
1	DTG	131256NOV05
2	Location of target (8-digit grid with identifier)	JL14593487
3	Type of target such as vehicle, building, or bridge	T-72 Tank
4	Description of target ***Physical Damage Assessment***--How much physical damage was inflicted by military force (munitions blast, fragmentation, or fire) on a particular target? This assessment is based on observed or interpreted damage. ***Functional Damage Assessment***--To what degree were the attack objectives achieved against a particular target? This assessment is based on the degree to which the application of military force degraded or destroyed the functional or operational ability of the targeted facility or objective to perform its intended mission.	Vehicle destruction is catastrophic. Vehicle is inoperable.
5	***BDA analysis*** The level of confidence in the accuracy of the assessment, and whether the reattack is necessary. ***Confirmed*** Data that has been confirmed visually or otherwise assured through IMINT, weapon system (aircraft cockpit) video, SIGINT, MASINT, or HUMINT (signal, measurement and signals, or human intelligence). • 95 percent confidence that assessment is accurate. • Data requires no additional intelligence. ***Probable*** • At least 50 percent confidence. • Data sources are reliable; data requires little additional intelligence. ***Possible*** • At most 50 percent confidence. • Data requires considerable additional intelligence.	Confirmed. No reattack necessary.

COMMUNICATIONS SECURITY

6-55. This function is management intensive for LRSU operations. The LRSC commander ensures the unit's COMSEC custodian keeps enough of the necessary materials on hand, both for training and contingency missions. Possible COMSEC considerations for LRSU operations include--

- JTF, corps or division nets.
- BFSB and R&S squadron nets.
- Internal company and team nets.
- Digital secure voice terminal key for MSE network.
- JTF, corps, division or BFSB UHF TACSAT keys.

Section V. ELECTRONIC WARFARE

Electronic warfare (EW) is any military action involving the use of electromagnetic and directed energy to control the electromagnetic spectrum or to attack the enemy (FM 1-02). There are three major subdivisions within electronic warfare: electronic attack, electronic warfare support, and electronic protection.

ELECTRONIC ATTACK

6-56. That division of EW involving the use of the electromagnetic energy, directed energy, or antiradiation weapons to attack personnel, facilities, or equipment with the intent of degrading, neutralizing, or destroying enemy combat capability and is considered a form of fires (FM 1-02).

ELECTRONIC WARFARE SUPPORT

6-57. That division of EW involving actions tasked by, or under the direct control of, an operational commander to search for, intercept, identify, locate or localize sources of intentional or unintentional radiated electromagnetic energy for the purpose of immediate threat recognition, targeting, planning and conduct of future operations (FM 1-02).

ELECTRONIC PROTECTION

6-58. That division of EW involving passive and active means taken to protect personnel, facilities, and equipment from any effects of friendly or enemy employment of EW that degrade, neutralize or destroy friendly combat capability (FM 1-02). LRSU are primarily concerned with electronic protection.

METHODS

6-59. These refer to anything a LRSU does to prevent or reduce the effectiveness of enemy EW and enhance electronic protection.

Security Tasks

6-60. *Emission security* includes--

- Using brevity lists.
- Masking antenna locations.
- Using directional antennas.
- Using the lowest possible output power.

6-61. *Transmission security* includes--

- Using voice communication only when essential.
- Developing and using brevity lists.
- Minimizing transmission time.
- Planning messages.
- Encrypting messages.

6-62. *Cryptograph security* includes--

- Exclusive use of *authorized* codes and key lists only. Only National Security Agency (NSA)-approved codes are authorized for encoding and decoding US Army message traffic. The same is true of mechanical cryptograph systems.
- *Physical security* of all cryptograph and equipment. This includes a comprehensive and workable plan for the destruction of material and equipment. It also includes the SOPs that identify to all team members where material and equipment are kept by the RTO. Table 6-10 shows the priority for destroying material and equipment.

Chapter 6

Table 6-10. Priority for destruction of communications devices.

1.	All superseded cryptographic keys.
2.	All current cryptographic keys.
3.	Zero all keyed devices.
4.	All future cryptographic keys.
5.	All cryptographic devices.
6.	Radios.
7.	Brevity list.
8.	Communications log.

Data-Burst Devices

6-63. To further reduce the chance that the enemy will use radio direction finding (RDF) equipment against them, the BRS and teams use data-burst devices. They can use the nonsecure OA-8990/P digital message device group, KL-43F, or portable computer. These devices shorten transmission times--they *do not* prevent the enemy from intercepting the radio traffic.

PROCEDURES

6-64. These procedures apply to interference, jamming, and deception. When someone at the BRS or on the team hears interference and suspects jamming, he should--

- Remain calm and continue to operate as if nothing unusual is happening.
- Prevent the enemy from knowing whether his jamming is successful or detected.
- Switch to a higher power on the radio.
- Reorient the antenna to the receiving station.
- Report the jamming using the meaconing, intrusion, jamming, and interference (MIJI) report format in the signal operating instructions supplemental instructions (Table 6-11). Send the report over a network free of jamming and interference to ensure that it reaches the intended recipient.
- Until communications can be established and maintain over the desired frequency, use an alternate one.

Table 6-11. Contents of a MIJI report.

Item	Contents
1	Type of report
2	Type of incident
3	Type of equipment affected
4	Frequency affected
5	Affected station call sign
6	Affected station coordinates

Section VI. ANTENNAS

This section discusses several concepts to help communications personnel select the best antenna.

WAVELENGTH AND FREQUENCY

6-65. A wavelength is the distance that an electromagnetic wave travels to complete one cycle at a particular frequency (Figure 6-6). In radio communication, the length of an antenna relates directly to the frequency's wavelength. This relationship is important to know when building antennae for a specific frequency or frequency range.

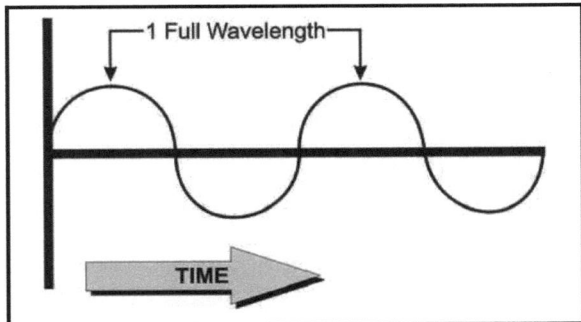

Figure 6-6. Measurement of a wavelength.

RESONANCE

6-66. Antennas are classified as either resonant or nonresonant, depending on their design. Both are commonly used on tactical circuits. However, if you can get a clear signal with a resonant antenna, that should be your first antenna choice rather than a nonresonant or standing-wave-ratio antenna.

RESONANT ANTENNAS

6-67. A resonant antenna matches the wavelength of one particular frequency.

- Advantage is efficiency; most radio signals sent to a resonant antenna radiate successfully.
- Disadvantage is lack of flexibility; a separate antenna must be built for each frequency used.

NONRESONANT ANTENNAS

6-68. These antennas match a range of frequencies.

- Advantage: This kind of antenna works with more than one frequency.
- Disadvantage: A nonresonant antenna, as the name implies, reduces resonance, which weakens the signal. The more frequencies the antenna resonates, the lower the resonance quality, which in turn reduces the efficiency of the signal.

STANDING WAVE RATIO

6-69. Signal energy resonates, or causes energy waves in a certain pattern on an antenna. These waves are measured and compared to the standard wave to determine if an antenna resonates at a particular frequency. Although a 1-to-1 ratio to a standing wave (standing wave ratio) is ideal, 1.1-to-1 ration is about the best ratio obtainable. When building wire antennas, the operator should adjust the length of the antenna until he obtains the lowest possible standing wave ratio. A 3-to-1 standing wave ratio is acceptable. Check the operator's manual for the particular radio in use to determine the maximum standing wave ratio that the radio can tolerate. Some radios automatically lower the power output of the transmitter if the standing wave ratio is too high.

POLARIZATION

6-70. Polarization is the relationship of radio energy radiated by an antenna to the earth. The most common polarizations are horizontal (parallel to the earth's surface) and vertical (perpendicular to the earth's surface). Others, such as circular and elliptical, also exist. A vertical antenna normally radiates a vertically polarized signal, and vice versa.

GROUND WAVES

6-71. For best communication with HF ground waves, both the sending and receiving antennas should have the same polarization. Vertical polarization works best for HF ground-wave propagation.

SKY WAVES

6-72. For HF sky-wave propagation, the sending and receiving antennas need not have the same polarization, because the ionosphere will bend the waves, thus randomly changing their polarization anyway. However, horizontal polarization works best for HF sky-wave propagation.

RADIO WAVE PROPAGATION

6-73. HF communications can be established using either ground- or sky-wave propagation. With low-powered, man-pack radios, ground-wave communication can be established out to about 30 km, depending on conditions. High-powered, vehicle-mounted equipment allows communication out to about 100 km. Sky-wave communications range from several to thousands of kilometers.

GROUND-WAVE PROPAGATION

6-74. Ground-wave propagation means sending a radio signal along or near the surface of the earth. The ground-wave signal has three parts: the *direct, reflected,* and *surface* waves (Figure 6-7).

Surface Wave

6-75. The surface wave travels, as the name implies, along the surface of the earth. It is the usual means of ground-wave communication. The surface wave depends on the type of surface that lies between the two antennas. With a good conducting surface, such as seawater, long ground-wave distances are possible. Poor surfaces, such as sand or frozen ground, shorten the distance the surface wave can travel. Heavy vegetation or mountainous terrain can do the same.

Direct Wave

6-76. The direct wave travels from one antenna to the other in what is called the line-of-sight mode. Maximum line-of-sight distance depends on the height of the antenna above ground. The higher the antenna, the longer the LOS. Because radio signals travel in the air, any obstruction between the antennas, such as a mountain, can block or reduce the signal. For an antenna 10 feet above the ground, the maximum LOS is 8 km (5 miles).

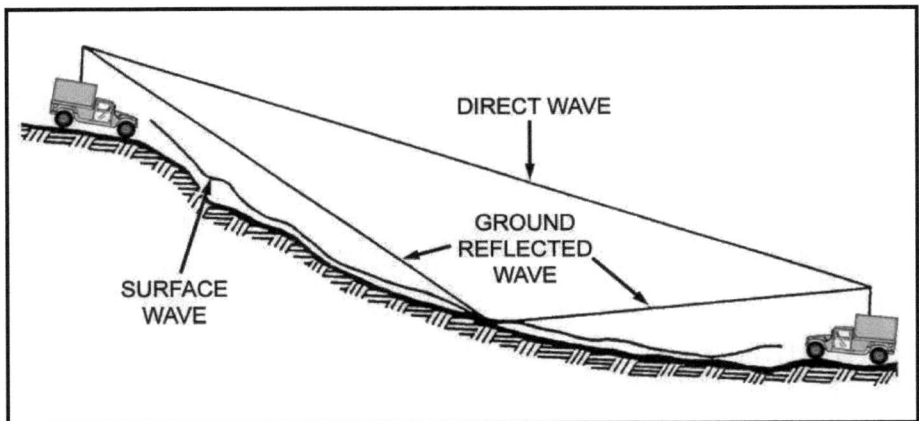

Figure 6-7. Components of ground wave.

Reflected Wave

6-77. This is a wave that bounces off the earth on its way to the receiving antenna.

Space Wave

6-78. This wave is the combination of a reflected wave and a direct wave.

SKY-WAVE PROPAGATION

6-79. HF signals travel much farther by sky-wave propagation than by ground-wave propagation. Sky-wave propagation is the bending (refraction) of the radio signal by a region of the atmosphere called the ionosphere.

6-80. The ionosphere is an electrically charged (ionized) region of the atmosphere that extends from an altitude of about 60 to 1,000 km (37 to 620 miles) above the earth's surface. Energy from the sun ionizes the atmosphere in this altitudinal range, and the electrical charge there refracts (bends) some radio signal that enters it, sending the signal back to the earth.

6-81. The area that affects HF communications the most lies between the altitudes of 48 km (29.6 miles, which lies below or inside the ionosphere) to 500 km (310 miles). This 440 km (273 mile) area is divided into four incremental altitudinal ranges: D, E, F1, and F2 (Table 6-12 and Figure 6-8).

Table 6-12. High frequency ranges in ionosphere.

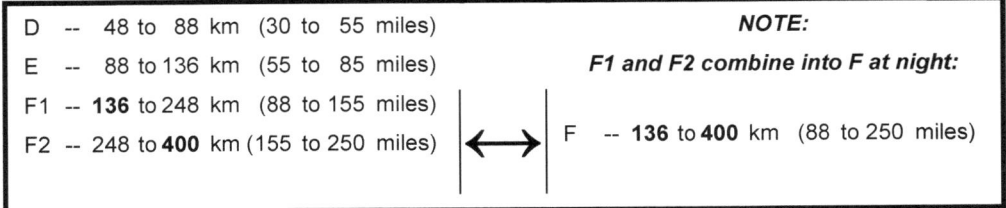

Chapter 6

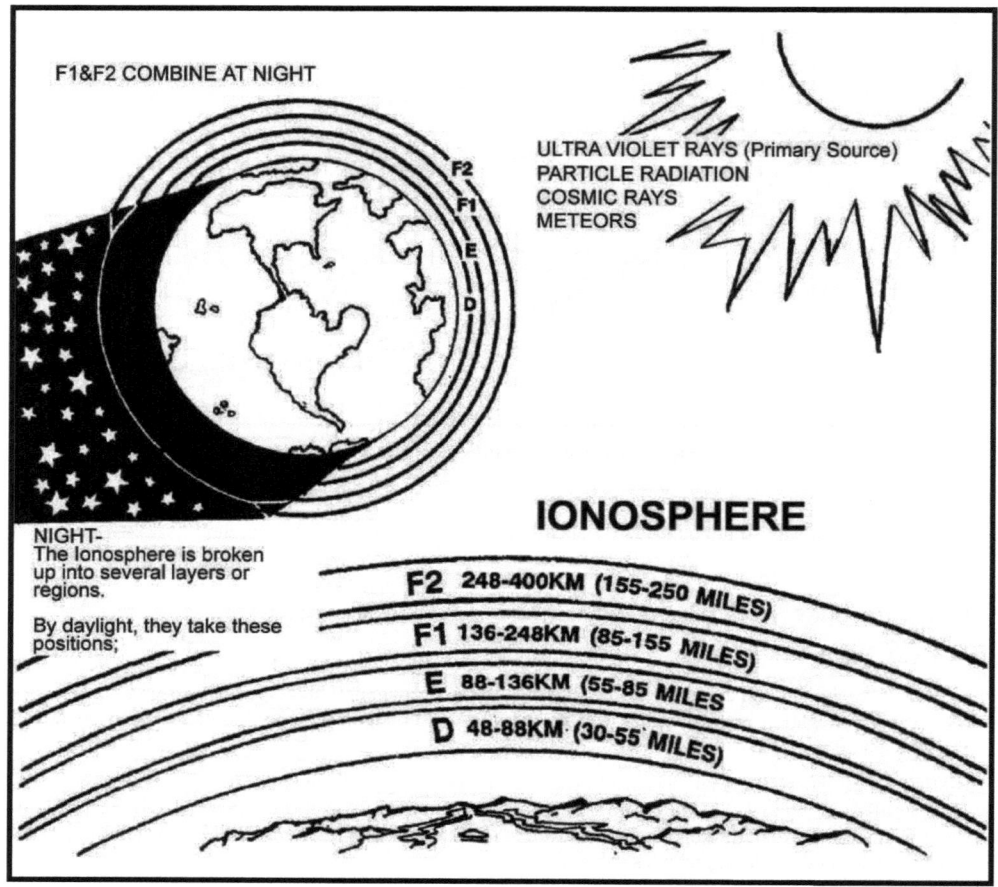

Figure 6-8. Structure of ionosphere.

6-82. The majority of HF sky-wave communications depend on the F1 and F2 regions. The F2 region is used the most for long-range daytime communications.

6-83. The E region is the next lower region. It is present 24 hours a day, although at night it is much weaker. The E region is the first region with enough charge to bend radio signals. At times, parts of the E-region become highly charged. This can either help or block HF communications. These highly charged areas are called "*sporadic E.*" They occur most often during the summer.

6-84. The D-region is closest to earth and only exists during the day. It cannot bend a radio signal back to earth, but it does play an important role in HF communication. The D-region absorbs energy from the radio signal passing through it, thereby reducing the strength of the signal.

6-85. The bending of the radio signal by the ionosphere depends on the frequency of the radio signal, the degree of ionization in the ionosphere, and the angle at which the radio signal strikes the ionosphere. At a vertical (straight up) angle, the highest frequency bent back to earth is called the *critical frequency*. Each region of the ionosphere (E, F1, and F2) has a separate critical frequency. For a vertical angle, signals above the highest critical frequency pass through all ionospheric regions and into outer space. Frequencies below the critical frequency of a region are bent back to the earth by that region; however, if the frequency is too low, the signal is absorbed by the D region. To have HF sky-wave communication, a radio signal must be a high enough frequency to pass through the D region, but not so high a frequency that it passes through the refracting region. Thus, radio operators must have current propagation charts from which to choose the most effective frequency during a given time period. To achieve an NVIS effect, the radio operator subtracts 20 percent from frequencies propagated on commercial computer propagation programs.

6-86. The angle at which a radio signal strikes the ionosphere plays an important part in sky-wave communication. As previously stated, any frequency above the critical frequency passes through the refracting region. If the radio signal having a frequency above the critical frequency is sent at an angle, the signal is bent back to earth instead of passing through the region. This can be compared to skipping stones across a pond. If a stone is thrown straight down at the water, it penetrates the surface. If a stone is thrown at an angle to the pond, the stone skips across the pond. For every circuit, there is an optimum angle above the horizon called the *takeoff angle*. This angle produces the strongest signal at the receiving station. This optimum takeoff angle is used to select the antenna for a specific circuit. By placing a dipole antenna between one-eighth and one-quarter wavelength above ground level, the radio operator achieves an NVIS effect, and he reduces or eliminates any skip zone (Figure 6-9).

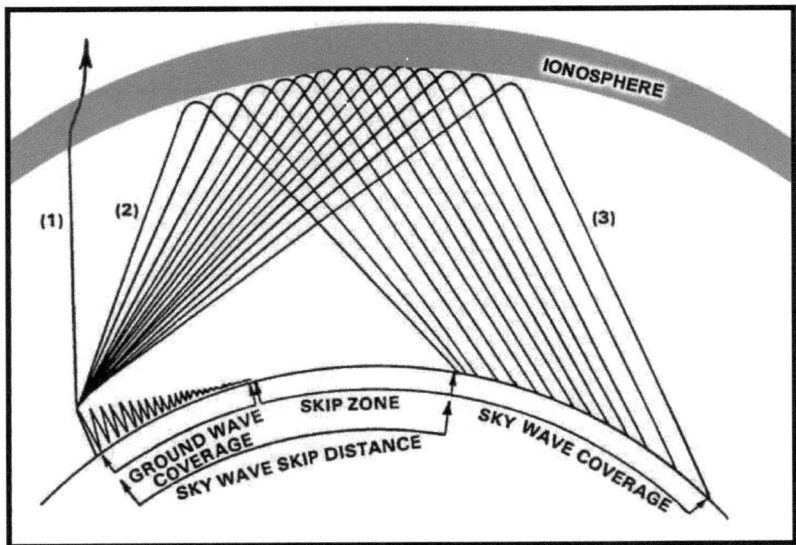

Figure 6-9. HF skip zone and distance.

6-87. Depending on the frequency, antenna, and other factors, an area may exist between the longest ground-wave range and the shortest sky-wave range where no signal exists. This is called the skip zone and no communication is possible. The NVIS effect can eliminate this problem.

6-88. Multiple frequencies are usually needed to maintain sky-wave communication. As a minimum, two frequencies, one for day and one for night are normally required.

CLASSIFICATION

6-89. Antennas are classified by the directions in which they can radiate energy. The three classifications include omnidirectional antennas (all directions), bidirectional antennas (two directions), or directional (one direction). A directional antenna is the best choice--if it works--because its signal is the most difficult for the enemy to locate.

DIRECTIONAL

6-90. This antenna's single lobe of energy sends a unidirectional signal (Figure 6-10). The width of the signal ranges from a narrow pencil beam to a 60-degree arc, depending on the type of directional antenna chosen.

Chapter 6

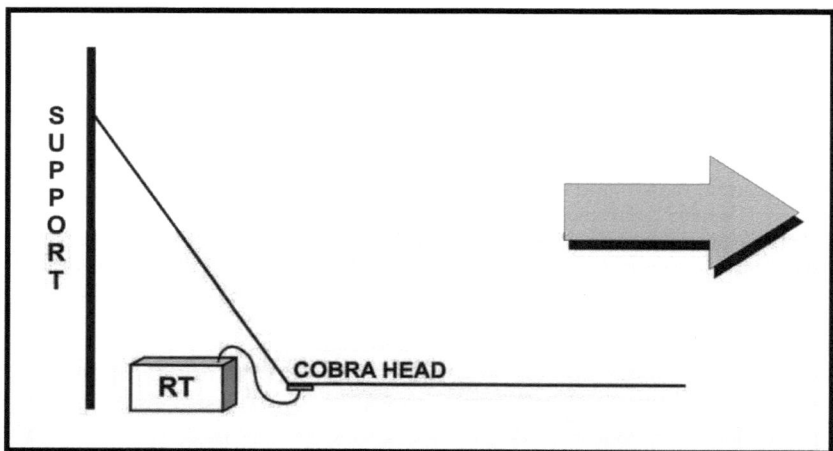

Figure 6-10. Unidirectional antenna pattern.

Application

6-91. Directional antennas are used on long-range, point-to-point circuits that need concentrated radio energy to ensure a reliable signal.

Orientation

6-92. A directional antenna concentrates most of its energy in one direction, so it requires careful orientation.

Detection

6-93. The enemy has a hard time determining the origin of directional antennas, which minimizes interference.

Adaptation of Bidirectional Antennas for Directional Use

6-94. Adding a terminating resistor to absorb the energy of the second lobe allows directional use of a bidirectional (long-wire or sloping "V") antenna. The terminating resistor must match the antenna. That is, it must be able to absorb one-half of the power output of the connected transmitter and provide 400 to 600 ohms of resistance.

BIDIRECTIONAL

6-95. A bidirectional antenna (Figure 6-11) has two opposite lobes of radio energy, with an area of null energy (no energy) between them. The lobes produce two strong signals in opposite directions, and weaker ones in all other directions.

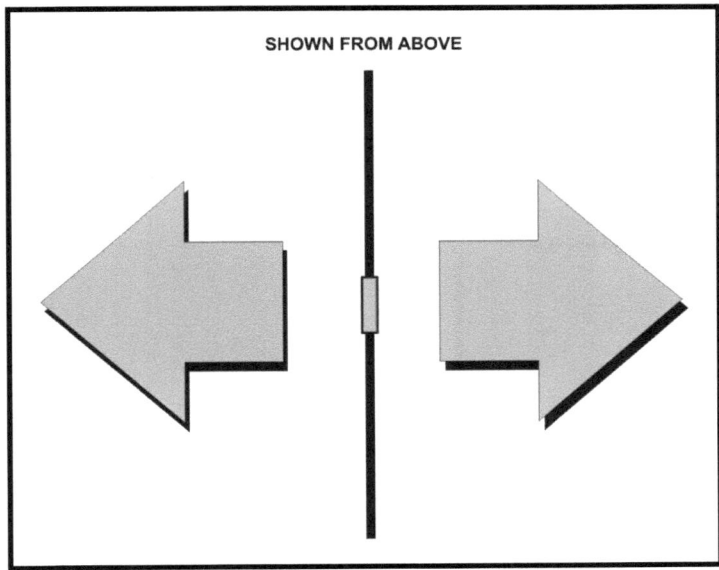

Figure 6-11. Bidirectional antenna pattern.

Application

6-96. Bidirectional antennas are usually used on point-to-point circuits and in situations where the antenna null can be positioned to reduce or block signals that could interfere with reception.

Orientation

6-97. To work properly (radiate in the desired directions), a bidirectional antenna must be oriented to the ground wave, and this is difficult to do. Lowering the antenna to create a near-vertical-incidence skywave (NVIS) effect makes this more difficult, because it increases the radiation pattern. A bidirectional antenna is best used near other antennas, which should be placed in its null to reduce interference and interaction between the antennas.

Examples

6-98. The bidirectional antennas most commonly used in tactical situations are the sloping-"V," random-length wire, and half-wave dipole.

Omni-Directional Antenna

6-99. An omnidirectional antenna (Figure 6-12) radiates and receives energy equally well in all compass directions. It is used when it is necessary to communicate in separate directions at once. However, it is also more susceptible to interference from all directions. The most common omnidirectional antenna is the whip. Some others are the quarter-wave vertical (RC-292 and OE-254) and the crossed dipole (AS-2259) antennas.

Chapter 6

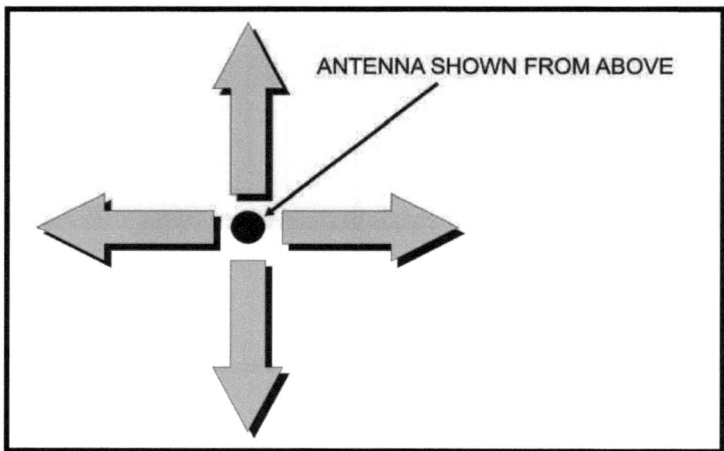

Figure 6-12. Omni-directional antenna pattern.

CONSTRUCTION AND SELECTION

6-100. Antenna construction is limited only by the imagination. There are many types and configurations. However, the operator must be careful not to construct an antenna that has a high standing wave ratio, which could damage radio equipment. He should use standing wave ratio meters when testing or using unfamiliar antennas. In selecting an antenna for an HF circuit, the operator must know the type of propagation.

GROUND-WAVE PROPAGATION

6-101. Ground-wave propagation requires low takeoff angles and vertically polarized antennas. The whip antenna provides good omnidirectional ground-wave radiation. If a directional antenna is needed, the operator selects one with a good low-angle vertical radiation.

SKY-WAVE PROPAGATION

6-102. Sky-wave propagation complicates antenna selection. After first finding the distance between radio stations, the radio operator can determine the required takeoff angle. The takeoff angle-to-distance tables give approximate takeoff angles for day and night sky-wave propagation. If the circuit distance is 966 kilometers (600 miles) during the day, the required takeoff angle is about 25 degrees. At night, it is 40 degrees. Therefore, the operator selects an antenna that has high gain from 25 to 40 degrees. He omits this step if the propagation predictions give the takeoff angles. For NVIS-constructed antennas and short-range HF communications, he subtracts 20 percent from these predictions and uses a planning range of 0 to 300 miles.

COVERAGE

6-103. The radio operator determines what type of coverage to use. If the radio circuit consists of mobile (vehicular) stations or of many stations at different directions from the transmitter, an omnidirectional antenna is required. If the circuit is point to point, he can use a directional or bidirectional antenna. Normally, the receiving station locations dictate this choice.

CONSTRUCTION

6-104. Before he can select an antenna, the operator must examine the materials available to build one. He needs two supports to build a horizontal dipole, and a third support in the middle for frequencies of 5 mega hertz (MHz) or less. If he has nothing he can use for a support, he cannot build a dipole antenna.

SITE

6-105. Another consideration is the site itself. The tactical situation usually determines the antenna positions. The ideal area is clear and flat with no trees, buildings, fences, power lines, or mountains. Unfortunately, the tactical communicator seldom finds such a perfect site, so he just tries to find one as flat and clear as possible. He will often have to settle for less ideal sites, and these sites usually interfere with the patterns and functioning of the antennas.

COMMON TYPES OF ANTENNAS

6-106. Common antenna types include the half-wave dipole, inverted "V," long wire, and the sloping "V."

HALF-WAVE DIPOLE

6-107. The half-wave dipole antenna is a balanced resonant antenna (Figure 6-13). It produces its maximum gain in a narrow range between 2 percent above and 2 percent below the design frequency. Since frequency assignments are normally several megahertz apart, the operator must build a separate dipole for each assigned frequency.

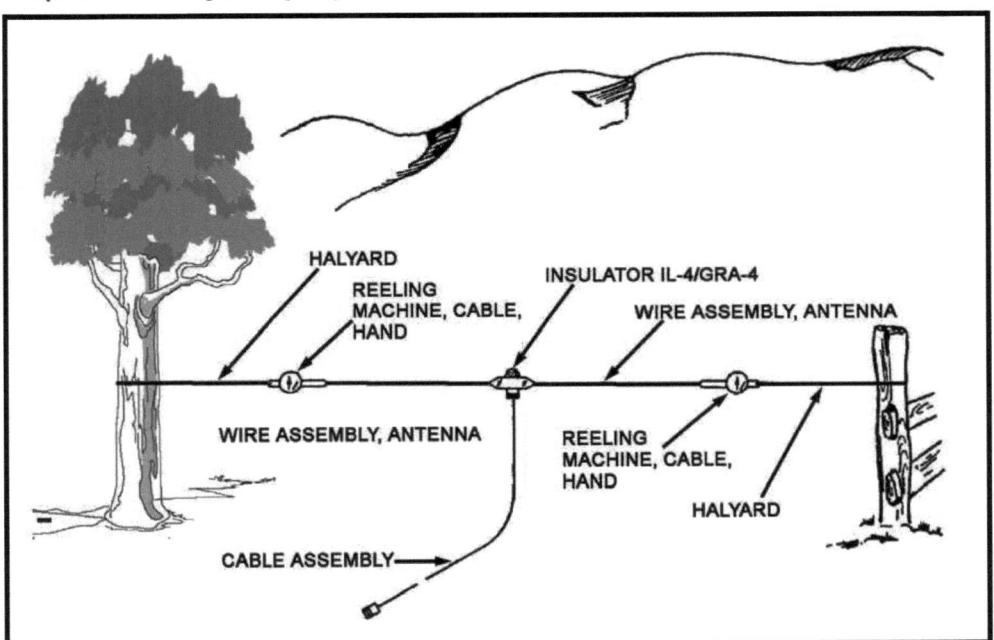

Figure 6-13. Half-wave dipole antenna.

Length of Dipole

6-108. The operator calculates the length of a half-wave dipole using the formula *length = 468/frequency*, as shown in Figure 6-14.

Height of Dipole

6-109. The operator normally keeps the height of a dipole between one-fourth and one-half wavelength above ground level for long-range sky wave. For NVIS of 0 to 300 miles, and for inverted and sloping "V" antennas, the operator raises the antenna between one-eighth and one-fourth wavelengths above ground level.

```
Length (in meters)   =   (150.00 x 0.95) = 142.50 Frequency in MHz
Length (in feet)     =   (492.00 x 0.95) = 468.00 Frequency in MHz

For harmonic operation, calculate the length of a long-wire antenna
(one wavelength or longer) as follows:

Length (in meters)   =   (150.00 x (N--0.05)) / Frequency in MHz
Length (in feet)     =   (492.00 x (N--0.05)) / Frequency in MHz
```
Where N = the number of half-wave lengths in the total length of the antenna.

For example, if the number of half-wavelengths is 3 and the frequency in MHz is 7, then--

```
If length (in meters)  =   (150.00 x (N--0.05)) / Frequency in MHz
Then                       (150.00 x (3--0.05)) / 7
                           (150.00 x (2.95)) / 7
                           (442.50 / 7)
so length in meters    =   63.20
```

Figure 6-14. Formula for calculating length of half-wave dipole antenna applied to example.

INVERTED "V"

6-110. The inverted "V" or "drooping dipole" antenna (Figure 6-15) is similar to a dipole antenna, except that it only requires one center support. Like a dipole, it is used for a specific frequency, and it has a bandwidth of plus or minus 2 percent of design frequency. Because of the inclined sides, the inverted "V" antenna produces a combination of horizontal and vertical radiation; vertical off the ends and horizontal broadside to the antenna. All the construction factors for a dipole also apply to the inverted "V." Although the inverted "V" has less gain than a dipole, the fact that it only requires one support makes it the preferred antenna in some tactical situations.

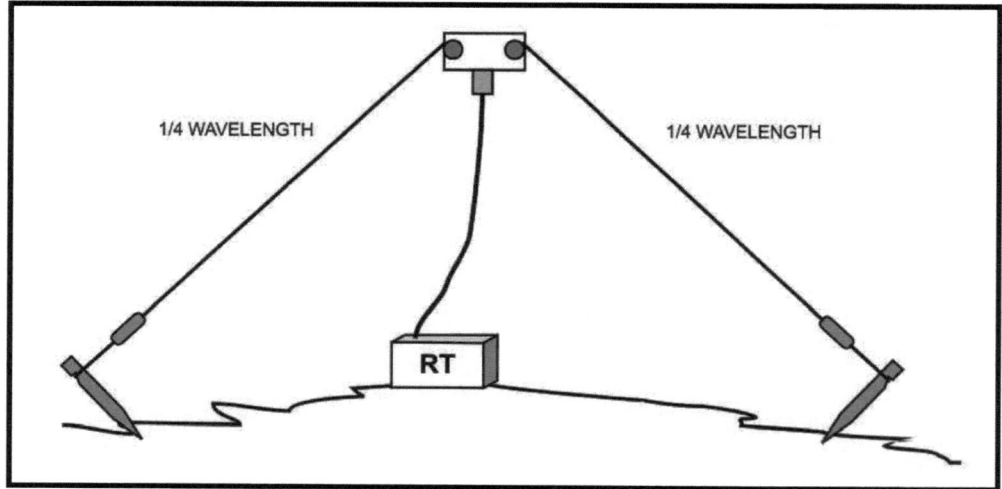

Figure 6-15. Inverted "V" antenna.

Communications

LONG-WIRE ANTENNA

6-111. A long-wire antenna is one that is at least as long as one wavelength (Figure 6-16). However, it should be longer to achieve good gain and directional characteristics. Constructing long-wire antennas is simple, but using the correct dimensions and making the correct adjustments are both critical to its success.

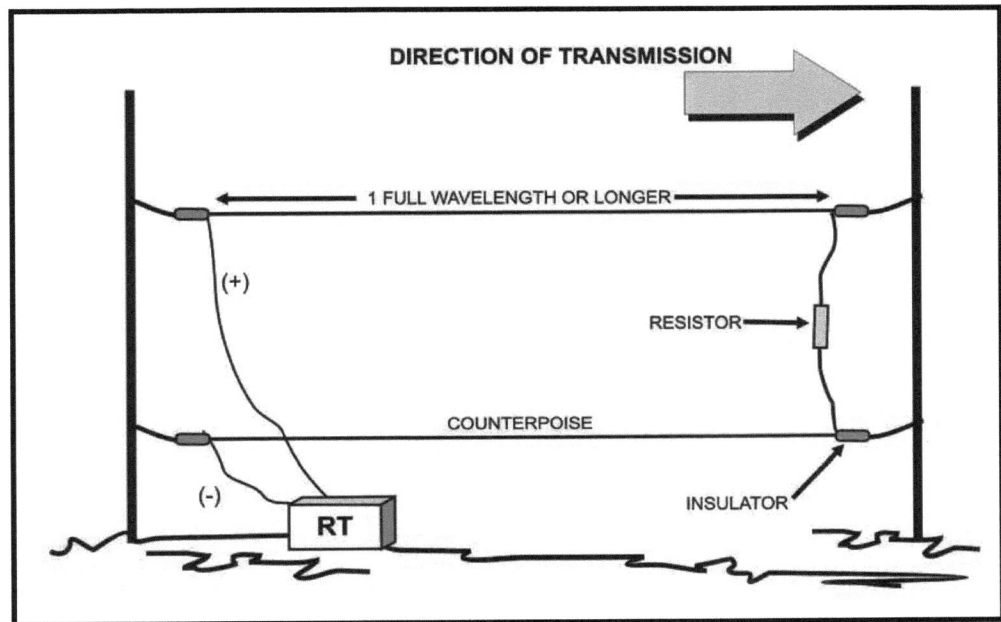

Figure 6-16. Long-wire antenna.

Direction

6-112. A long-wire antenna is made directional by placing a terminating resistor at the distant station end of the antenna. The terminating resistor should be a 600-ohm, noninductive resistor that can absorb at least one-half of the transmitter power. Terminating resistors are part of some radio sets, but can be made locally using a 100-watt, 106-ohm resistor (NSN 5905-00-764-5573).

Construction

6-113. Building a long-wire antenna only requires wire, support poles, insulators, and a terminating resistor (if directionality is desired). The only other requirement is that the operator string the antenna in as straight a line as the situation permits. Because the antenna is less than 20 feet tall, it requires no tall support structures.

SLOPING WIRE

6-114. If an HF circuit is only a single point-to-point ground link or a short skywave link, and if all other stations are oriented in the same direction, then the team can use a sloping wire antenna (Figure 6-17). The radiating wire is normally one quarter of the wavelength. (Antenna length is measured from the radio equipment.) The far end of the antenna should be connected to a rope whose other end is tied to a nonconductive weight such as a stone or brick. The weighted end is then thrown over a tree so that the antenna forms a 30- to 45-degree angle to the ground. Angles greater than 45 degrees are used for ground waves, and less than 30 degrees for sky waves. The angle formed by the wire should point in the direction opposite that of the intended receiver.

Chapter 6

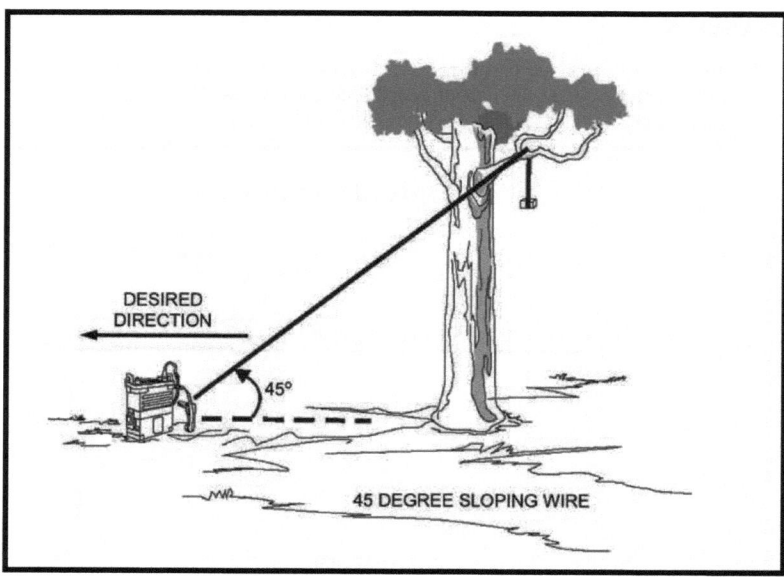

Figure 6-17. Sloping wire antenna.

TERMINATED SLOPING "V"

6-115. The sloping "V" antenna is a short- to long-range sky-wave antenna that the radio operator can build in the field (Figure 6-18). Gain and directivity depend on leg length. For reasonable performance, the antenna should be at least one-half wavelength long. To make the antenna directional, the operator puts terminating resistors on each leg on the open part of the "V." The terminating resistors should be 300 ohms and be capable of absorbing one-half of the transmitter's power output. These terminating resistors are either procured or locally made. Using the terminating resistors, the operator aims the antenna so that the line cutting the "V" in half points at the distant station.

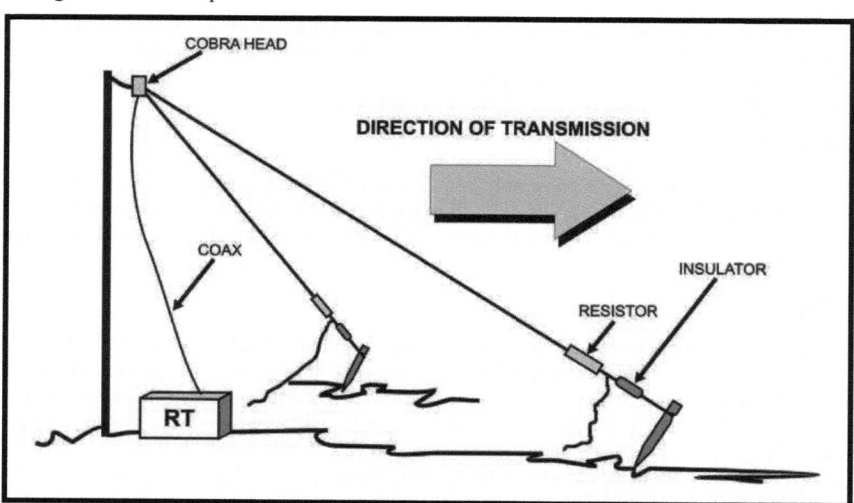

Figure 6-18. Terminated sloping "V" antenna.

Communications

FIELD-EXPEDIENT ANTENNAS

6-116. Operators must know the importance of field-expedient antennas. The operator will have to construct field-expedient antennas if conventional ones are damaged or missing parts.

REPAIR OF DAMAGED ANTENNA

6-117. A broken whip antenna can be temporarily repaired (Figure 6-19)--

- If the whip is broken in two sections, the operator can join the sections. First, he removes the paint and cleans the sections where they join. This ensures a good electrical connection. Then, he places the sections together and secures them using bare wire or tape.

- If the whip is badly damaged, the radio operator can use field wire (WD1/TT) of the same length as the original antenna. The radio operator removes the insulation from the lower end of the field wire antenna, twists the conductors together, sticks them into the antenna base connector, and secures the conductors with a wooden block. He supports the antenna wire with a tree or a pole.

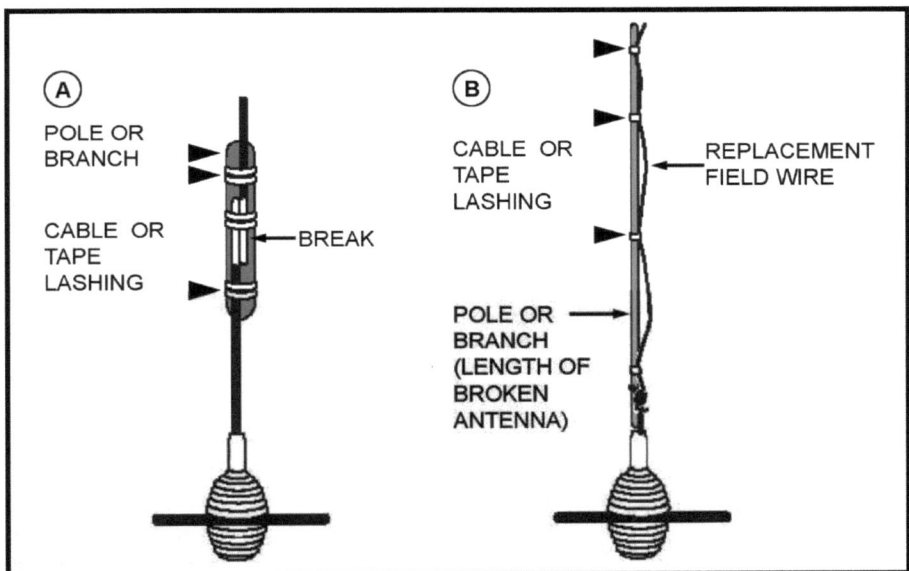

Figure 6-19. Repair procedure, whip antenna.

INSULATORS

6-118. The radio operator can make these from items that are readily available (Figure 6-20). He should choose materials that do not absorb water, as those that do, such as rope or cloth, will lose their insulating characteristics and become conductors themselves should they get wet.

Chapter 6

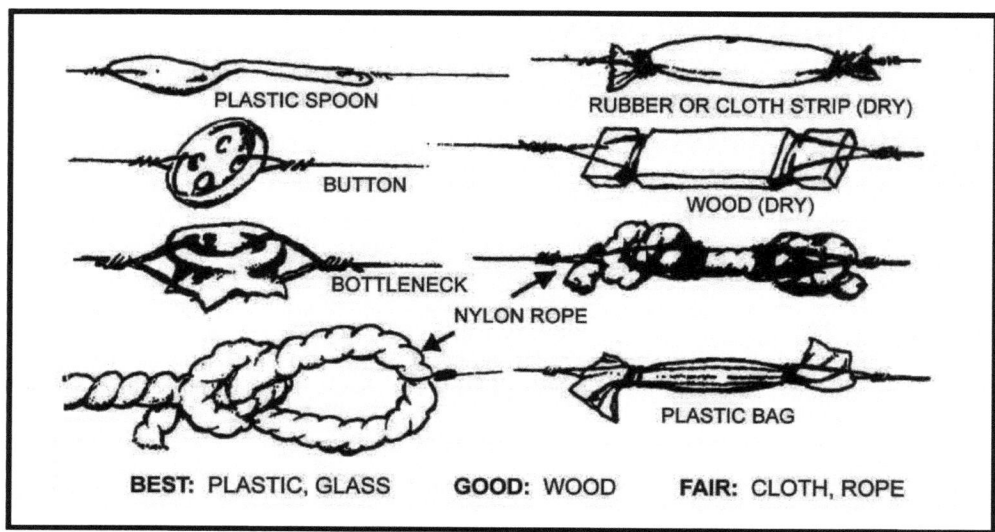

Figure 6-20. Expedient insulators.

SUPPORTS

6-119. Many expedient antennas require support. The most common support is a strong tree that can survive heavy windstorms. However, even the largest tree sways enough in the wind to break a wire antenna. The operator attaches a spring or piece of old inner tube to one end of the antenna to keep it taut while preventing it from breaking or stretching as the tree sways. If a small pulley is available, he attaches that to the tree. He attaches one end of a rope to the end of the antenna, passes the rope through the pulley on the tree, then attaches a heavy weight to the free end of rope. This lets the tree sway without straining the antenna.

TERMINATING RESISTORS

6-120. Resistors for low-power (man-pack) HF radios are readily available from commercial radio supply stores. However, carbon resistors that can dissipate more than 5 watts are hard to find.

6-121. As a field-expedient technique, the radio operator can connect the low-power (5-watt) resistors in parallel to enable a terminator to handle greater power. For example, eight 5-watt, 4,000-ohm resistors connected in parallel become a 500-ohm, 40-watt terminator. Unfortunately, this is still too small to work with a high-power, HF terminator. A terminator for a 1,000-watt transmitter requires 100 5-watt resistors. However, a series of 100-watt, 106-ohm resistors (NSN 5905-00-764-5573) may be mounted on a single insulating board to serve as a terminator for a high-powered transmitter.

FIELD-EXPEDIENT WIRE

6-122. If regular antenna wire is unavailable, the radio operator can use field telephone wire (WD1/TT) to build antennas. Field wire consists of two insulated wires, and each of those has four copper and three steel strands.

- When making electrical connections with field wire, the operator uses the copper strands. To identify them he removes about 1 inch of insulation from one end of the insulated wire. He holds it where the insulation ends and bends the strands to the side. When he releases the pressure, the steel strands snap back to their original positions, but the copper strands remain bent. He can then wrap these copper strands around the steel strands for a good electrical connection.

- If field wire is used as the radiating element of an antenna, the two insulated wires in the twisted pair must be connected together at the ends so that electrically the two wires act as one.
  -- First, the radio operator tightly twists together all six steel strands from the two wires (for strength).
  -- Second, he twists the eight copper strands together (to connect them electrically).
  -- Third, he twists the copper strands *around* the steel strands.
- When using them as a feed line for a dipole antenna, the radio operator connects each of the two insulated wires of the twisted pair to a separate leg of the dipole. At the radio, he connects one wire (any wire) to the center connector of the radio antenna terminal and the second wire to a screw on the antenna case.
- In an emergency, any wire of sufficient length can be used for an antenna, for example, barbed wire, electrical wire, fence wire, or metal-cored clothesline. Communication has even been successful using metal house gutters and metal bed springs. A radio operator's mission is incomplete until he establishes communication.

GROUND

6-123. All radio equipment should be grounded to prevent shock and damage to equipment during electrical storms. This protects the operator and his equipment. Also, some antennas must have a radio-frequency ground before they will function. Most radio sets come with a ground rod that should provide enough ground if used properly in good soil. The radio operator checks to ensure that the ground rod is neither oily nor corroded. He drives the rod into the ground so that the top of the rod is below surface. To ensure a good electrical connection, he makes sure that the top of the ground rod and the end of the ground strap are both clean and bright. Then, he uses a clamp or a nut and bolt to make a good mechanical and electrical connection at the ground rod.

Alternative Materials

6-124. If he has no ground rod, he can use water pipes, concrete reinforcing rods, metal fence posts (protective paint coating removed), or any length of metal. If a water system has metal pipes, he can make a good ground by clamping the ground strap to a water pipe. He can also use underground pipes, tanks, and metal building foundations.

> **WARNING**
>
> Never ground on any piping or underground tanks that contain flammable materials such as natural gas or gasoline.

Soil Additives

6-125. The operator can improve the conductivity of dry soil by adding water and chemicals such as table or Epsom salt to it (Epsom salt is less corrosive than table salt). First, the radio operator digs a hole around the ground rod. Then, he mixes and pours into it one pound of the chemical and one gallon of water. He should periodically add water to keep the ground damp. He can use urine in place of water, if needed.

Multiple Ground Rods

6-126. Using multiple ground rods can also improve the electrical ground. If he has enough ground rods, the operator can build a "star ground." He drives a single rod into the center of a circle that measures about 20 feet in diameter. Then, he drives additional ground rods around the outside of the circle. He connects the ground strap from the radio to the center rod, which he in turn connects to the rods along the outside of the circle. Finally, he connects the rods around the circle.

HIGH FREQUENCY, DIRECTIONAL, FIELD-EXPEDIENT ANTENNAS

6-127. The long-wire (Figure 6-21) and vertical half-rhombic (Figure 6-22) are two field-expedient, directional antennas. These antennas consist of a single wire, preferably two or more wavelengths long, supported on poles at a height of 3 to 7 meters (10 to 20 feet) above the ground. The antennas will, however, operate satisfactorily at less than 1 meter (about 3 feet) aboveground. The far end of the wire connects to ground through a noninductive resistor of 500 to 600 ohms. The resistor should have a rating of at least one-half the wattage output of the transmitter. This ensures that the output power of the transmitter does not burn out the resistor. A reasonably good ground, such as a number of ground rods or a counterpoise, should be used at each end of the antenna. The radiation pattern is directional. The antennas are used primarily to transmit or receive HF signals. The "V" antenna is another field-expedient, directional antenna (Figure 6-23). It consists of two wires forming a "V" with the open area of the "V" pointing toward the desired direction of transmission or reception. To make construction easier, the legs may slope downward from the apex of the "V." This is called a sloping "V" antenna (Figure 6-24). The angle between the legs varies with the length of the legs in order to achieve maximum performance. To make the antenna radiate in only one direction, add noninductive terminating resistors from the end of each leg (not at the apex) to ground. The resistors should be approximately 500 ohms and have a power rating at least one-half that of the output power of the transmitter being used. Without the resistors, the antenna radiates bidirectionally, both front and back. The antenna must be fed by a balanced transmission line.

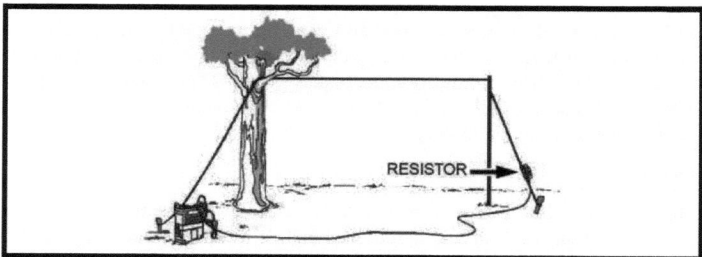

Figure 6-21. High frequency antenna, long-wire type.

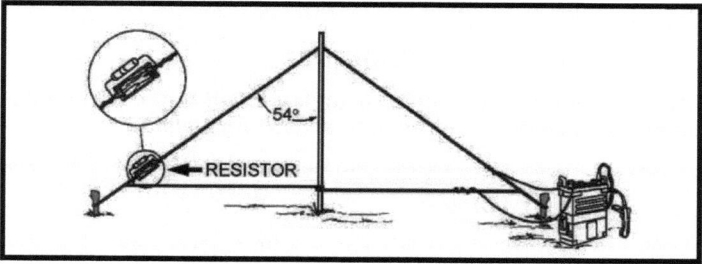

Figure 6-22. High frequency antenna, half-rhombic type.

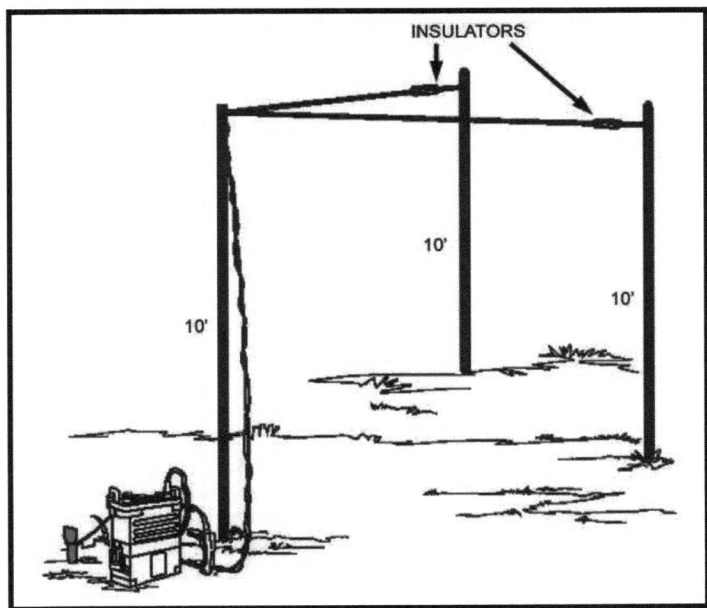

Figure 6-23. High frequency antenna, "V" type.

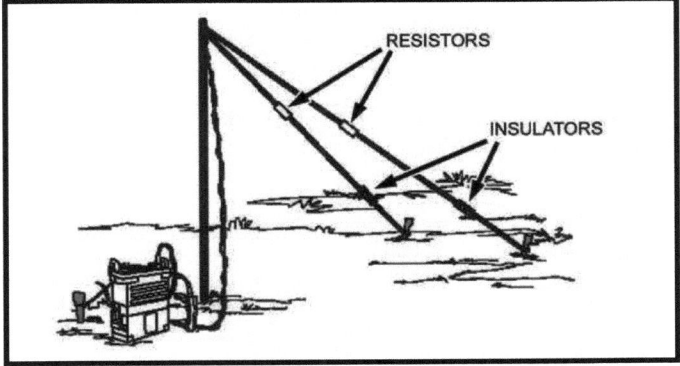

Figure 6-24. Sloping antenna, "V" type.

ANTENNA, AN/GRA-50

6-128. The AN/GRA-50 antenna is a component of the AN/PRC-104, AN/GRC-213, and AN/GRC-193 and can be used with modern radio systems. It is a half-wave, dipole-type antenna. This means that it can be used for long- (2,500 miles) or short-range, near-vertical-incidence, sky wave (0 to 300 miles) messages. (See TM 11-5820-467-15 for more detail.)

ANTENNA, AS-2259/GR

6-129. The AS-2259/GR antenna has two inverted "V" dipoles positioned at right angles. It uses a foam-electric center pole as its coaxial. It is used for short-range (NVIS, from 0 to 300 mile) messages. (See TM 11-5985-379-14&P for more detail.)

Section VII. UNUSUAL ENVIRONMENTS

Climatic variations such as temperature, electricity, humidity, and wind require variations in the ways radios and antennas are set up, used, and maintained.

DESERT OPERATIONS

6-130. The factors that most affect radios and antennas in the desert are poor electrical grounding, temperature and humidity extremes, and wind-blown contaminants.

ELECTRICITY

6-131. For the best operation in the desert, locate radio antennas on the highest terrain available. Poor electrical grounding in the desert reduces the ranges of whip antennas by one-fifth to one-third. For this reason, complete antenna systems such as horizontal dipoles and vertical antennas with adequate counterpoises are generally more effective.

STATIC ELECTRICITY

6-132. Static electricity is caused by many factors present in the desert, including wind-blown dust particles. Also, dry air increases static discharges between charged particles. When operating in fixed positions, ground communications equipment properly to prevent such discharges. Tape all sharp edges and tips of antennas to reduce wind-caused static discharges and the accompanying noise. Since static-caused noise diminishes at higher frequencies, use the highest frequencies authorized for use.

HEAT

6-133. Radio sets can overheat and fail, so turn them on only when necessary. Keep air vents and filters clean to prevent overheating. High temperature conditions will degrade radio wave propagation. A station that can be reached at night may not be reachable during the day.

MOISTURE

6-134. In deserts with high humidity, condensation occurs wherever a surface is cooler than the air. To keep contacts dry, tape electrical plugs, jacks, and connectors. To prevent arcing, make sure these are dry before connecting.

DUST AND DIRT

6-135. Wind-blown particles (dust, dirt, sand, and grit) damages everything it contacts. They cause the most damage to items with moving or electrical parts, or with vents, grids, or grilles. Because radios with servomechanisms are particularly vulnerable, extra cleaning and precautions are required. For example, use dustproof covers to protect communications equipment such as HF amplifiers. Regularly check items such as receiver-transmitter units that have ventilating ports or channels. Keep these openings clear of dust so the equipment remains within operating temperatures. Keep an extra supply of dry batteries on hand, because heat drains batteries at a higher rate and produces a higher failure rate. Protect electrical wire insulation with tape while it is still intact. Use an old toothbrush or other brush to clean electrical contacts and plugs before connecting, then seal the connection with tape. Conduct preventive maintenance checks often. Check parts that require lubrication often. Dust and dirt can collect in lubricants and damage moving parts.

JUNGLE OPERATIONS

6-136. Communications in jungle areas must also be carefully planned. Thick jungle growth vertically polarizes RF energy, which reduces transmission ranges. Heat and humidity increase maintenance problems. Siting is the most important consideration in the jungle, followed closely by maintenance.

ANTENNA SITE

6-137. Complete antenna systems, such as ground planes and dipoles, work better than fractional, wave-length, whip antennas. To further improve communications--

- Locate antennas in clearings. Place them on the edge farthest from the distant station, and as high as possible.
- Keep all cables and connectors--antenna, power, and telephone--off the ground to reduce damage from moisture, fungus, and insects.
- If possible, clear vegetation from antenna sites. Foliage touching an antenna will ground its signal.
- If vegetation cannot be removed, especially dense or wet vegetation, always horizontally polarize the antennas.

HUMIDITY

6-138. The high humidity of jungle environments condenses moisture on equipment. This encourages rust and fungus, complicating maintenance. Operators and maintenance personnel should check their TMs for special requirements but mainly they must--

- Keep the equipment as dry as possible and in lighted areas to retard fungus growth.
- Keep air vents clear to help keep equipment cool and dry.
- Keep connectors, cables, and bare metal parts as free of fungus as possible. After repairs or damage, paint all surfaces of equipment with moisture-fungus-proofing paint.

FIELD-EXPEDIENTS

6-139. LRSU can greatly improve their ability to communicate in the jungle by using expedient antennas. Moving units are generally restricted to using the short and long antennas that come with the radios. However, when not moving, field-expedient antennas increase range and improve reception.

COLD WEATHER OPERATIONS

6-140. In very cold weather, ionospheric storms and night lights, such as the Aurora Borealis, can degrade sky-wave propagation and disable radio communications. Static can block frequencies for extended periods; changes in the density and height of the ionosphere can fade a signal for weeks. When these disturbances occur is difficult to predict. However, when they do, radio operators must be ready to use alternate frequencies, or other means of communication. Put radios in vehicles, if possible. This simplifies transport and provides shelter for radio operators. It also helps alleviate grounding and antenna installation problems caused by the cold: permafrost and deep snow limit grounding. Frozen ground conducts electricity too poorly to propagate ground waves well. To improve ground wave transmission, install a counterpoise far enough aboveground to prevent the snow from covering it. When installing antennas--

- Handle the mast sections and the antenna cables carefully; they become brittle at very low temperatures.
- Run antenna cables overhead to avoid damage from heavy snow and frost. Use nylon rope for guy wires rather than cotton or hemp. Nylon absorbs less moisture, so it is less likely to freeze and break.
- Use extra guy wires, supports, and anchors to help antennas withstand heavy ice and wind loading.

- Allow radios to warm up for several minutes before use. Since extreme cold lowers the voltage output of a dry battery, try warming the battery before operating the radio set. This minimizes frequency drift. Flakes or pellets of highly electrically charged snow have been reported in northern regions. When these particles strike the antenna, the resulting electrical discharge causes a high-pitched static roar that can blanket all frequencies. To prevent this, cover all antenna elements with polystyrene tape and shellac.

6-141. Protect radios from blowing snow. Snow can freeze to dials and knobs and blow into the wiring, causing shorts and grounds. Handle cords and cables carefully because they lose their flexibility in extreme cold. Properly winterize all radio equipment and power units. Check the appropriate technical manuals (TM) for winterization procedures.

Power Units

6-142. As temperature decrease, operating and maintaining generators becomes increasingly difficult. Protect them from weather as much as possible.

Batteries

6-143. The effect of cold weather on wet and dry cell batteries depends on the type and kind of battery, the load on the battery, the use of the battery, and the degree of exposure to cold temperatures.

Shock Damage

6-144. In extreme cold, most synthetic shock mounts get brittle and fail to cushion the equipment. The jolting of a vehicle during movement can damage radios. Check the shock mounts often, and change them when needed.

Microphones

6-145. Moisture from a Soldiers breath can freeze on the perforated cover plate on the microphone. Use standard microphone covers to prevent this. If no standard covers are available, improvise one from rubber or cellophane membranes or from rayon or nylon cloth.

Breathing and Sweating

6-146. A radio generates heat when it is operated. When it is turned off, the air inside cools and contracts, drawing cold air in. This "breathing" can bring still-hot parts into contact with subzero air. This can cool the glass, plastic, and ceramic parts too quickly, and cause them to break. If the cold equipment is brought suddenly into contact with warm air, moisture will condense on its parts. This is called sweating. Before cold equipment is brought into a heated area, wrap it in a blanket or parka to ensure that it will warm gradually to reduce sweating. Thoroughly dry all equipment before taking it back out into the cold. Otherwise, moisture caused by sweating will freeze equipment.

MOUNTAIN OPERATIONS

6-147. Operation in mountainous areas presents many of the same problems as operation in northern or cold weather areas. It also makes selecting transmission sites a critical task. Terrain restrictions often make relay stations necessary for good communications. Terrain obstacles often make line-of-sight transmission necessary. Also, the dirt in mountainous areas seldom conducts electricity well. Use a complete antenna system such as a dipole or ground-plane antenna with a counterpoise. Maintenance requirements in mountainous areas resemble those in northern or cold weather areas. Tricky mountain climates require flexible maintenance planning.

URBAN OPERATIONS

6-148. Communications in urbanized terrain pose special problems. Some problems are similar to those encountered in mountainous areas. Obstacles can block transmission paths. Pavement surfaces conduct electricity poorly. Commercial power lines cause electrical interfere. VHF radios are generally less effective in urban terrain. Due to their power output and operating frequencies, VHF radios require a LOS between antennas. Urban areas sometimes prohibit the establishment of a street-level LOS. HF radios require and rely on LOS less than VHF radios, because they use lower operating frequencies and transmit at higher powers. The antenna should be hidden or blended into its surroundings to prevent discovery. Antennas can be concealed by blending them with existing structures such as: water towers, existing civilian antennas, or steeples. In urban areas, the LRSU should--

- Park radio-equipped vehicles inside buildings for cover and concealment and then remote the antennas outside the buildings.
- Dismount radio equipment and install it inside buildings, ideally, in basements.
- Conceal generators against buildings or under sheds; this also decreases noise. Provide adequate ventilation to prevent heat buildup.

Chapter 7
Intelligence Preparation of the Battlefield

The IPB process applies to all types of operations, from stability operations to war. It helps leaders reduce uncertainty. IPB is conducted as part of the MDMP. The MDMP is generally used at battalion or squadron level and above--units with staffs. LRSUs do not have staffs and as a result use TLP to plan operations. However, because LRSUs make extensive use of IPB products, it is imperative LRSU Soldiers have an in-depth understanding of the continuous IPB process. FM 2-01.3 defines IPB as, *"The staff planning activity undertaken by the entire staff to define and understand the operational environment and the advantages and disadvantages presented to friendly and threat forces."* IPB has four steps, which this chapter will treat as sections:

- *Define the Operational Environment* (Section I).
- *Describe Environmental Effects on Operations* (Section II).
- *Evaluate the Threat* (Section III).
- *Determine Threat Courses of Action* (Section IV).

SECTION I. DEFINE THE OPERATIONAL ENVIRONMENT

This first of the four steps in the IPB process identifies for further analysis specific features of the environment or activities within it and the physical space where they exist that may influence available COAs or the commander's decision. To conduct this step, the staff performs five substeps:

- Identify significant characteristics of the environment.
- Identify the limits of the command's AO.
- Establish limits of the area of influence and the area of interest (AOI).
- Evaluate existing databases and identify intelligence gaps.
- Initiate collection of information required to complete IPB.

IDENTIFY SIGNIFICANT CHARACTERISTICS OF THE ENVIRONMENT

7-1. The Army uses METT-TC as the framework for analysis in IPB. Much of the information on environmental characteristics can be obtained from existing databases. The BFSB S-2 GI&S section and USAF combat weather team contribute much of the commander's and staff's information and analysis.

TERRAIN

7-2. In this first substep of defining the operational environment, commanders and staffs use OAKOC to develop the military aspects of terrain. These are examples of terrain characteristics:

- Hydrological data.
- Elevation data.
- Soil composition.
- Vegetation.

CLIMATE AND WEATHER

7-3. Climate and weather can significantly impact military operations. Climate is the prevailing pattern of temperature, wind velocity, and precipitation in a specific area measured over a period of years. Weather describes the conditions at a specific place and time, and is only somewhat predictable. The following are military aspects of weather:

- Visibility.
- Wind.
- Precipitation.
- Cloud cover.
- Temperature.
- Humidity.

CIVIL CONSIDERATIONS

7-4. In urban terrain, manmade infrastructure, civilian institutions, attitudes and activities of civilian leaders, populations, and organizations also affect the environment (Figure 7-1).

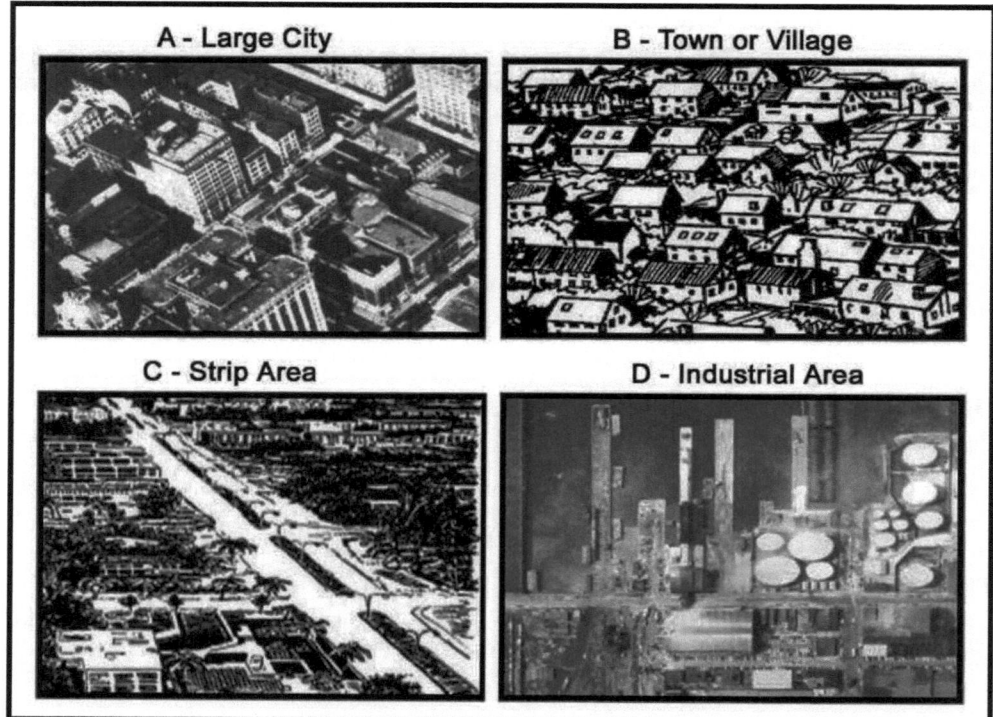

Figure 7-1. Classification of urban area by size.

POPULATION

7-5. Most urban environments have a significant population in both number and density. Civilians in the urban environment may pose significant threats, obstacles, and logistical support problems such as medical. They can also provide both support and information. Therefore, it is important to gain as much insight into the population as possible. This should start with the basics of demographic information: ethnic background, religion, age structure, growth rate, birth and death rates, net migration rates, communicable disease issues, languages spoken, and literacy rate. Any information on friendly, enemy, or neutral personnel will aid in planning.

POLITICAL OR SOCIOECONOMIC CONDITIONS

7-6. Knowing the political affiliations of the populace may help the team determine whether they are likely to be for, against, or neutral towards the U.S.. Clans, tribes, and gangs may not only influence area politics and economic conditions, they may control them. Cultivation of and traffic in illicit drugs can also impact the political and economic conditions.

INFRASTRUCTURE

7-7. Infrastructure refers to all basic resources, support systems, communications, and vital industries. The physical and social attributes of an infrastructure operate interdependently. Understanding the functions and interrelationships of these components allow an assessment of how disruption or restoration of the infrastructure affects the population and ultimately the mission.

Physical

7-8. Lines of communication such as roads, rails, trails, and waterways, as well as telecommunications means, are key aspects of the physical environment to consider.

Social

7-9. The social attributes of infrastructure include such things as the medical system, the commerce system, and the transportation system. Each of these needs people in order to operate efficiently. For example, the doctors, nurses and support staff (pharmacists, ambulance drivers, clerks and janitors) interact with the physical infrastructure. The transportation system allows the people of the medical system to get to and from work. The commerce system provides the goods to supply the needs of the medical system.

IDENTIFY THE LIMITS OF THE COMMAND'S AREA OF OPERATIONS

7-10. In this second of five subtasks, the unit commander defines the AO. The AO is one of the most basic and important control measures. It is where he has the responsibility and authority to conduct military operations. AOs may be contiguous or non-contiguous. Contiguous means adjacent, touching, sharing a single boundary. Thus, contiguous AOs share a boundary, and noncontiguous AOs do not. Higher headquarters controls the unassigned areas between noncontiguous AOs.

ESTABLISH THE LIMITS OF THE AREA OF INFLUENCE AND THE AREA OF INTEREST

7-11. In this third of five tasks, the commander defines the area of influence and the AOI. The area of influence is larger than and includes the AO:

AREA OF INFLUENCE

7-12. The area of influence (no acronym) is a geographic area--

- Where the commander can directly influence operations by maneuver or fires support systems normally under the commander's C2.
- An area that includes terrain inside and outside the AO.
- An area determined by both the G-2/S-2 and the G-3/S-3.

AREA OF INTEREST

7-13. The AOI is the commander's area of concern (interest). It includes the area of influence and adjacent areas, and it extends into enemy territory to the objectives of current or planned operations. It also includes areas occupied by enemy forces who could jeopardize the accomplishment of the mission. The

Chapter 7

AOI is established by the commander with input from the G-2/S-2 and G-3/S-3. It is normally larger than the area of influence, and may therefore require more intelligence assets to monitor. Time is the key consideration in establishing the limits of the AOI. Time limits consider two primary factors: the mobility of the threat and the time needed for the friendly unit to accomplish the mission.

EVALUATE EXISTING DATABASES AND IDENTIFY INTELLIGENCE GAPS

7-14. In this fourth of five subtasks, national, multinational, joint, and higher echelon databases are examined to determine if the information required already exists and is available. Not all needed information will be available, which causes gaps in the available information. Identifying the gaps early allows actions to be initiated to collect the intelligence required to fill them. The commander's initial intelligence requirements and intent also allows these gaps to be prioritized. Any gaps not expected to be filled within the available time must be substituted with reasonable assumptions.

INITIATE COLLECTION OF INFORMATION REQUIRED TO COMPLETE IPB

7-15. In this last of five subtasks, the G-2/S-2 identifies and prioritizes the gaps in current holdings (Table 7-1). Based on this analysis, collection activities are initiated or RFIs are submitted to fill intelligence gaps to the level of detail needed to conduct IPB. IPB products are constantly updated as the staff receives new information.

Table 7-1. Identification of gaps in existing databases.

Who ?	What ?	Why or When ?
Leader	Evaluates LRS and SOF debrief and patrol/surveillance log archives	To identify and prioritize gaps in current information
Teams	Evaluates the MPF	To find information gaps
Leader	Reviews the MPF	To thoroughly identify intelligence or information gaps
Leader	Loses planning time and manpower	If he fails to thoroughly research information in the MPF
Team	Formulates requests for RFI and RII	After intelligence gaps are identified

7-16. Once the LRSC and LRS team receives their mission, they conduct a similar analysis. The leaders and team members review and evaluate the MPF, OPORD, annexes, briefings and debriefings to identify intelligence or information gaps. Once this is complete, they formulate RFIs and requests for intelligence information (RIIs).

SECTION II. DESCRIBE ENVIRONMENTAL EFFECTS ON OPERATIONS

At this second step in the IPB process, staffs determine how the environment affects both the friendly and threat operations and the friendly and threat COAs. Performing this step in a determined and thorough manner may prevent the unit from being surprised by an unexpected enemy COA.

ANALYZE THE ENVIRONMENT

7-17. An AO evaluation is generally more detailed than an AOI evaluation. For each, the leader considers the areas that might favor one type of military operation, such as attack or defend, or those associated with stability operations such as peace enforcement, peacekeeping, and arms control.

ANALYZE TERRAIN

7-18. Terrain analysis is the study and interpretation of natural and manmade features of an area, their effects on military operations, and the effects of weather and climate on these features. Terrain analysis is a continuous process. Changes in the operational environment may change the analysis of its effects on the operation or on threat COA.

7-19. Ideally, analysis of the military aspects of terrain is based on reconnaissance of the AO and AOI. This starts with a map and imagery reconnaissance. If METT-TC permits, the LRS team can conduct an aerial or vehicular reconnaissance of the AO and of the objective AO. Automated digital terrain tools can assist in the analysis of environmental factors and can display data over maps. Multispectral imaging processors (MSIPs) and other digital tools can also help. The BFSB S-2's GI&S team can provide these and other tools, as available. Automated tools supplement ground, air, map, or imagery reconnaissance and products include--

- Cross-country mobility.
- Lines of communication.
- Vegetation types and distributions.
- Surface drainages and configurations.
- Surface materials.
- Subsurface (bedrock) materials.
- Obstacles.
- Infrastructures.
- Flood zones.
- Potential helicopter landing zones.
- Potential amphibious landing zones.

Note: Lines of communication include transportation, communications, and power.

Analyze the Military Aspects of Terrain

7-20. The military aspects of terrain--OAKOC--follow:

Observation and Fields of Fire

7-21. Observation refers to the ability to see the threat, either aided by surveillance devices or unaided. This includes observation through electronic and optical LOS systems, thermal imaging devices, laser range finders, jamming devices, radars, and radios as well as observation from overhead platforms.

7-22. A field of fire is an area that a weapon or group of weapons can effectively cover with fire from a given position. A field of fire is evaluated for threat and friendly indirect- and direct-fire weapons. Even if a clear opening offers the best observation, it might have poor fields of fire.

7-23. The evaluation of observation and fields of fire allows identification of--

- Potential engagements areas.
- Defensive terrain and specific equipment or equipment positions.
- Areas where friendly forces are most vulnerable to observation and fires.
- Areas of visual dead space.

7-24. Intervisibility and line of sight have a close relationship. Intervisibility is the condition of being able to see one point from the other. This condition may be altered or interrupted by adverse weather, dusk,

Chapter 7

terrain masking, and smoke. Line of sight is an unobstructed path from a Soldier weapon, weapon sight, electronic sending and receiving antennas, or reconnaissance equipment from one point to another.

7-25. Observation and field of fire require special consideration in urban environments such as observation and weapons effects. Urban situations create a lot of dead space. However, high structures generally offer excellent observation. Line of sight distances may decrease in urban settings.

Avenue of Approach

7-26. This is an air or ground route of an attacking force (friendly or threat) leading to its objective or key terrain. Avenues of approach normally show the size of unit that can use them.

Key Terrain

7-27. This refers to any place whose seizure, retention, or control affords a marked advantage to either combatant. In an urban environment, key terrain can include tall structures, choke points, intersections, bridges, industrial complexes, or other facilities, for example. High ground can serve as key terrain, because it dominates an area with good observation and fields of fire. In an open or arid environment, a draw or wadi could serve as key terrain.

7-28. Decisive terrain is key terrain that has an extraordinary impact on the mission. The successful accomplishment of the mission depends on seizing, retaining, or denying decisive terrain to the threat. Note that key terrain is not necessarily decisive. The commander designates decisive terrain to show his staff and subordinate commanders how important that terrain is to his concept of the operation.

7-29. Other services emphasize the importance of the population, and include groups of people such as ethnic groups, the media, or political parties as terrain, and in some cases key terrain. However, the US Army does *not* consider people to be terrain.

Obstacles

7-30. An obstacle is any obstruction designed or employed to disrupt, fix, turn, or block the movement of a threat, and to impose additional losses in personnel, time, and equipment on the threat. Obstacles can be natural, manmade, or a combination of both. Some examples are--

- Buildings.
- Mountains.
- Steep slopes.
- Dense forests.
- Rivers.
- Lakes.
- Urban areas.
- Minefields.
- Certain religious and cultural sites.
- Wire obstacles such as concertina wire, barbed wire.

7-31. Obstacles could affect certain types of movement differently. As an example, obstacles such as rivers, lakes, swamps, densely forested areas, road craters, rubble in streets, or dense populations in urban areas may have a greater effect on mounted movement than on dismounted movement. Mine fields, concertina wire, or steep slopes may have a greater effect against dismounted movement. Obstacles that can affect air mobility include terrain features that are higher than an aircraft's service ceiling, that restrict nap-of-the-earth flight, or that force the aircraft to use a particular flight profile. Examples include tall buildings (skyscrapers), cell phone towers, phone and power lines, rapidly rising terrain features, mountains, smoke, and other obscurants. High mountains can impact rotary- and fixed-wing aircraft lift capabilities.

7-32. Leaders combine the several factor overlays into a single product known as the combined obstacle overlay (COO). They integrate these overlays with the evaluations of various other factors, for example, into a single product, the modified COO, or MCOO, that shows the effect of the operational environment on mobility.

7-33. The MCOO provides the basis for identifying air and ground AA and mobility corridors. It integrates all obstacles to movement including, but not limited to, built-up areas, slopes, soil, vegetation, and transportation systems (bridge classification, road characteristics) into one overlay. It is important that the MCOO be tailored to operational METT-TC factors. It is a collaborative effort involving input from the entire staff. The MCOO shows the terrain according to mobility classification. These classifications are severely restricted, restricted, and unrestricted.

- *Unrestricted*--This terrain is free of any restrictions to movement. Examples include gently sloping terrain with scattered or widely spaced obstacles such as trees or rocks.
- *Restricted*--This terrain hinders movement to some degree. It is represented as "/////////" on overlays. Restricted terrain includes "zigzagging" or frequent detours; swamp or rugged terrain for LRS teams or dismounts; and moderately to densely spaced obstacles for armor or mechanized forces.
- *Severely Restricted*--This terrain severely impedes or redirects movement. It is represented as "XXXXXXXX" (cross-hatching) on overlays. Examples include minefields; unfordable rivers; and road, railroad, and stream embankments.

Cover and Concealment

7-34. This aids in identification of defensible terrain, approach routes, assembly areas, or deployment and dispersal areas. Cover and concealment is evaluated the same as observation and fields of fire. Each factor is combined onto a single product such as a cross-hatched overlay.

Cover

7-35. This means protection from bullets, fragments of exploding rounds, flame, nuclear effects, and biological and chemical agents. Cover does not necessarily provide concealment.

Concealment

7-36. This is protection from observation such as that provided by woods, underbrush, snowdrifts, and tall grass. Concealment considerations for urban operations include using NSTVs or wearing the types of clothing worn by the populace. Both may offer some concealment and help the LRS teams blend in. Concealment and cover *are not the same thing.* Concealment hides, cover protects.

Evaluate the Terrain's Effect on Military Operations

7-37. The BFSB and R&S squadron staffs evaluate how terrain will affect military operations. They disseminate the results of this analysis in the intelligence annex or estimate. The staff uses any of four basic techniques to evaluate and graphically show the results of the analysis:

- *Concentric Ring Technique*--This technique establishes concentric rings around US forces that start from the unit's base of operation and work out.
- *Belt Technique*--This technique divides the AO in belts (areas) that run the width of the AO. The shape of the belt is based on METT-TC analysis.
- *Avenue-In-Depth Technique*--This technique focuses on one avenue of approach. It is good for offensive COAs or for defense when canalized terrain inhibits mutual support.
- *Box Technique*--This technique requires a detailed analysis of a critical area such as an engagement area, river-crossing site, or LZ. It is most useful when time is short for operations in noncontiguous AOs.

Chapter 7

WEATHER ANALYSIS

7-38. The BFSB USAF combat weather team and the S-2 section work closely during much of the analysis process. The weather team analyzes the weather's direct effects and its effects on terrain and other aspects of the environment that integrates climate, forecasts, and current weather data with terrain analysis and with the overall analysis of the environment. The weather team describes in detail how the weather will affect each equipment system and subsystem.

7-39. Terrain and weather aspects of the environment are inseparable. During terrain analysis, the analyst determines how the weather will affect terrain. In this substep, the analyst also evaluates how the weather will directly affect operations.

7-40. The Integrated Meteorological System (IMETS) produces the Integrated Weather Effects Decision Aid (IWEDA). The IWEDA shows the LRSU how the weather will affect current and planned operations. IMETS forecasts wind turbulence, surface temperatures, cloud ceilings, humidity, visibility, and ice.

Military Aspects of Weather

7-41. The military aspects of weather are visibility, wind, precipitation, cloud cover, temperature, and humidity.

Visibility is the greatest distance from which prominent objects can be seen and identified by the unaided, normal eye.

Wind of sufficient speed from any direction can blow dust, smoke, sand, or precipitation, reducing the combat effectiveness of a force.

Precipitation is any moisture that falls from a cloud in frozen or liquid form. Rain, snow, hail, drizzle, sleet, and freezing rain are examples.

Cloud Cover affects ground operations by limiting illumination. It can also reduce the thermal signature of targets.

Temperature extremes can reduce effectiveness of troops and equipment.

Humidity is the amount of water vapor suspended in the atmosphere.

Additional Weather Considerations

Thermal Crossover

7-42. Temperature of targets and objects on the ground is important for the use of thermal sights and forward-looking infrared (FLIR). Thermal crossover, which is an additional weather consideration, is a natural phenomenon that normally occurs twice daily when temperature conditions reduce thermal contrast between adjacent objects.

Direct and Indirect Effects

7-43. Weather has both direct and indirect effects on military operations. The following are examples of direct effects and indirect effects on military operations:

- Temperature inversion can increase the risk of contamination by chemical agents.
- Low visibility, such as that caused by fog, obviously affects the observation capabilities of both friendly and threat forces.
- Hot, dry weather might force friendly and threat forces to consider water sources key terrain.

Intelligence Preparation of the Battlefield

Civil Considerations

7-44. An appreciation of civil considerations—the ability to analyze their impact on operations—enhances several aspects of operations: among them, the selection of objectives; location, movement, and control of forces; use of weapons; and protection measures. Civil considerations comprise six characteristics, expressed in the memory aid ASCOPE:

- Areas.
- Structures.
- Capabilities.
- Organizations.
- People.
- Events.

DESCRIBE ENVIRONMENTAL EFFECTS

7-45. Combine the evaluations of the effects of terrain, weather, and civil considerations into a product that best suits the LRS team's needs. *Avoid guessing or assuming.* Focus on the total environment's effects on the COAs available to both the LRS team and threat forces.

7-46. On request from the LRSC, the BFSB S-2 and R&S squadron S-2 will provide all the products previously described. The LRSC may only need some of these products as is; others they may be able to use, with some adaptation, in LRS team-planning operations. Training and close coordination between the LRSC, the BFSB, and the R&S squadron staffs will produce useful products for team planning.

SECTION III. EVALUATE THE THREAT

In step 3, the G-2/S-2 and staff analyze the command's intelligence holdings, which they identified in step 1, to determine how the threat normally conducts operations under similar circumstances. Every threat can be analyzed, understood and, to some extent, predicted. Threat doctrine may be simple or even nonexistent. However, a threat will usually, at some level of command, act based on some set of ad hoc or established procedures. This third step in the IPB process begins with analyzing the threat, after which two substeps are performed: 1) update or create threat models, and 2), identify threat capabilities.

ANALYZE THREAT FACTORS

7-47. When operating against a new or less defined threat, the G-2/S-2 may need to develop or expand intelligence databases and threat models concurrently. In order to accomplish this, the G-2/S-2 should conduct threat characteristic order of battle (OB) analysis for each group identified in step 1. To do this, the staffs analyze--

- Composition.
- Disposition.
- Tactics.
- Training.
- Logistics.
- Operational effectiveness.
- Communications.
- Intelligence.
- Recruitment.
- Support.
- Finance.
- National agencies.
- Law enforcement agencies.
- International organizations and nongovernmental organizations.
- Personality.
- Other threats such as CBRN, diseases, or toxins.

Chapter 7

UPDATE OR CREATE THREAT MODELS

7-48. Creating or updating a threat model lets the analyst piece together information, identify gaps, predict threat activities or COAs, and plan ISR. There will always be information gaps in the threat model, so the analyst will always have some uncertainty. Threat models have three parts:

- Convert threat doctrine or patterns of operation to graphics.
- Describe the threat's tactics and options.
- Identify HVTs and HPTs.

CONVERT THREAT DOCTRINE OR PATTERNS OF OPERATION TO GRAPHICS

7-49. Threat templates graphically portray how the threat might use its capabilities to perform the functions required to accomplish its objectives. Construct threat templates by analyzing the intelligence database and by evaluating the threat's past operations. Determine how the threat normally organizes for combat, and how he deploys and employs his forces and assets. Look for patterns in how the threats organize their forces, timing, distances, relative locations, groupings, or use of the terrain and weather. Threat templates are tailored to the needs of the unit or staff section creating them.

7-50. Threat templates for a LRS team are tailored to the team's mission. If the LRS team is tasked to locate the threat's regimental or division reconnaissance, the LRSU needs the threat template showing how the threat reconnaissance units are deployed. In some OEs, threat templating can be more difficult and unpredictable. The LRS team must consider the enemy situation and mission. Ask this question: "If I had to accomplish the same mission, and had no terrain constraints, where would I place my assets?" Analyze patterns and associations, even though these require frequent updates and are somewhat reactive.

DESCRIBE THE THREAT'S TACTICS AND OPTIONS

7-51. The threat model includes a description of the threat's preferred tactics. A description is still needed, even if the threat's preferred tactics are shown in graphic form. The description--

- Lists the threat's available options.
- Is not a "snapshot in time"--should portray actions as events unfold.
- Aids in war-gaming and in developing threat COAs, and situational templates.
- Addresses timelines, phases, WFFs.

7-52. Describe and determine the threat's goal(s). Threat objectives are often, but not always, what the unit's mission is trying to prevent. Threat objectives are also often actions taken by the threat to prevent unit mission accomplishment. Describe them in terms of purpose and endstate.

IDENTIFY HVTS AND HPTS

7-53. An HVT is the asset the threat commander requires for successful completion of a COA. It is shown and described on the template. Examples include--

Fires, for example, regimental artillery group (RAG), division artillery group (DAG), IV13, IV14, SNAR-10, or individual artillery and mortars.

Protection, for example, ZSU-23-4, 2S6, straight flush radar, fan song radar, man-portable air defense systems, and heavy machine guns.

Command and Control, for example, IV13, IV14, TOCs, and vehicles with multiple antennas.

Intelligence, for example, twin box DF, dog-ear radar, scanners and local populace.

Movement and Maneuver, for example, IMR, MT-55, PMM-2, GSP, MDK-2M, and BTM.

Sustainment, for example, a list of assets key to the threat commander's execution of the primary mission is recorded. Then, the assets are ranked by their relative worth to the threat's operation. Throughout the course of an operation, the HVTs will change. The HVTs for the phase of each operation are recorded and annotated on the threat model. The LRS team should identify HVTs for their mission with the help of the unit staff.

7-54. An HPT is a target whose loss to the threat commander, will contribute to the success of the friendly COA. HVTs and HPTs may be one in the same. In some OEs, HPTs might be key personalities. However, the loss of the enemy key personality might support the friendly COA, but may fail to deter the enemy from completing the COA.

Link Analysis

7-55. This tool identifies HVTs and HPTs in the OE. It is used to show contacts, associations and relationships between persons, events activities and organizations in an unconventional setting. Link analysis tools use link diagrams, association matrixes, relationship matrixes, activities matrixes and time-event charts. All help in identifying HVTs and HPTs in the OE.

Link Diagram

7-56. This tool seeks to graphically show relationships between people, locations, or other factors deemed significant in any given situation. It reflects information from both association and activities matrixes, is easy to read and interpret and is generally an effective briefing tool. Link diagrams show participants in activities, personal and nonpersonal links, internal and external contacts, structures and lines of C2 (Figure 7-2).

Chapter 7

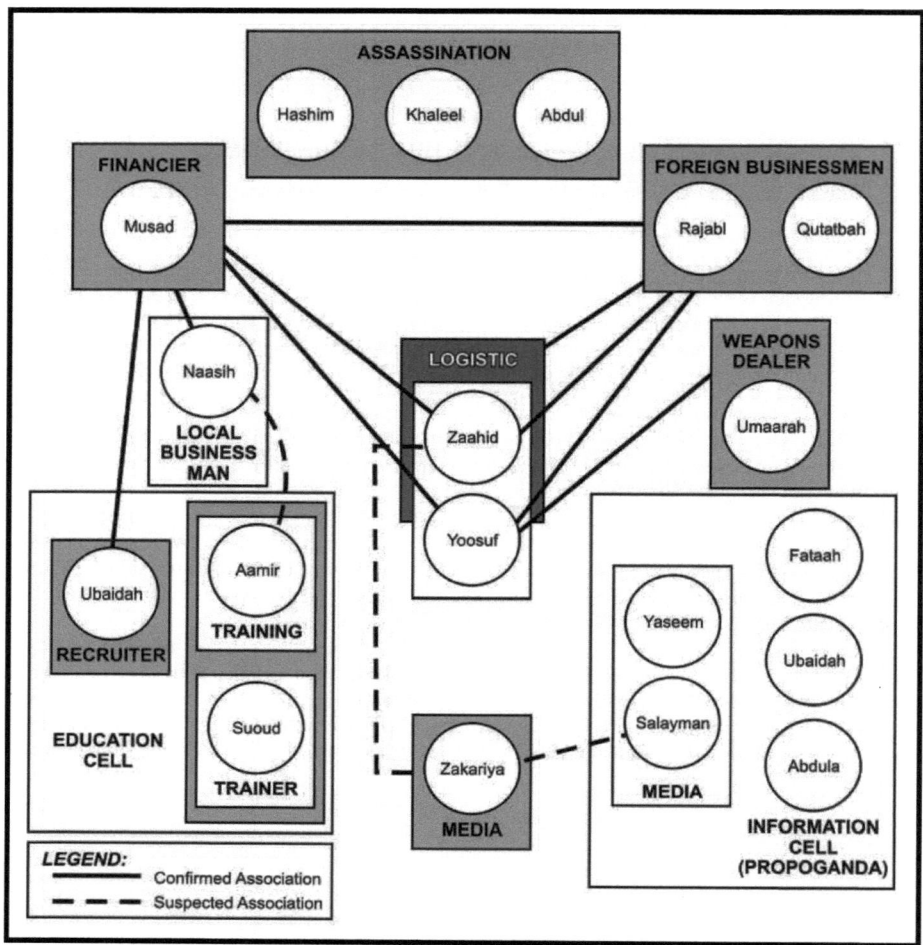

Figure 7-2. Example link diagram.

Association Matrix

7-57. The association matrix is used to establish the existence of an association between individuals. Analysts can use association matrices to identify those personalities and associations needing a more in-depth analysis in order to determine the degree of relationship, contacts, or knowledge between the individuals. The structure of a threat organization is formed as connections between personalities are made (Figure 7-3).

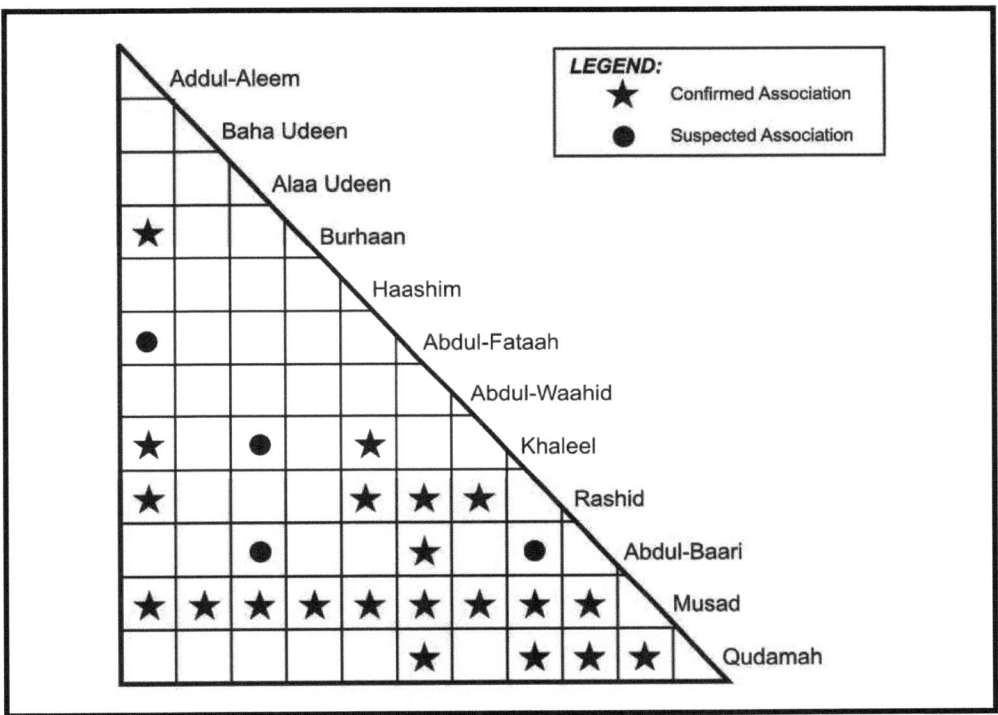

Figure 7-3. Example association matrix.

Relationship Matrix

7-58. Relationship matrices are intended to show the nature of relationships between elements of the AO. The elements can include members from the noncombatant population, the friendly force, international organizations and adversarial groups. Utility infrastructures, significant buildings, media and activities might also be included. The nature of the relationship between two or more components includes measures of contention, collusion or dependency. The purpose of this tool is to demonstrate graphically how each component of AO interacts with others and whether these interactions promote or degrade the likelihood of mission success. The relationships represented in the matrix can assist the analysts in deciphering how best to use the relationship to shape the environment (Figure 7-4).

Chapter 7

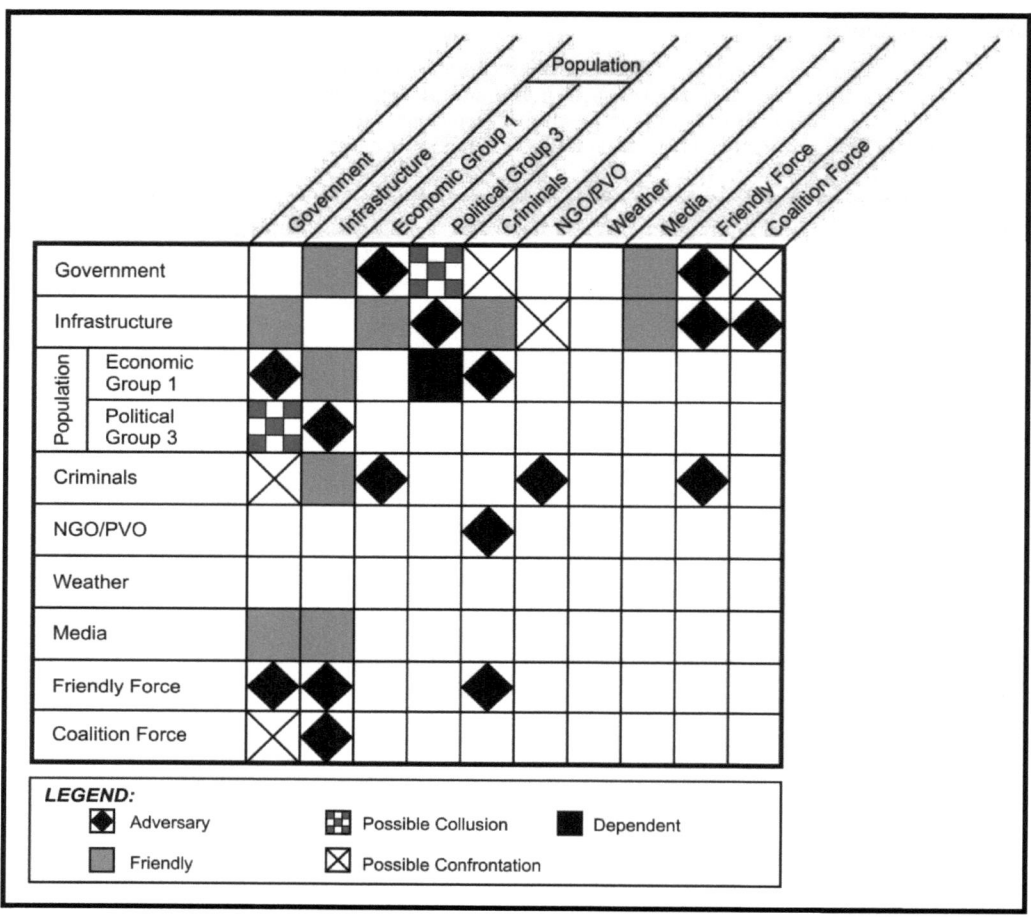

Figure 7-4. Example relationship matrix.

Activities Matrix

7-59. Activities matrices help analysts connect individuals (such as those in the association matrices) to organizations, events, entities, addresses and activities--anything other than people. Information from this matrix, combined with information from association matrices, can assist analysts in linking personalities as well (Figure 7-5).

Intelligence Preparation of the Battlefield

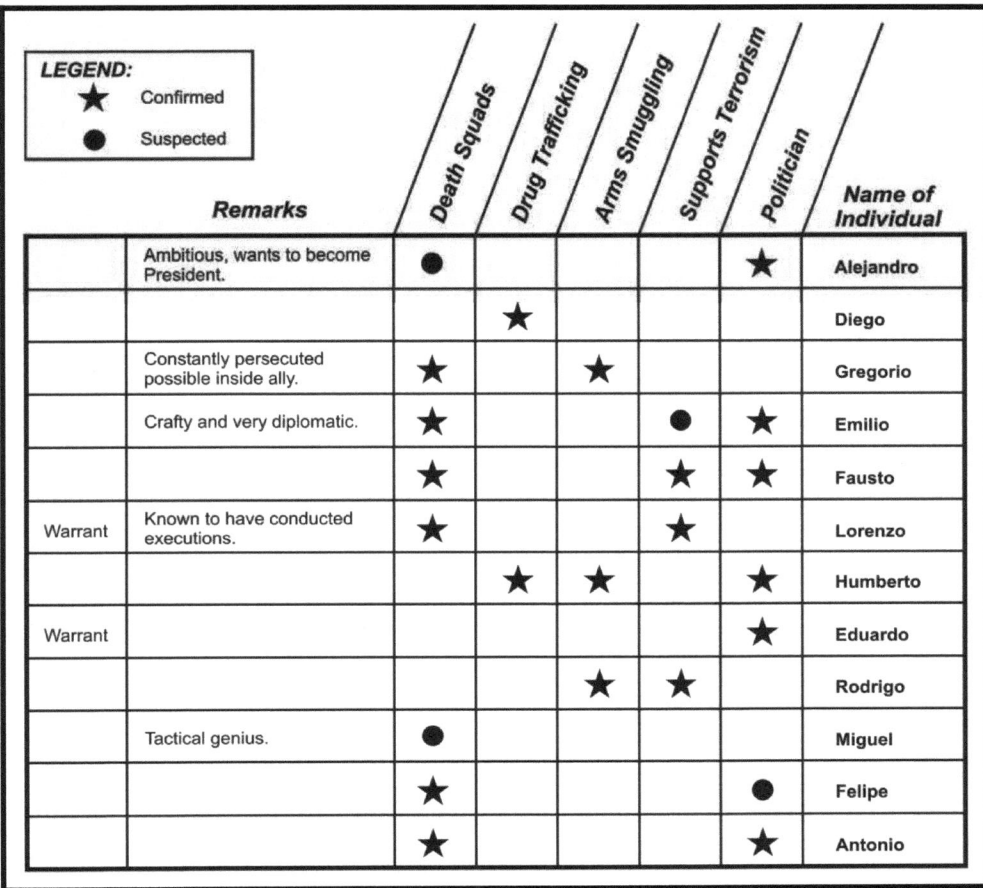

Figure 7-5. Example activities matrix.

Time Event Chart

7-60. Time event charts are chronological records of individual or group activities designed to store and display large amounts of information in a small space. Analysts can use time event charts to help analyze, for example, large-scale patterns of activity and relationships (Figure 7-6).

Chapter 7

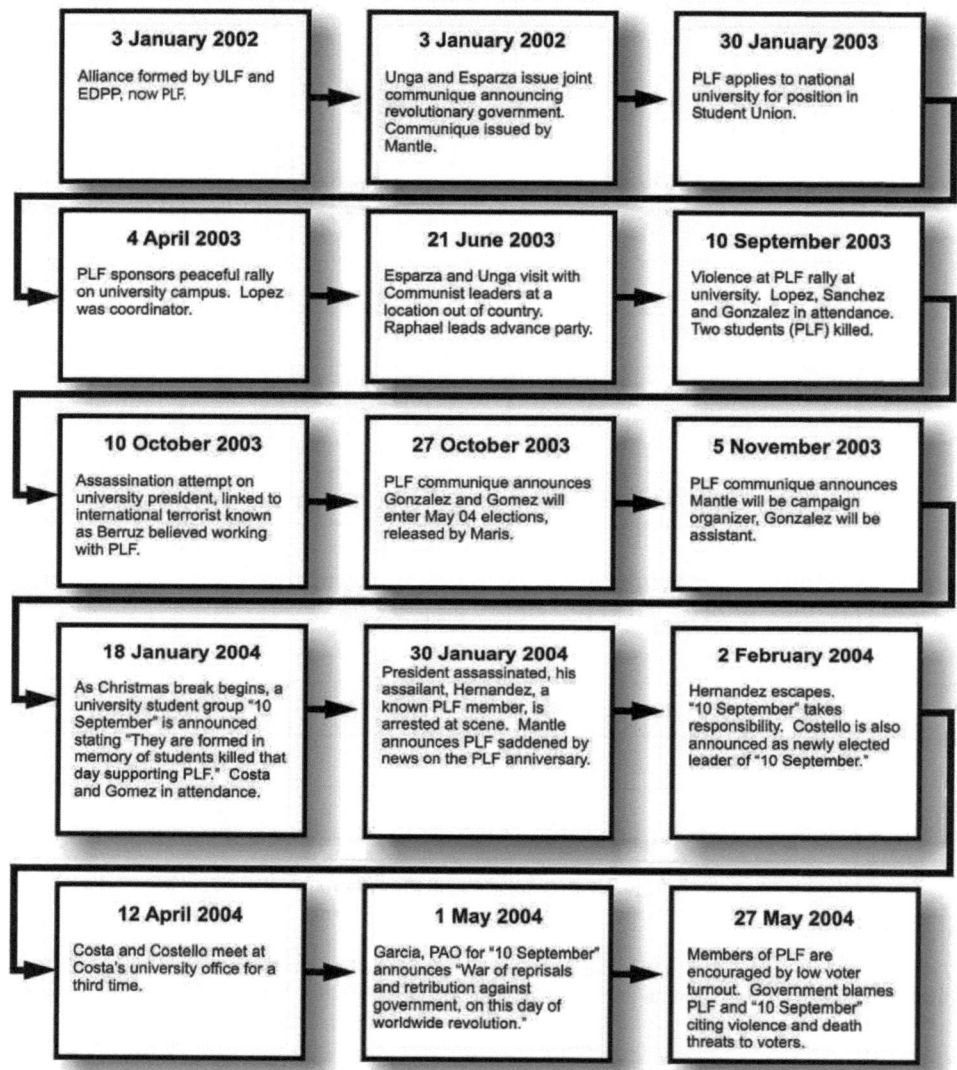

Figure 7-6. Example time event chart.

IDENTIFY THREAT CAPABILITIES

7-61. Leaders identify broad COAs and supporting operations that the threat can take to influence the accomplishment of the friendly mission. Use statements to define the capabilities, for example--

"The threat has the capability to attack with up to eight divisions supported by 150 daily sorties of fixed-wing aircraft."
"The criminal organization has the ability to pay off local law enforcement agencies (LEAs)."
"The threat can establish a prepared defense by 14 May."
"The terrorists have the ability to conduct multiple car bombings simultaneously."

Intelligence Preparation of the Battlefield

7-62. In a stability operation, the lack of, or a variety of, uniforms and equipment makes it more difficult to identify the threat. Varying levels of organization increases the range of options afforded the enemy threat. This makes it easier for the threat to conduct large-scale demolitions (car bombs and rigged buildings) against friendly forces. The ability of the threat to collect information and conduct covert operations is greatly enhanced in a stability operations. Sniper employment and booby traps are capabilities that are often unpredictable and very dangerous to friendly forces.

SECTION IV. DETERMINE THREAT COURSES OF ACTION

This is the fourth and final step in the IPB process. Staffs identify and develop threat COAs, along with other facts and assumptions about the OE, that drive the friendly COAs analysis (wargaming). This leads to development of the friendly COA development. The LRS team uses these threat COAs to drive their own war-gaming process, and to develop the LRS team COA. Developing threat COAs has five substeps:

- Identify the threat's likely objectives and desired endstate.
- Identify the full set of COAs available to the threat.
- Evaluate and prioritize each COA.
- Develop each COA.
- Identify initial ISR requirements.

IDENTIFY THE THREAT'S LIKELY OBJECTIVES AND DESIRED ENDSTATE

7-63. At a minimum, the staff determines likely objectives and desired endstate. Against a conventional threat, the analysis should start at more than one level above the friendly echelon unit and work down. For those threats that are asymmetrical in nature, the analysis should start at the lowest level possible.

7-64. The LRS leader starts with the threat command one level above that of the threat he expects the LRS team to encounter. He then ensures the objectives of the threat commands at each level are identified. The BFSB S-2 fusion element advises the LRSU on the threat's overall objective. A LRS team should break down the threat objective to the level and detail appropriate for the team mission.

IDENTIFY THE FULL SET OF COAs AVAILABLE TO THE THREAT

7-65. The staff and the LRS team consider possible threat COAs.

CONSIDERATIONS

7-66. At a minimum they consider--

- The COAs that the threat's doctrine assumes that the threat will use in the current situation, and the threat's most likely objectives.
- The threat COAs that could greatly influence the friendly unit's mission.
- The threat COA that may exceed the boundaries of known threat doctrine or TTP, even if threat doctrine generally considers them unfeasible.
- The threat COAs indicated by recent activities and events.

CRITERIA

7-67. Each threat COA should meet five criteria:

Suitability--If the COA is successful, it will accomplish the threat's objective.

Feasibility--The threat has the time, resources, and space available to execute the COA.

Acceptability--The threat's tactical or operational advantage gained by executing the COA must justify the cost. The threat might undertake an unfavorable COA if he believes there are no other choices.

Distinguishability--Each threat COA must differ greatly from the others.

Completeness--The COA must show how the decisive operation accomplishes the mission, and how shaping and sustaining operations support the decisive operation.

EVALUATE AND PRIORITIZE EACH COURSE OF ACTION

7-68. The commander and staff need to develop a plan that is optimized to one of the COAs, while allowing for contingency options should the threat choose another COA. Therefore, the staff must evaluate each threat COA and prioritize it according to how likely the threat will adopt that option. Use judgment to rank the threat COAs in their likely order of adoption. Modify the list as needed as the current situation changes.

DEVELOP COURSES OF ACTION

7-69. Once a COA set is complete, develop each COA into as much detail as time allows. To ensure completeness, each COA must answer six basic questions: who, what, when, where, how and why. Each developed COA has three parts--

- Situation templates.
- Threat course of action and options.
- High value targets.

SITUATION TEMPLATES

7-70. Situation templates are developed on the threat's current situation, for example, training and experience levels, logistical status, losses, dispositions, the environment, threat doctrine. or patterns of operations. The SITEMP is a graphic that shows the expected threat dispositions should the threat adopt a particular COA.

Construction

7-71. To construct a SITEMP, start with the threat template and lay it over the MCOO. Adjust the disposition of the arrayed force on the threat template to account for battlefield effects such as weather and terrain. Try to array forces as the threat commander might. Construct the SITEMP in as much detail as time allows:

- *Conventional Threat*--evaluate time and space factors to develop time phaselines showing threat movement.

- *Unconventional Threat*--other tools such as pattern analysis. Pattern analysis uses multiple map overlays and text assessing military, terrorist, or other threat activity in an urban area. These events can be related by any of several factors to include location and time. These events can be analyzed by plotting them on maps over time, using multiple historical overlays (analog or digital) that can be compared to one another over time, and using a time-event wheel or other analysis tools (Figure 7-7).

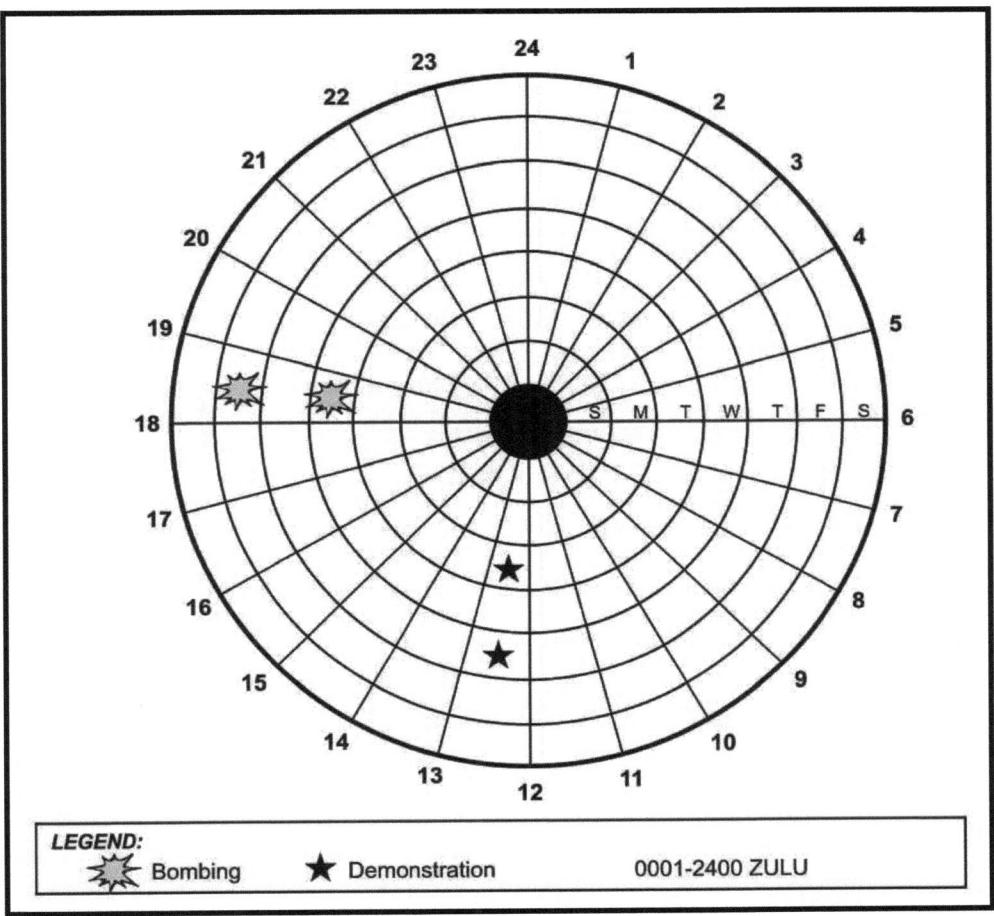

Figure 7-7. Example pattern analysis plot sheet.

Pattern Analysis

7-72. The pattern analysis is a technique that is based on certain characteristic patterns that can be identified and interpreted. Pattern analysis can be critically important when facing an enemy whose doctrine is undeveloped or unknown. It is necessary for the intelligence analyst to create or update the threat model and threat templates.

7-73. A pattern analysis requires the organization and recording of large volumes of incoming information and adding it to existing information so meaningful relationships are clarified. Often, indicators can initially be ambiguous and incomplete. but over time patterns of activity or tip-offs that characterize specific threat emerge. Expect to deal with a thinking threat that learns from previous success or failures and will change or vary patterns of operations.

7-74. Once all related items of information from the intelligence files, sources, and agencies are obtained, assemble the available information to form as many logical solutions or hypothesis as possible. Assembly of information to develop logical hypothesis requires good judgment and considerable area expertise. When developing hypothesis, avoid reaching conclusions based on prejudices or preconceived notions.

Chapter 7

7-75. Pattern analysis tools consist of automation, situation map (SITMAP), incident map, IPB templates, working files, hot files, resource files, coordinate register and time-event charts.

Automation--Automation provides immediate access to situational updates and execution information and allow transmittal of situational understanding and execution orders. Automation enhances situational awareness and enables friendly forces to share a common picture while communicating and targeting in real or near-real time.

Incident Map--Working incident maps and overlays are considered informal. Post current information on the incident map as received. This map is a working aid that graphically shows threat activity in both AO and AOI considered important to the operation. Information on the incident map can provide a good foundation for the SITMAP.

Situation Map--The SITMAP is built from the incident overlay. The SITMAP adds current intelligence and activities indicating movement, resupply operations, or attacks.

Working Files--Working files are critical to properly store the research material generated. Whether done manually or with automation, ensure the filing system is easily understood, information is easy to retrieve, and includes cross-referencing.

Hot Files--Hot files are the most important working file. The hot file contains material pertaining to a specific incident, as well as information from related incidents. Reports of planned demonstrations, sabotage, or attack all initiate hot files. A hot file becomes inactive when the event occurs, does not occur, or when the unit's priorities change.

Resource Files--Resource files includes all material which is important but not of immediate value. It includes hot files that are obsolete, inactive incident files, inactive personality and organization files, and photographs.

Level of Command, Operational Type, and Named Areas of Interest

7-76. The level of command and type of operation bears directly on the level of detail that goes into a situation template.

Tactical Level--Templates at this level sometimes show individual vehicles in threat dispositions. The NAIs are often pinpoint locations such as road junctions or small-unit battle positions.

Operational Level--Templates at this level focus on large reserve formations, major staging bases and LOCs. The NAIs are often large dispersal areas, reserve assembly areas, or logistical support areas.

Strategic Level--Templates at this level might focus on the shift of large forces from one theater to another as well as political and economic developments. The NAIs can sometimes encompass large regions.

THREAT COURSES OF ACTION AND OPTIONS

7-77. This describes the activities of the forces shown on the situation template. It can range from a narrative description to a detailed synchronization matrix showing activities of each unit, WFF, or asymmetrical activity in detail. It should address the earliest time the COA can be executed, timelines and phases associated with the COA, and decisions the threat commander will make during and after execution of the COA. Use the COA description to support staff wargaming and to develop the event template and supporting indicators

HIGH-VALUE TARGET

7-78. As the SITEMP is prepared, mentally war-game and note how and where each of the WFF provide critical support to the COA. This leads to identification of the HVTs. Note on the SITEMP where the HVTs must appear or be employed to make the operation successful. Focus on times just before or when the HVTs are most valuable. These HVTs lead to potential TAIs, engagement areas, and decision points.

IDENTIFY INITIAL ISR REQUIREMENTS

7-79. After identifying the set of potential COAs, focus on which one the threat will most likely adopt. This is crucial to identify ISR requirements for an effective ISR plan. At the LRSC and LRS team level, they identify the COA RFIs and RIIs to send to the R&S squadron S-2 and the BFSB S-2 ISR fusion element to "fill in the blanks" in the COA. The team identifies the areas where they expect key *events* to occur. These are called "named areas of interest" (NAIs). Activities or events that reveal a COA are called "*indicators.*"

EVENT TEMPLATE

7-80. The event template is a guide for ISR synchronization and ISR planning. It shows the NAIs where activity or lack of activity will indicate which threat COA the threat has adopted. It shows who, what, where, and when to collect information that reveal the threat's chosen COA. Comparing and contrasting the various NAIs and indicators associated with each COA helps identify the differences. These differences are markers that help to recognize which COA the threat has chosen to execute. The selected NAIs are marked on the event template. This initial event template focuses only on identifying which of the predicted COAs the threat has adopted.

EVENT MATRIX

7-81. The *event matrix* complements the *event template*. The matrix provides details on the type of *activity* is expected to occur at each NAI, when it is expected to be active, and how those *activities* relate to other *events* (*indicators*) on the battlefield. The matrix contains the *event* (*indicator*) associated with each NAI. It also includes the team phase lines from the SITEMP, and the LTIOV time line. Its main uses are in collecting intelligence and aiding in situation development. The elements of the event matrix follow:

- *Information Requirement*—IR includes all of the information elements that the commander and staff need to successfully conduct operations, that is, all elements necessary to address the factors of METT-TC.

- *Priority Intelligence Requirement*—PIR includes all intelligence requirements for which a commander has an anticipated and stated priority in his task of planning and decision-making.

- *Intelligence Requirement*—Intelligence requirements include knowledge of lesser importance than the PIR.

- *Specific Information Requirement*—SIR consists of indicators that will answer all or part of a PIR or IR.

- *ISR Tasks and RFIs*—ISR tasks and RFIs are the orders or requests that generate planning and execution of a collection mission or analysis of database information.

- *Named Area of Interest*—An NAI is a geographic area where indicators can be collected.

- *Target Area of Interest*—A TAI is a geographic area where high-value targets (HVTs) can be acquired and engaged.

- *Decision Point*—A DP is a point in space and time where the commander or staff anticipates making a decision concerning a specific friendly COA. A DP is usually associated with a specific TAI, and is located in time and space to permit the commander sufficient lead time to engage the enemy in the TAI.

PREPARE DECISION SUPPORT TEMPLATE

7-82. The decision support template is a graphic version of a war game. From this template, the BFSB and R&S squadron S-2 and S-3 prepare a detailed ISR plan. This plan shows where and when R&S elements such as LRS should look for the enemy. The plan directs specific tasks and priorities to LRSC, which are then assigned as missions to individual LRS teams.

Chapter 8

Evasion and Recovery

Evasion is the process whereby people isolated in hostile or unfriendly territory avoid capture and return to areas under friendly control. Recovery is the return of such evaders to friendly control, either with or without aid, as the result of plans, operations, and individual actions by recovery planners, conventional or unconventional forces, and sometimes the evaders themselves. Evasion is considered the highest form of resistance. Both E&R are integral to military operations (Appendix K).

FUNDAMENTALS

8-1. Conduct of E&R operations requires Soldiers understand each kind of operation and the laws that govern them. Soldiers participating in an evasion operation are classified as one of the following:

EVADER

8-2. An evader is considered a lawful combatant for the duration of the evasion, since evasion is an extension of combat and a refusal to capitulate to the enemy. Evaders are obliged by the Code of Conduct to do all they can to avoid capture and rejoin friendly forces. If needed, they may commit acts of violence against legitimate military targets, without being prosecuted by the enemy for violating local criminal laws.

> **Code of Conduct. Article II:**
> I will never surrender of my own free will. If in command, I will never surrender the members of my command while they still have the means to resist.

ESCAPEE

8-3. An escapee is someone who has escaped from a confinement facility. Escapees are noncombatants, no longer able to commit hostile acts, who may be charged under the laws of the detaining power for certain acts committed against its military or civilian population during escape or avoidance of recapture. Such an escapee may carry no arms of any kind, nor may he try to arm himself. If he commits any crime(s) of no specific military significance, he may be tried and punished for war crimes, not only by the detaining country, but also by his own. The *Code of Conduct* and *Law of Land Warfare* both list the responsibilities of the evader and escapee in an evasion. However, under Geneva Convention Articles 91-94, *Geneva Convention Relative to the Treatment of Prisoners of War*, POWs have a national obligation to escape and rejoin their own forces. Those apprehended during or after an attempted escape are subject to disciplinary punishment only. This assumes that any offences they may have committed were for the sole purpose of escaping, and that the offences entail no violence against life or limb, or the goal of self-enrichment.

DETENTION OF EVADERS BY NEUTRAL COUNTRIES

8-4. The evader who crosses into a neutral country is subject to detention by that country for the duration of the war. A neutral country that receives escapees may leave them at liberty. If it allows them to remain in its territory it may assign them a place of residence. The neutral country is also authorized to confiscate all equipment of the evader/escapee.

TIMES TO INITIATE E&R

8-5. These include—
- On order of commander.
- When considered "isolated personnel" by individual unit standards or SOP.

CHAIN OF COMMAND

8-6. The agencies, units and individuals within the E&R chain of command are responsible for the successful planning for and execution of E&R operations.

JOINT PERSONNEL RECOVERY AGENCY

8-7. The JPRA develops joint E&R tactics, techniques, and procedures; E&R aids; tools; and specialized equipment for E&R. They also provide expertise on E&R and survival to all services.

JOINT SEARCH-AND-RESCUE CENTER

8-8. The JSRC represents two or more services or countries. They coordinate recovery efforts among joint services. They develop and distribute the Air Tasking Order Special Instructions (ATOSPINS). They maintain ISOPREP cards for missing personnel.

RESCUE COORDINATION CENTER

8-9. A RCC is located at each service in the theatre of operation. The RCC notifies the JSRC of isolated personnel. They help individual units develop their own EPA. They maintain a copy of ISOPREP cards for isolated personnel. They coordinate recovery efforts for that service.

INDIVIDUAL UNITS

8-10. Each unit is responsible for developing an EPA, and for ensuring that all personnel properly fill out their ISOPREP cards, after which the unit keeps the cards updated, stores them in a secure location, and forwards a copy to the RCC.

INDIVIDUAL SOLDIER

8-11. Responsible for filling out the EPA and the ISOPREP card.

PLANS

8-12. Personnel assigned to LRSU are considered high-risk-of-capture and subject to isolation in hostile territory. Therefore, they should prepare for the possibility of being in an evasion situation. Successful evasion is dependent on detailed planning, as well as peacetime training and proficiency in survival and E&R tactics, techniques and procedures.

8-13. The LRSC commander, with assistance from the RCC and JSRC, is responsible for and prepared to conduct personnel recovery operations in support of their own operations. The commander also coordinates with the RCC when elements are preparing to enter a possible evasion situation. He relays information such as ISOPREP cards and EPA with overlays of the AO. After the LRSC commander coordinates with evasion planning agencies, he may determine the unit must make independent evasion plans. The LRSC commander starts by identifying the team's AO and formulating and evasion annex with the assistance of the JSRC.

8-14. LRS team evasion planning begins with receiving the E&R annex to the OPORD along with any evasion aids that will assist them in the planning phase. After receiving area briefings and examining E&R area studies, the team formulates an EPA. The EPA is normally an annex to the team OPORD and briefed to the commander. All available evasion aids are requested to assist the team if evasion is required. Finally, the team reviews their ISOPREP cards before leaving the planning facility to start infiltration.

TYPES OF RECOVERY

8-15. The two types of combat recovery of isolated personnel are conventional and unconventional (Figure 8-1):

Evasion and Recovery

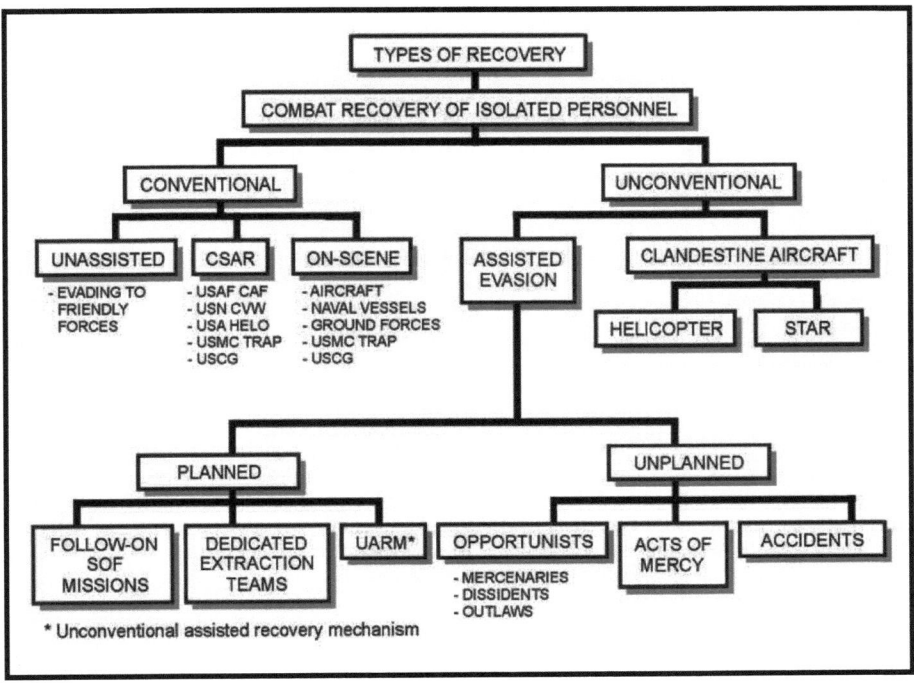

Figure 8-1. Types of recovery.

CONVENTIONAL

8-16. There are three types of conventional recovery:

Unassisted--Unassisted recovery by isolated personnel requires evasion back to friendly lines.

CSAR--Available PR assets, controlled and initiated by the JPRA, can recover the isolated personnel.

On-Scene--Assets already in the AO, such as aviation, ground or naval forces can recover the isolated personnel.

UNCONVENTIONAL

8-17. There are two types of unconventional recovery: assisted and clandestine (aircraft). Assisted evasion is the most likely of the two types that LRSU will have available. Planned recovery by SOF with a dedicated recovery mission, or clandestine units may already be in place in the AO. Opportunists sometimes perform unplanned recoveries in hope of receiving a reward for their actions. Civilians may perform acts of mercy when evaders need medical attention. At other times, those sympathetic to the cause will also provide accidental assistance.

CLASSIFICATIONS OF EVASION

8-18. In a short-range evasion, the evader is close to the main battle area and becomes isolated. The evader usually has the means to return to the unit within a few days. In an extended evasion, which involves greater distances behind enemy lines, the evader might have to travel an extended distance through foreign terrain, possibly with little food and equipment. LRS teams generally fit into this group. Characteristics of a successful extended evasion include—

Chapter 8

KNOWLEDGE OF THE AO

8-19. Study of the AO and detailed E&R are critical to success.

KNOWLEDGE OF SURVIVAL

8-20. These skills should be reinforced in the unit with thorough survival and evasion classes. Like most perishable skills, survival training should be a regular training task.

SUPPLY ECONOMY

8-21. Limiting use and making the best use of available supplies.

ENDURANCE

8-22. The physical and mental conditioning of individuals and the team will largely determine the ability to successfully execute E&R.

PATIENCE AND DISCIPLINE

8-23. Recovery may not be immediate--individuals and teams should be prepared to evade for days, weeks, months or even years. Patience and discipline are often traits acquired though tough and realistic training. Soldiers in an E&R and survival situation will be subjected to many physical and mental hardships. These hardships are compounded if detailed area studies are not conducted during the planning phase and Soldiers have limited survival skills. One major factor that has lead to more successful evasions than any other is the individuals will to survive. Many factors affect evaders, they include—

- Family and home ties.
- Self-preservation.
- Material items.
- Religion.
- Sense of responsibility—to your family, to your team, and to your country.

> **Code of Conduct. Article III:**
> If I am captured, I will continue to resist by all means available. I will make every effort to escape and aid others to escape. I will accept neither parole nor special favors from the enemy.

MOVEMENT

8-24. The team must consider the following movement factors in order to ensure successful evasion:

TECHNIQUES

8-25. The movement techniques used during evasion are the same as with normal patrolling, as well as the SOPs for movement.

ROUTE SELECTION

8-26. Route selection is an extremely important factor when conducting evasion. Avoid all major built up areas and areas with high traffic.

CAMOUFLAGE AND CONCEALMENT

8-27. Camouflage and concealment are crucial when moving, and should never be sacrificed.

Rate of Movement

8-28. A rate of movement is established that allows the team to achieve goals established in the E&R plan. Moving as fast as possible is seldom the preferred technique. Some of the goals and reasons for them are--

Progress

8-29. The team moves to the terrain feature/location goals established in the E&R plan. Measure progress on the ground by the hole-up areas that are reached. Be flexible--make adjustments to the E&R plan based on METT-TC considerations.

Stealth

8-30. Although progress is extremely important, care should be taken so that the team is not rushed. Speed should not be substituted for security.

Energy Conservation

8-31. Along with assisting in stealth, the slower you move, the more energy you conserve, and the longer you will be able to evade.

Physical Condition

8-32. Physical condition is another deciding factor in how successful the team will be in the evasion. Equipment cross loading may be necessary if some Soldiers are caring a heavier load than others, or if there are wounded personnel on the team. The team is only as strong and fast as their weakest or slowest Soldier. Since the goal of the team is to be recovered by friendly forces, at times it may be necessary to cache equipment in order to assist the team in evading. This technique is highly discouraged since the team should have a goal of extracting with 100 percent of Soldiers, weapons, and equipment.

Team Movement

8-33. The team may evade by whatever technique they have for a SOP, such as moving as a whole element or by breaking up into two to three groups. There are advantages and disadvantages for both and they must be considered when deciding how the team will move during E&R.

Countertracking

8-34. The team should use counter tracking techniques throughout the E&R. During the course of the evasion there are certain considerations that will greatly affect the routes that are taken and the way the evasion is conducted.

Obstacles

8-35. Obstacles will be encountered throughout the evasion corridor. The two types of obstacles are natural and man-made. Obstacles not only restrict movement, but they may also funnel movement. However, some obstacles may serve as navigation aids for the evader, as well as aid in movement:

Mountains and Hilltops

8-36. Use these for navigation, but avoid traveling on them due to silhouetting.

Rivers and Streams

8-37. You can also use rivers and streams as guides and for water resupply, if the water is potable, but remember that the local populace may also use it.

Roads or Trails

8-38. Avoid roads and trails, and *never* use them for movement.

Chapter 8

Bridges

8-39. Avoid these, too, because, although they look like the easiest and fastest way across a water obstacle, they are also used for checkpoints and roadblocks.

Populated Areas

8-40. Avoid entering populated areas, though if you are being tracked by dogs, traveling near a populated area can throw off your scent.

SPECIAL CONSIDERATIONS

8-41. Here are some things to considered while evading:

- Know how the locals feel about your cause. If you are compromised during the evasion, this information can improve your chances of survival.
- Blend in with the natives. Knowing and following their customs and habits will aid in your survival during the evasion.
- Remember that pets and livestock will compromise your situation.
- Know local curfews. They may determine your movement in an evasion.

DISGUISES

8-42. The use of disguises, have been discouraged due to the belief that if captured you could possibly be shot as a spy for treason. That is not to far from the truth, however there are ways that you may use a disguise to your advantage.

WEARING OF ENEMY UNIFORMS

8-43. It is a violation of international law to make improper use of the uniform of the enemy. It is, however, permissible for military personnel isolated in hostile territory to use the enemy's uniform to assist in evading capture, as long as no other military operations are carried out while so attired.

WEARING OF CIVILIAN CLOTHING

8-44. It is a violation of international law to kill, injure, or capture the enemy by feigning civilian status. As with the military disguise you should avoid conducting operations that might be defined as either military or espionage. Even if the evader is within his legal rights to use the clothing to help him evade, he might, if captured, have to identify himself as an American. If partisans are aiding his evasion, then he could be treated as one of them and tried for acts of treason. Every Soldier must keep his ID card and blood chit on his person.

PROTECTED EMBLEMS

8-45. The only disguises unauthorized for use in escape or evasion are protected emblems such as those of medical, religious, or relief agencies.

IDENTIFICATION

8-46. According to the Geneva Convention, all Soldiers involved in armed conflict must keep control of his ID card and ID tag at all times.

LOCAL CURRENCY

8-47. Soldiers are strongly discouraged from carrying any type of payment for helpers, partisan or otherwise sympathetic to your cause, since this requires contact, which would compromise the team's status. Higher echelons control payment to partisans; teams should not be concerned with this.

UNIFORMS

8-48. It has traditionally been the practice of LRS teams to go to the field wearing sterile uniforms. The Geneva Convention only requires them to wear a nametag and US Army tags. The sole purpose of the sterile uniform is to avoid giving away the major supported unit or command. Wearing of rank is permitted. If captured, a Soldier must provide name, rank, SSN, and date of birth anyway.

EVASION AIDS

8-49. Evasion aids are easily accessible to the teams as long as the team knows where to request them. Some of the evasion aids available and methods of requesting them are--

ISOLATED PERSONNEL REPORT (ISOPREP) (DD FORM 1833 TEST (V2))

8-50. This is the most important source of authentication data in the US and in some coalition countries. When filled out, it becomes a classified document. JP 3-50.2 provides instructions for completing the ISOPREP card. When an individual enters an evasion mode the ISOPREP card should be forwarded to the RCC to assist in recovery (Appendix K).

BLOOD CHITS

8-51. The blood chit is a small piece of material imprinted with an American flag, control numbers in each corner, and a statement in English and in several other languages (spoken by the indigenous populace in the AO). This chit identifies the bearer as an American and promises a reward to any one or more people who help him return to friendly control. The presenter of the blood chit may either tear the control number off one of the corners, or he can let them write down the control number. He may also give them his name, rank and SSN, to assist in identification. When the blood chit number is presented to friendly authorities and the claim has been properly validated, it represents an obligation by the US government to provide compensation to the claimant for services rendered to the evader. Blood chits are available through JPRA, the office of primary responsibility for policy and authorizing the production, distribution, and use of blood chits. The National Imagery Mapping Agency (NIMA) maintains the capability to produce and reprint blood chits at the request of the combatant commander as coordinated through the JPRA.

POINTIE TALKIE

8-52. This is similar to language guides used by invasion forces in WWII. They are distributed by the JPRA. Each English phrase has the same phrase in various languages to the side. Simply point to the phrase you wish to use. One of the disadvantages of this are the same as the blood chit, which are you must make contact with people in order to use it. Literacy is also a problem that might arise among the local populace.

EVASION CHARTS

8-53. The National Imagery Mapping Agency (NIMA) distributes evasion charts. Each evasion chart is actually a series of eight 1:250,000 scale joint operations graphic (JOG) charts. The National Imagery Mapping Agency (NIMA) distributes evasion charts. Each evasion chart is actually a series of eight 1:250,000 scale joint operations graphic (JOG) charts. The JOG charts are usually printed four to a side. The chart is overprinted with a camouflage pattern suitable for area terrain, and also includes an American flag that allows the evader to identify himself. An evasion chart combines standard navigation charts and includes evasion and survival information in the margins. In addition, it typically provides information on local navigation, survival medicine, environmental hazards, personal protection, and water and food procurement, plus photos of edible and poisonous plants and wildlife. The evasion chart is waterproof, and can provide a make-shift shelter in an emergency. It folds up small enough to fit into a cargo pocket or flight suit. If evasion charts are unavailable, tactical pilotage charts (1:500,000 scale) may be substituted.

INTELLIGENCE DESCRIPTION FOR SELECTED EVASION AREA

8-54. The term usually used for this is SAID, which means "SAFE area intelligence description" or, spelled out completely, "Selected Area for Evasion Intelligence Description." The "SAID" (say each letter) is distributed by the Defense Intelligence Agency for training and real-world contingencies. The SAID is an in-depth study of all-source evasion designated to help in recovering military personnel from a SAFE under hostile conditions. The SAID includes a 1-to-1:50,000 scale map of the AO, colored photographs of contact points within the SAFE, possible LZs and PZs, survival information, terrain surveys, a chart that shows average rainfall by month, high and low temperatures, and any other information about the AO that the evader might find useful.

AIR-TASKING-ORDER SPECIAL INSTRUCTIONS

8-55. The JSRC develops the ATOSPIN using information that will allow recovery to be conducted with the least amount of problems possible. These instructions include--

- Point of contact for PR incident report.
- Communications report.
- Color of the day.
- Number of the day.
- Codeword of the day.
- Letter of the day.
- Search and rescue point (SARDOT).

EVASION PLAN OF ACTION

8-56. The individual units, not the team, develop the EPA. The RCC helps. Speed of recovery depends on how closely everyone follows the EPA. All units operating in or over hostile territory should develop an EPA or review their existing EPA each time a designated target or AO changes. Responsibility for properly preparing and planning the evasion rests with the potential evaders. The headquarters element supports planning, the team develops the plan, and joint theatre assets support the recovery effort. Successful evasion depends on detailed planning, including contingency plans, initiation mechanisms, and incorporates information from available reference sources. The EPA provides critical information to the recovery force, such as the scheme of maneuver. The EPA format has six mandatory and five optional components (Appendix K).

Mandatory Components

- Identification.
  — Name and rank for each team member.
  — Mission number.
  — Team call sign or identifier.
  — Team position.
  — Call sign suffix.
  — Other.
- Planned flight or travel routes.
  — Describe routes for both ingress and egress.
  — In-flight emergency plans for each leg of the mission.
- Evasion actions and intentions for first 48 hours, uninjured.
  — Compare evasion to resupply and continuation of the mission.
  — Plan for evading alone, in small groups, or with entire team.

- Plan travel, including such factors as distance, duration, and speed.
- Plan intended actions and length of stay at initial hole-up location(s).
- Evasion actions and intentions for first 48 hours, injured.
  - Plan for treatment of the injured or self-aid.
  - Considerations for movement techniques.
  - Litters, canes, crutches, and so on.
  - Rate of march.
- Evasion actions and intentions after 48 hours.
  - Routes, plans to destination.
  - Actions and intentions at potential contact or recovery locations.
  - Contact and recovery point signals, signs or procedures.
  - Contingency plans.
- Communication and authentication.
  - Code words.
  - Bona fides.
  - Color or letter of the day, month, or quarter.
  - Challenge or password and any number combinations.
  - Available communications and signaling devices (day or night, near or far).
  - Primary communication schedule procedures and frequencies.
  - Alternate communication schedule procedures and frequencies.

Optional Components

- Weapons and ammunition carried.
- Personal evasion kit items.
- Listing of issued survival kit items.
- Mission evasion preparation checklist.
- Signature of reviewing officer.

EVASION AREAS

8-57. Evasion areas within the theatre or area of operation are decisive to mission accomplishment should the team or individual become isolated from his unit. Using the evasion area within the study region increases the evader's chance to reunite to friendly control.

SAFE

8-58. A SAFE is an area within a potentially hostile region where an individual may become isolated and must evade to avoid capture by the enemy. This area will be pre-determined to have the best conditions for evasion and survival opportunities. Within the SAFE will be contact points and usually a recovery site. An area study of the SAFE will be found in the SAID and should be used when planning for a possible evasion situation.

Chapter 8

DESIGNATED AREAS FOR RECOVERY

8-59. The DAR can be any size or shape within a potentially hostile region where individuals may become isolated and must evade to avoid capture. The DAR is issued when no SAFE is available in the AO or can supplement the SAFE in the area of recovery. An area study will be used similar to the SAID during the planning process. If a DAR is used, the contact points are called 'recovery points.' (For more information on E&R see JP 3-50.2, JP 3-50.21, and JP 3-50.3.)

Appendix A
Recruitment, Assessment, and Selection Program

The physical and psychological demands on the LRSU Soldier generally exceed those on conventional Infantry and reconnaissance Soldiers. Instead of operating near friendly supporting units, they operate deep in the enemy's rear, surrounded by hostile forces. They operate independently for long periods of time without support. They must rely only on each other and on what they can carry. The pressure of having to perform under such extreme circumstances requires more than extraordinary physical capabilities--it also requires specific character traits. Ideally, LRSU Soldiers are selected in three stages: recruitment, assessment, and selection. This appendix covers the purpose, organization, and elements of the recruitment, assessment, and selection program (RASP) as well as reassignment during and after the 90-day probationary period. The purpose of the RASP is to ensure quality personnel are assigned and maintained to meet the demands of the LRSU mission.

PURPOSE AND ORGANIZATION

A-1. The purpose of a LRSU RASP is to identify and gauge the potential of individual Soldiers to meet LRSU standards within a reasonable training period.

Memorandum of Understanding—The BFSB S-1, the R&S squadron S-1 and the LRSC should develop a RASP *Memorandum of Understanding* (MOU). The RASP MOU should have the approval of the BFSB and R&S squadron commanders.

Three-Time Volunteers—Potential new Soldiers for a LRS team should be at least three time volunteers: they have joined the Army, they are Airborne qualified, and they agree to be assigned to a LRSC.

Assessment and Selection Board—The LRSU assessment and selection board consists of senior LRSC leaders, R&S squadron and BFSB commanders and Command Sergeants Majors (or their designated representatives). The board centrally reviews the candidate's qualifications and results of the assessment process. Assignment to the LRSC ideally occurs only after the Soldier successfully completes the assessment phase and receives a recommendation for assignment from the selection board.

Phases—A LRSU RASP normally has three phases: recruitment, assessment, and selection.

RECRUITMENT

A-2. LRSU recruitment should not be used to reward or promote particular Soldiers. Neither should other units reassign Soldiers with "problems" or marginal performance to the LRSC. For the LRSU to properly perform its mission, Soldiers should volunteer or be chosen based on their proven ability or potential to perform and thrive under the demanding conditions of LRS missions. LRSU recruitment should be viewed as an extension of the Army's personnel-management program. It identifies Soldiers who meet the requirements for specific LRSU needs.

A-3. Privates through specialists will be assigned to the BFSB by the Human Resources Command. With very few exceptions, 11B1P Soldiers assigned to the BFSB are intended to fill slots in the LRSC. It is unlikely that these Soldiers will initially all meet the unique characteristics of a LRS Soldier. The BFSB S-1 needs to establish a relationship with units on the host installation that require 11B10s and institute a reassignment policy so Soldiers not meeting the RASP standards can be transferred. A Soldier not meeting the RASP standards does not necessarily mean he is a substandard performer. Likewise, RASP standards should not be so stringent that the failure rate of potential new LRSU Soldiers puts an undue burden on the BFSB, the host installation and the US Army Human Resources Command.

Appendix A

A-4. 11B2V through 11B4V Soldiers will be assigned to the LRSC or the R&S squadron insertion and extraction section from the Human Resources Command, Ranger assignments NCO.

A-5. Officer RASP candidates are normally volunteers from other Infantry organizations. Experience proves that for LRSC commanders to be successful, it must be a second or third company command, the previous command(s) being very successful. LRSD platoon leaders are normally Captains and should have had highly successful previous Infantry rifle platoon leader time including time on an Infantry unit staff. The best candidates also having highly successful scout/reconnaissance platoon or company XO experience.

ASSESSMENT

A-6. To be qualified for selection consideration, the recruit must meet all of the requirements and none of the disqualifiers. The board assesses recruits to determine whether they qualify for consideration for assessment.

QUALIFIERS

A-7. Recruits must meet all of the following criteria:

- Airborne qualified (privates first class and specialists).
- Airborne and Ranger qualified (sergeants and above).[1,2]
- Earned a GT score of 110 or above (all).
- Willing to volunteer for Ranger School.
- Willing to volunteer for the RSLC.
- Has at least two years retainability in unit upon selection.
- Are eligible for Secret security clearance.

Notes: [1] MOS 11B sergeants and above should not be automatically disqualified if they are not Ranger qualified.

[2] Non-MOS 11B sergeants and above are not required to be Ranger qualified.

PHYSICAL SCREENS

A-8. A LRS Soldier normally carries over 100 pounds of equipment during an operation. This far exceeds the normal Soldier's load, so the LRS Soldier needs extraordinary physical capabilities to meet the physical requirements:

- Pass the Army Physical Fitness Test + (Ranger school standard for push-ups, sit-ups and chin-ups).
- Pass the Combat Water Survival Test.

A-9. The LRS Soldier must have the potential to become experts in the following skills:

- Map reading and land navigation.
- Communications.
- Patrolling.
- Field craft.
- Threat equipment recognition.

- Medical.
- Planning and orders.

PSYCHOLOGICAL EVALUATION

A-10. The success of the LRS Soldier is essential to success of the overall mission. LRS leaders must understand the psychological characteristics of those most likely to succeed so that team composition will provide commanders with the best possible resources. Studies such as that conducted by LTC (Dr.) John C. Chin, United States Army Infantry Center command psychologist in 2002 indicate successful LRS Soldiers have certain identifiable psychological characteristics. While not resource supportable in all cases, it is recommended a psychological evaluation be included in the assessment phase of RASP.

Desirable

A-11. The Soldier is evaluated on the basis of the following mental and psychological criteria:

- Effective, functional intelligence.
- Conceptual complexity.
- Emotional stability and high stress tolerance.
- Trainability and situational awareness.
- Humor.
- Learned optimism.
- Self-efficacy.
- Moral reasoning and integrity.

Undesirable

A-12. The desirable mental and psychological criteria can also be contrasted with unacceptable psychological criteria for LRSU Soldiers which include the following:

- Anxiety or mood disorders.
- Antisocial personality.
- Serious phobias.
- High substance abuse potential.
- Undue sensitivity.
- Impulsivity.
- Chronic relationship problems.
- Uncooperative family.
- Chronic financial problems.
- Low trainability.
- Rambo syndrome.
- Low motivation.

Appendix A

DISQUALIFIERS

A-13. To be assessed as eligible for consideration, the LRS recruit's record must not indicate a history of any one of the following:

- Disciplinary problems.
- Drug use.
- Alcohol abuse.
- Financial irresponsibility.
- Emotional instability.

CONDUCT OF ASSESSMENT

A-14. The LRSC normally conducts the RASP with the support of the R&S squadron. Techniques to conduct the RASP include—

- Training by consolidating cadre/trainers at the LRSC,
- Training conducted by LRSD, or
- Training by consolidating cadre/trainers with the program administered by the insertion and extraction section in the R&S squadron S-3 under the supervision of the LRSC.

A-15. To maintain test standards, the LRSC should not have individual LRS teams conduct the RASP. The RASP is conducted quarterly. It is normally three weeks long, divided into three phases. Phase 1 includes both classroom instruction and physical training. Phase 2 is field instruction on critical skills. Phase 3 is testing, evaluation, and conduct of the selection board.

SELECTION

A-16. Soldiers selected for consideration as a LRS Soldier must meet—

- Earn at least 70 percent on the Army Physical Fitness Test in each event, within age groups.
- Meet height and weight or body fat standards in AR 600-9.
- Pass the Combat Water Survival Test.
- Complete 5-mile run within 40 minutes.
- Complete 12-mile foot march with LCE, weapon, and rucksack (with 35-pound load) within 3 hours.
- Pass a written test on map reading and land navigation.
- Complete practical exercises in day and night land navigation.
- Demonstrate proficiency following training in basic LRS team skills, including--
    -- Communications.
    -- Threat-equipment recognition.
    -- Fieldcraft.
    -- Patrolling.
    -- Medical aid.
    -- Planning and orders production.
- Pass psychological screening.
- Pass a comprehensive examination given by the unit selection review board.

REASSIGNMENT

A-17. Once the Soldier meets the RASP criteria for assignment to the LRSC, he must perform, and continue to perform, his duties to LRSU standards.

90-Day Probation

A-18. During the first 90 days of the Soldier's assignment to the LRSC, the commander should be allowed to reassign him at any time, with adequate documentation, for failure to meet unit standards.

A-19. During the first 90 days it is recommended each Soldier attend the RSLC. Attendance and graduation of the RSLC will greatly enhance the LRS Soldier skill set and provide a base for LRSU-specific training.

Subsequent Release Authority

A-20. After the 90-day probationary period, the LRS commander should be allowed to reassign any Soldier whose performance or personal situation degrades the ability of the LRSU to accomplish its mission.

Appendix B
Orders and Briefs

This appendix helps LRSU prepare orders (WARNO, OPORD, and FRAGO) and briefs (confirmation, mission analysis, decision, and mission concept briefs; backbriefs; and debriefs).

Brief and order types are distinguished by--

- Time--When the order or brief occurs.
- Contents--What the order or brief includes.

Briefs are also distinguished by--

- Briefer.
- Briefee.

Section I. ORDERS

This section discusses the three types of orders (FM 1-02). A WARNO is a preliminary notice of an order or action that is to follow. A OPORD is a directive issued by a commander to subordinate commanders for the purpose of effecting the coordinated execution of an operation. A FRAGO is issued after an OPORD to change, modify, or execute a branch or sequel to that order. WARNOs follow the five-paragraph OPORD format. FRAGOs also follow the OPORD format, but seldom include all five paragraphs. For extensive changes, the leader normally issues a new OPORD.

WARNING ORDER

B-1. Leaders issue WARNOs as soon as they complete their initial assessment of the situation and available time. Leaders do not wait for more information. They issue the best WARNO possible with the information at hand and update it as needed with additional WARNOs. The WARNO follows the five-paragraph OPORD format (Figure B-1 (page B-3) shows an example WARNO). The WARNO contains as much information as possible. Normally an initial WARNO includes--

- Mission or nature of the operation.
- Time and place for issuing the OPORD.
- Units or elements participating in the operation.
- Specific tasks not addressed by unit SOP.
- Time line for the operation.

B-2. The WARNO may also include--

- The higher unit's mission statement.
- Commander's intent.
- Commander's WARNO.
- Commander's operations brief or order.
- Task organization changes.
- Attachments/detachments.
- The unit AO.
- Initial intelligence requirements or CCIR.

Appendix B

- Risk guidance.
- ISR tasks and RFIs.
- Initial movement instructions.
- Security requirements.
- Security measures.
- Specific priorities.
- Movement time to planning site.
- Strength figures (for the XO and first sergeant) to support planning for movement and Class I.
- Class I planning.
- Coordination actions for communications.
- Isolated Personnel Report (ISOPREP) DD Form 1833 TEST (V2).
- Times for--
    - -- For personnel and equipment attachments (communications, transportation, and medic).
    - -- For communication exercise.
    - -- For vehicle inspections and dispatches.
- Times and locations for--
    - -- For issue and turn-in of classified material.
    - -- For air-mission brief and coordination.
    - -- For issue of equipment.
    - -- For test-firing and zeroing of equipment (including night-vision devices).
    - -- For rehearsals (day or night, with or without equipment).
    - -- For distribution of ammunition.
    - -- For initial or final inspections.
    - -- For religious services.
    - -- For take-off and time on target.

(CLASSIFICATION--WHEN FILLED OUT)

WARNING ORDER

1. SITUATION.
Team Leader briefly states enemy and friendly situations, and lists attachments and detachments. When stating the enemy situation, he includes where the enemy is operating and what he is doing.
He shows this information on the Enemy Situational Template.

2. MISSION.
Team Leader--
   a. Answers the five W's twice:
     (1) Who?
     (2) What?
     (3) When?
     (4) Where?
     (5) Why?
   b. Gives the expected duration of the mission:
   c. Avoids real unit names: "Detachment 6, 52nd LRS…"
   d. Avoids giving other real details as well: "…will conduct a zone reconnaissance to determine the disposition of enemy forces in AO Darby NLT 161300z Mar 03 to facilitate follow-on offensive operations of Task Force Ranger."
   e. For this example, says instead, "Detachment Alpha, [Unit] Kilo will conduct a zone reconnaissance…"

3. GENERAL INSTRUCTIONS.
Team Leader gives the team basic instructions for preparing for the mission. Include--

(CLASSIFICATION--WHEN FILLED OUT)

Figure B-1. Example warning order.

Appendix B

```
(CLASSIFICATION--WHEN FILLED OUT)
    a. Chain of command.
    b. Special teams and the organization within each.
    c. Uniform and equipment common to all.
    d. Weapons, ammunition, and special equipment.
    e. Schedule.
4. SPECIFIC INSTRUCTIONS.
Team Leader gives instructions and checklists to individual team
members (subordinates) to help them prepare for the mission.
    a. Assistant Team Leader (ATL).
        • Setup of planning area.
        • Paragraph 4.
        • Supplies.
        • Ammunition.
        • Food.
        • E&R annex.
    b. RATELO.
    c. Senior Scout Observer (SSO).
        • Routes.
        • Terrain model.
        • Overlays.
        • Intelligence annex.
        • Paragraph 5.
        • Communications equipment.
        • Direct support (DX).
    d. Scout Observer.
        • Route assistance.
        • Paragraph 1.
        • Terrain model.
        • Other annexes.
    e. Assistant RATELO. Assist RATELO and help with other annexes.
(CLASSIFICATION--WHEN FILLED OUT)
```

Figure B-1. Example warning order (continued).

OPERATION ORDER

B-3. Figure B-2 shows the format used for operation orders. After briefing the OPORD body, the leader issues annexes and schedules, then asks for questions.

Orders and Briefs

```
           (CLASSIFICATION--WHEN FILLED OUT)
OPERATION ORDER

1. SITUATION.
Team Leader gives information about the following:
   a. Enemy Forces.
      (1) General road to war (big picture).
      (2) Activity (big picture).
      (3) Probable courses of action (big picture).
   b. Friendly Forces.
      (1) Mission of next higher unit.
      (2) Location of planned actions of units in the area.
   c. Attachments and Detachments.

2. MISSION.
Team Leader states and restates who, what (operation and task),
when, where, and why (purpose).

3. EXECUTION.
   a. Commander's Intent.
   b. Concept of the Operation. Cover in general from start to
finish. Use maps, sketches, or terrain models.
   c. Task Organization and Responsibilities.
      (1) Subunit instructions (during movement and actions on the
objective, who, what, when, where and why).
      (2) Special teams such as surveillance, hide, EPW and search,
aid and litter, and R&S teams.
      (3) Key individuals such as the ATL, RATELO, compassman,
paceman, and en route recorder.
   d. Maneuver.
      (1) Infiltration.
         (a) Formations and order of movement.
         (b) Actions at halts (short and long).
         (c) Routes (use terrain model).
            1) Primary routes (azimuths, distance, phase lines,
TRPs, description of terrain along routes, and location of
tentative initial entry report).
            2) Alternate routes (same as primary routes).
         (d) Actions at planned rally points/en route RV's, location
and terrain reference.
         (e) Actions on enemy contact.
         (f) Actions at planned, suspected, linear, and open
danger areas.
      (2) Actions on the objective (use terrain model or sketch).
           (CLASSIFICATION--WHEN FILLED OUT)
```

Figure B-2. Example operation order.

Appendix B

> (CLASSIFICATION--WHEN FILLED OUT)
>
>   1) Location and terrain reference.
>   2) Plan for occupation or SLLS.
>   (b) Leaders reconnaissance for hide or surveillance site. Team leader must pinpoint objective.
>   1) Five-point contingency plan.
>   2) Communications checks prior to departure, en route, and upon arrival at the tentative hide or surveillance positions.
>   3) Planned routes to hide or surveillance site. Team leader should physically designate the team internal linkup site during the movement to the tentative surveillance site.
>   4) Order of movement, and list of equipment to take.
>   5) Plan for approach and reconnaissance of surveillance site.
>   6) Countertracking instructions.
>   (c) Tentative surveillance or hide site.
>   1) Construction plan for security, site construction, or configuration.
>   2) Specific instructions to surveillance or hide team such as reporting, sketches, actions on contact, security, alert, evacuation, RVs, rest, departure, priorities of work, actions on enemy contact, and so on.
>   3) Plan actions if no communications between sites.
>   (d) Plan for reconnoitering alternate surveillance or hide site.
>   (e) Locations of planned TRPs in the objective area.
>   (f) Withdrawal plan.
>   1) Plan for team internal linkup for withdrawing from the objective.
>   2) Plan for dissemination of information. Conduct this a safe distance from the objective.
>   (3) Exfiltration. Use terrain model.
>   (a) Routes. State primary and alternate routes, to include azimuth, distance, phase lines, TRPs and description of terrain along the route.
>   (b) Plan for reconnaissance of PZ and actions at PZ.
>   (c) Fires.
>   1) Fire-support plan. Use fire-support overlay with all planned targets, no-fire areas, and restrictive fire areas.
>   2) Unit providing support.
>   3) Purpose and priority of fires.
>   4) Types of fire support available.
>   5) Method of requesting fire support.
>   6) Time requirements for fire mission requests.
>   (d) Coordinating instructions.
>   1) Time of departure and time of return.
>   2) MOPP level.
>   3) Rules of Engagement.
>   4) Abort criteria.
>
> (CLASSIFICATION--WHEN FILLED OUT)

Figure B-2. Example operations order (continued).

```
(CLASSIFICATION--WHEN FILLED OUT)
            5) Time schedule, with time, event, place, attendees,
and uniform or equipment.
                  - Rehearsals and their priority(ies).
                  - Briefback time, place, and uniform.
                  - Final inspections.
                  - Debrief time and place.
            6) Annexes to OPORD.
```

4. **SERVICE AND SUPPORT.**
  a. **Materials and Services.**
    (1) Supply.
      (a) Uniform and equipment common to all.
      (b) Rations and water.
      (c) Weapons and ammunition.
      (d) Pyrotechnics and demolition.
      (e) Medical.
        1) Supplies and breakdown by personnel.
        2) Mission hazards such as disease or weather effects.
        3) Preventive measures and hygiene plan.
        4) Medical qualifications of team members.
      (f) Itemized list of special carried equipment.
    (2) Services. Methods of handling--
      (a) Friendly dead and wounded.
      (b) Enemy dead and wounded.
      (c) Enemy prisoners of war.
      (d) Detained civilians.
  b. **Maintenance Plan.**
  c. **Resupply Plan.**

5. **COMMAND AND SIGNAL.**
  a. **Command.**
    (1) Chain of command.
    (2) Locations of key leaders during all phases.
    (3) Locations of the COB (or MSS) and AOB.
  b. **Signal.**
    (1) Hand and arm signals.
    (2) FM communications (include here or in communications annex).
    (3) Long range communications (include here or in communications annex).
      (a) HF.
      (b) TACSAT.
    (4) Challenge and password.
    (5) Running password or number combination.
    (6) Code words.
    (7) Destruction plan (method and priority).
    (8) Brevity plan.
    (9) Special communications procedures and equipment.
    (10) Team no-communications plan with COB (or MSS) or AOB.

(CLASSIFICATION--WHEN FILLED OUT)

Figure B-2. Example operations order (continued).

Appendix B

FRAGMENTARY ORDER

B-4.   A FRAGO provides timely changes to existing orders. A FRAGO includes only the items that have changed since the last OPORD. The items in the FRAGO follow the five-paragraph OPORD format. However, a significantly changed mission or a brand new mission requires a new OPORD. Figure B-3 shows an example FRAGO.

```
                  (CLASSIFICATION--WHEN FILLED OUT)
FRAGMENTARY ORDER

TASK ORGANIZATION
Team Leader includes this only if changed.
  1. SITUATION.
     a. Enemy Forces.
     b. Friendly Forces.
  2. MISSION. Team Leader states mission twice.
  3. EXECUTION.
     a. Commander's Intent.
     b. Maneuver.
     c. Fires.
     d. Intelligence and electronic warfare.
     e. Individual Tasks.
     f. Coordinating Instructions.
  4. SERVICE SUPPORT. Include only if changed.
  5. COMMAND AND SIGNAL. Include only if changed.

                  (CLASSIFICATION--WHEN FILLED OUT)
```

Figure B-3. Example fragmentary order.

Section II. BRIEFS

Briefs are presentations of information from one leader to another, either up or down the chain of command.

TYPES

B-5.   Figure B-4 shows the types of briefs, and Table B-1 compares them based on who gives them to whom and why, and on what elements each type must, should, or might include--

- Confirmation
- Mission Analysis
- COA Decision Brief
- Backbrief
- Initial GO/NO-GO Brief
- Mission Concept Brief
- Final GO/NO-GO Brief
- Debrief

Figure B-4. Brief types.

Table B-1. Comparison of brief types.

Type	Purpose	Conditions	Briefer	Audience	Time	Elements
Confirmation	Ensure team understands OPORD before they analyze mission	NA	Team leader	Company commander (or unit rep)	NLT 30 minutes after receipt of Company OPORD	• Restatement of mission (if specified in OPORD) • Restatement of commander's intent • Initial issues or concerns • Next key hard time
Mission Analysis	Extract all critical and pertinent data from OPORD and conduct detailed MPF before developing team COAs	Given a time-- or an operational paced-constrained environment-- and situational awareness	Team leader	Commander or his representative (IAW unit SOP, number of unit teams, and operational pace), or team only	NLT 2 hours after receipt of Company OPORD	• Intent two levels up • Specified tasks • Implied tasks • Facts about friendly, enemy, terrain, time • Assumptions about friendly, enemy, terrain, and time • Facts that could hinder execution • Restatement of mission
	Prepare team leader to develop COAs		Commander (or rep)	Team leader	After team leader's Mission Analysis (MA) brief	• Assumed risk by phase • Criteria to evaluate (some or all) • Number of COAs to develop • Detailed instructions on higher coordinations for insertion and extraction • Approval of mission statement
COA Decision Brief	Determine best possible COA for team execution		Team leader	Commander or rep (PL, PSG, operations officer, or 1SG, depending on operational pace and number of teams)		• Intent of higher • Restated mission • Updated IPB (enemy SITEMP) • COA table or matrix with criteria* * Weighted criteria are allowed
				Team, if team leader does not brief higher; team then votes on COA.	After team completes MA--developing COA and preparing for COA decision brief takes about 2 hrs	• Residual risk by phase of the operation • Strengths and weaknesses of each COA • Team's recommendation • Issues and concerns
			Commander or representative			• Commander or rep chooses or refines COA

Appendix B

Table B-1. Comparison of brief types (cont'd).

Type	Purpose	Conditions	Briefer	Audience	Time	Elements
Backbrief: initial GO/NO-GO brief	To instill confidence in audience. To assure them that team understands plan thoroughly, and has applied contingencies and proper risk mitigation to ensure mission success and team preservation.	Audience generally understands the mission and knows the presenters.	Team leader	Company commander and other guests such as R&S Squadron or BFSB Commander, or other VIPs.	NLT 4 hours before insertion. This gives the detachment or company commander time to deliver the final GO/NO-GO brief to his higher before insertion.	• Initial GO/NO-GO brief • Team introductions (if needed). • Orientation to facility. • Announcement of who briefs what topic. • Mission statement • Commander's intent • Concept of the operation (insertion, infiltration, execution, fires, exfiltration, abort criteria, extraction info, and EPA) • Risk mitigation (by phase) • Team's recommendation (issues and GO/NO-GO) • Commander's decision (GO/NO-GO)
Mission concept brief Final GO/NO-GO brief	To obtain a GO/NO-GO decision		Company or detachment commander	Decision maker is normally one of the following: • LRSC Cdr • R&S Squadron Cdr • BFSB Cdr	NLT 2 hours prior to insertion. This gives the commander time to return to the team's location and inform them of MIssion Concept Brief results before insertion	• Overall recommendation • Purpose of NAI coverage (BFSB S-2 ISR fusion element Warrant Officer) • Enemy situation (BFSB S-2 ISR fusion element Warrant Officer) • Weather (BFBS S-2 weather team) • Mission statement (LRS commander) • Mission statement (if applicable for aviation commander supporting insertion) • Concept of the operation (intelligence, movement and maneuver, fire support, protection, sustainment, C2, risk mitigation) • Decision (GO/NO-GO)

Table B-1. Comparison of brief types (cont'd).

Type	Purpose	Conditions	Briefer	Audience	Time	Elements
Debrief	To gather any unreported intelligence from the team, and to create an historical record of the mission		Debriefer is officer from LRSU or external supporting agency (R&S Squadron S-2 or BFSB ISR fusion element member)	Team members	NLT 4 hours after extraction, so team can collate patrol, R&S, and communication logs, and to allow debriefing while information might still remain fresh in the minds of team members	• Appropriate team logs, cameras, recorded messages turned in • Map • Recorder (manual or recording device) • Communications representative • Debriefer in charge

CONFIRMATION BRIEF

B-6. The leader gives a confirmation brief to ensure that he understands the company OPORD and to focus the team's mission planning in the right direction. The confirmation brief should occur NLT 30 minutes after receipt of the OPORD. The confirmation brief should include three elements:

- Restated mission (team's mission statement).
- Restated commander's intent.
- Initial issues or concerns.

MISSION ANALYSIS BRIEF

B-7. Figure B-5 through Figure B-9 (this page through B-15) show example formats for mission analysis worksheets.

Appendix B

MISSION ANALYSIS WORKSHEET
MISSION, INTENT, AND PIR
Mission and Intent (two levels up)
PIR
Mission and Intent (one level up):
PIR

Figure B-5. Analysis of mission, intent, and priority intelligence requirements.

Orders and Briefs

MISSION ANALYSIS WORKSHEET
SPECIFIED AND IMPLIED TASKS

Specified Tasks: (mission, task to maneuver, coordinating instructions, annexes, and so on)

* Mission-Essential Task:

Implied Tasks:

Figure B-6. Analysis of specified and implied tasks.

Appendix B

MISSION ANALYSIS WORKSHEET
FACTS AND ASSUMPTIONS

FACTS

   Friendly - Unit locations, fire-support ranges and types, partisans, equipment, classes of supply…

   Enemy - DOCTEMP, weapons, uniforms, fire support, equipment…

   Terrain - MCOO, soil composition, line of sight, time of camouflage…

   Time - Crucial times, time to infiltrate, duration of mission, time to exfiltrate, light data…

Figure B-7. Analysis of facts.

Orders and Briefs

MISSION ANALYSIS WORKSHEET
FACTS AND ASSUMPTIONS

ASSUMPTIONS

Friendly - CAS, FS, NGF, evacuation support, partisans…

Enemy - SITTEMP, EVENTEMP…

Terrain - Stand-off, camouflage, support of movement, effects of weather…

Time - Estimated times of movement, tentative timeline…

Figure B-8. Analysis of assumptions.

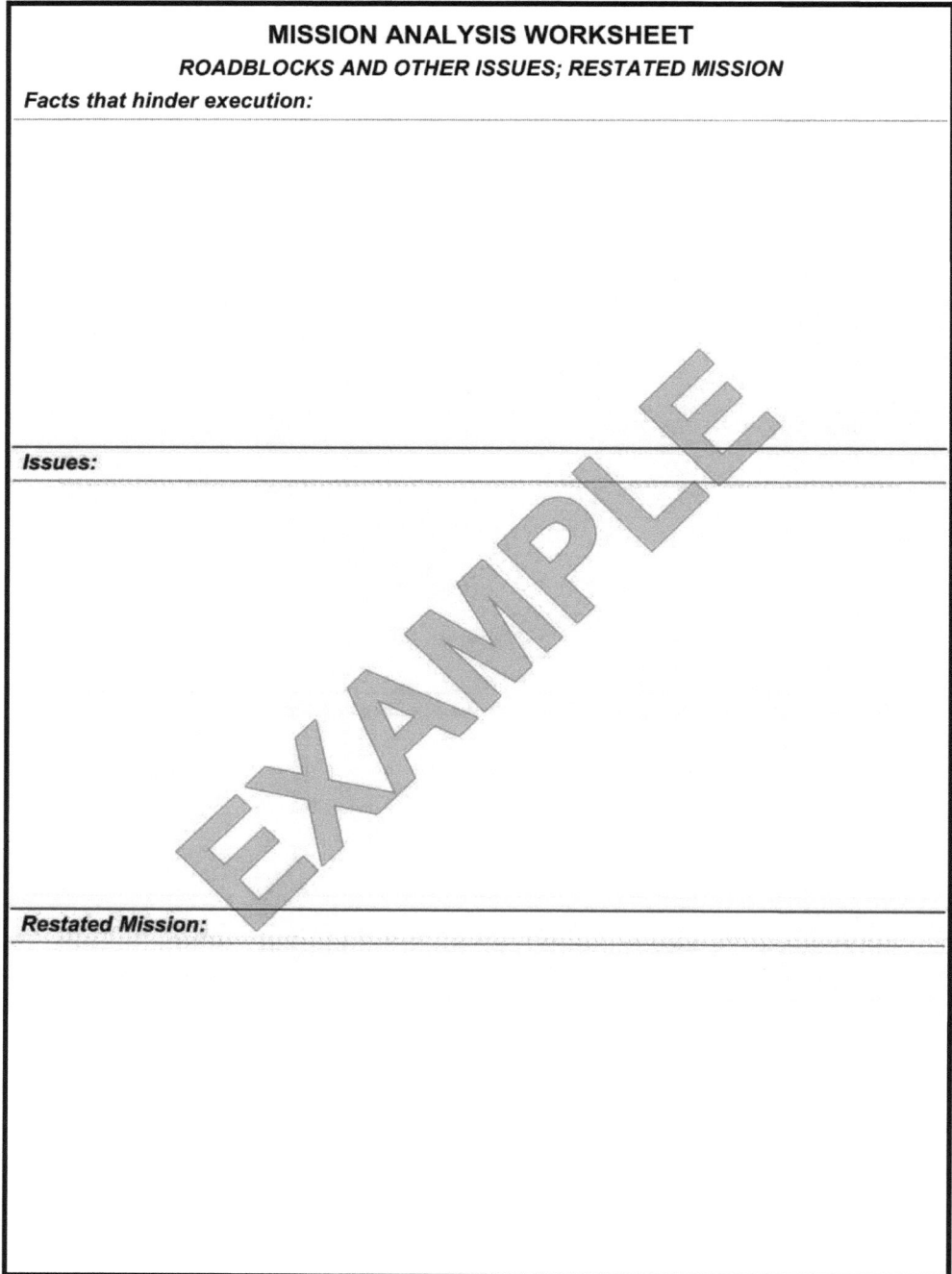

Figure B-9. Analysis of mission roadblocks, issues, and restated mission.

DECISION BRIEF

B-8. The decision brief occurs after the team has completed its initial mission planning. Team members present possible courses of action (COAs) and the criteria for evaluating each. The team leader chooses one COA, explains why he chose it, and then presents all of the COAs to the commander. The commander can approve the selected COA or provide additional guidance to the team leader for continued COA development. The decision brief should include these six elements:

- Restated mission (the team's mission statement).
- Restated commander's intent.
- Two to three courses of action on the objective.
- The team's recommended COA.
- Limitations and constraints.
- Any issues or concerns.

BACKBRIEF

B-9. A backbrief is an informal transfer of information to one or more leaders, usually the commander and his guests, about the impending and completed missions, respectively. The amount of information briefed is usually set by SOP. All present should understand the mission before they arrive. The purpose of the backbrief is only to demonstrate that the team understands the plan thoroughly and has applied contingencies and proper risk mitigation procedures to ensure mission success and team preservation. Figure B-10 shows an example backbrief format, with explanations.

(CLASSIFICATION--WHEN FILLED OUT)

BRIEFBACK

1. **INTRODUCTIONS.**
Team Leader introduces the person being briefed to the team, if needed. Shows the person being briefed around the room and explains briefing products. Unless he is the only presenter, the team leader also announces the topic that each team member will brief:
    a. **Team Leader.**
    b. **Assistant Team Leader.**
    c. **Senior Scout Observer.**
    d. **Radio Telephone Operator.**
    e. **Assistant Radio Telephone Operator.**
    f. **Scout Observer.**
    g. **Attachments as Necessary.** This might include the LNO or Scout Observer No. 2, for example.

2. **MISSION STATEMENT.**
Team Leader briefs the mission statement twice.*
    a. **Who?**   Team number.
    b. **What?**   Type of operation and mission-essential task.
    c. **When?**   Date-time group (DTG) and duration of mission.
    d. **Where?** Grid coordinates for named area of interest (NAI).
    e. **Why?**   Purpose of mission.

3. **COMMANDER'S INTENT.**
Team Leader briefs the commander's intent.*
    a. **Purpose.**
    b. **Key Task.**
    c. **End State.**
    d. **Named and Target Areas of Interest (NAI and TAI).** The Team Leader explains the intelligence requirement, indicators, and ISR tasks.
        (1) PIR:
        (2) ISR Tasks:
        (3) LTIOV:
    e. **Enemy Situation.** The Team Leader uses the SITTEMP, enemy in AO, on objective. He might also discuss the MPCOA and MDCOA.
    f. **Enemy Strengths and Weaknesses.** The TL explains how the team will exploit the enemy weaknesses and counter his strengths.

(CLASSIFICATION--WHEN FILLED OUT)

Figure B-10. Example backbrief.

(CLASSIFICATION--WHEN FILLED OUT)

4. **CONCEPT OF THE OPERATION.**
Senior Scout Observer briefs the concept of the operation.*
  a. **Insertion.**
    (1) Primary platform.
      (a) Primary means of insertion and location.
      (b) Alternate means of insertion and location, and what might cause the team to use alternate means.
    (2) Abort criteria for insertion.
  b. **Infiltration.** The SSO briefs general routes into the objective area.
    (1) Fire support for infiltration.
    (2) Abort criteria during infiltration.

5. **EXECUTION.**
  a. **Intelligence Collection Plan.**
    (1) Tentative surveillance and vantage points.
    (2) Occupation plan for ORP or construction of hide or surveillance site.
    (3) Abort criteria (during execution).
  b. **Communication Plans.** The RATELO or Assistant RATELO briefs.*
    (1) Team external HF, TACSAT, and FM windows.
    (2) Antenna used (primary and alternate).
    (3) Team internal. FM-type windows.
    (4) No-communications plan (internal).
    (5) No-communications plan (external).
  c. **Fire Support Plan.** Scout Observer briefs the fire plan.*
    (1) Planned target locations.
- Insertion (SEAD).
- Infiltration.
- Execution.
- Exfiltration.
- Extraction.

    (2) Purpose of fires, that is, to isolate, suppress, or whatever.
    (3) Assets available.
    (4) Locations of supporting fires.
    (5) Minimum safe distance for munitions.
  d. **Departure Plan.** The senior scout observer briefs the departure plan.*
    (1) Sterilization.
    (2) Linkup.
    (3) Abort criteria.
  e. **Exfiltration.** Senior Scout Observer briefs the exfiltration plan.*
    (1) General routes to the extraction zone.
    (2) Fire support for exfiltration.
    (3) Abort criteria.

(CLASSIFICATION--WHEN FILLED OUT)

Figure B-10. Example backbrief (continued).

Appendix B

```
                    (CLASSIFICATION--WHEN FILLED OUT)
     g. Evasion Plan of Action. Assistant Team Leader briefs the EPA.*
        (1) Reason for initiation.
        (2) Rendezvous points (2, 4, 24 hr RVs), and locations and
actions at each.
        (3) Contact point locations.
     h. Risk Mitigation. The Assistant Team Leader briefs the controls
implemented, by phase.*
        (1) Planning Phase.
            (a) Risks.
            (b) Controls implemented.
        (2) Insertion and Infiltration.
            (a) Risks.
            (b) Controls implemented.
        (3) Execution.
            (a) Risks.
            (b) Controls implemented.
        (4) Exfiltration or Extraction.
            (a) Risks.
            (b) Controls implemented.
        (5) Recovery.
            (a) Risks.
            (b) Controls implemented.
6. TEAM RECOMMENDATION.
Team Leader gives overall recommendation and clarifies any issues.*
     a. Issues and Concerns.
        (1) Equipment.
        (2) Additional coordinations required.
     b. GO or NO GO. The team recommends either a GO or NO GO on the
mission briefed.
7. REQUEST DECISION.
Team Leader asks the commander if the mission is a GO or a NO GO.
                    (CLASSIFICATION--WHEN FILLED OUT)
```

Figure B-10. Example backbrief (continued).

MISSION CONCEPT BRIEF

B-10. Generally 2 to 12 hours before the planned insertion time, the company commander briefs the appropriate decision makers on the mission concept to obtain a final "GO" or "NO GO" on the mission. This gives the commander time to conduct movement back to the team to inform them of the decision and any modified guidance before the planned insert time. The decision maker is normally the R&S squadron or BFSB commander. The mission concept brief includes the following elements, and an example brief is shown in Figure B-11 through Figure B-38 (this page through page B-33).

- Overall recommendation.
- Purpose of NAI coverage (S-2 or ISR fusion element chief).
- Enemy situation (S-2).
- Weather (Air Force battlefield weather team).
- Mission statement (LRS commander).

- Mission statement (if applicable for the aviation commander supporting the insertion).
- Concept of the operation (intelligence, movement and maneuver, fire support, protection, sustainment, C2, risk mitigation).
- Request for decision (GO or NO GO).

LRSC
MICON
BRIEF
Coordinated with –
___ S3
___ S2
___ ISR FUSION ELEMENT
___ AVN
___ ARTY

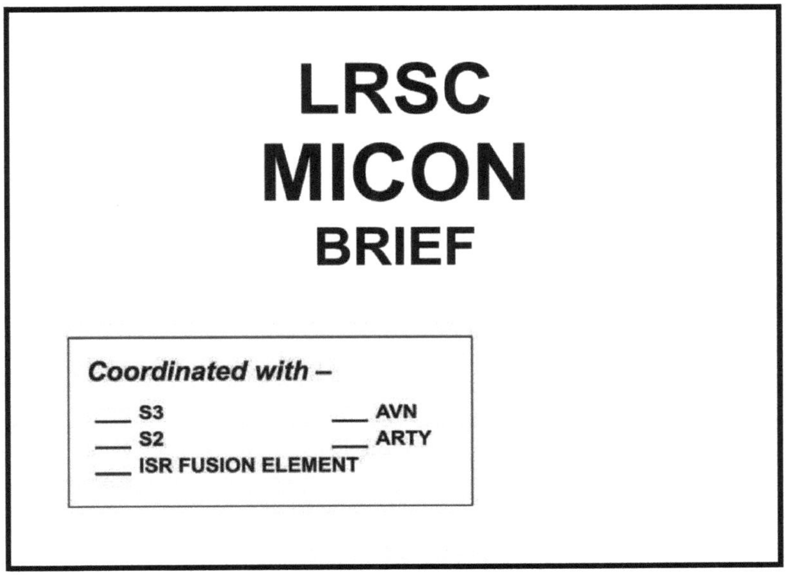

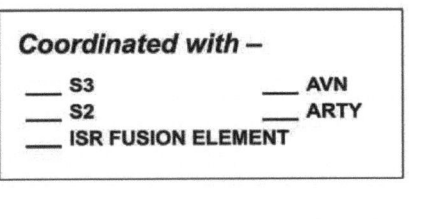

Figure B-11. Slide 1, LRSC mission concept brief.

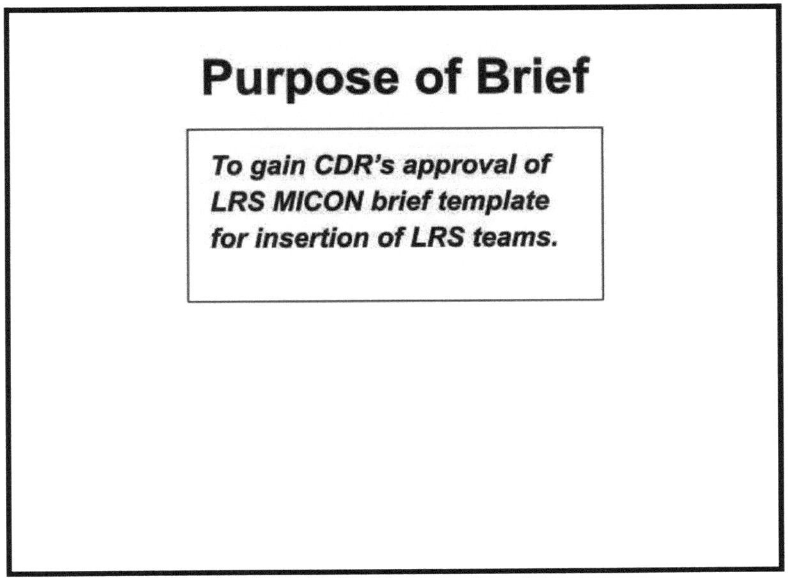

Purpose of Brief
To gain CDR's approval of LRS MICON brief template for insertion of LRS teams.

Figure B-12. Slide 2, statement of purpose.

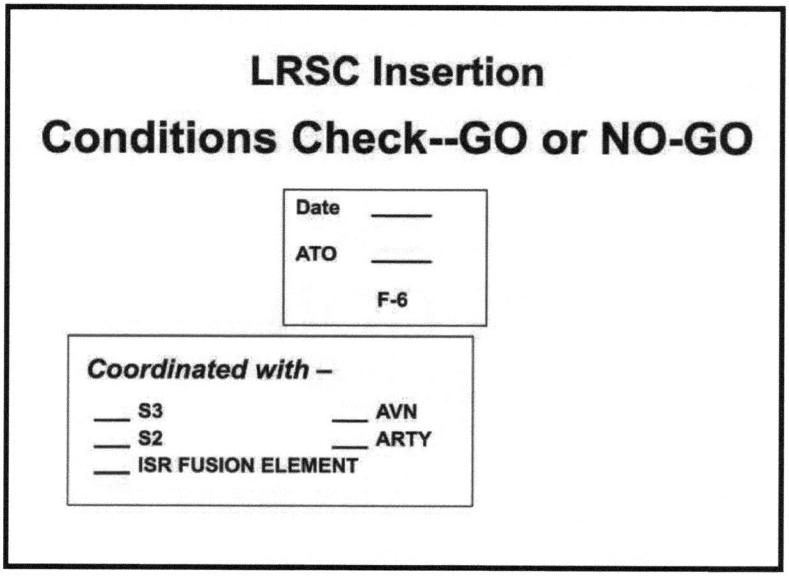

Appendix B

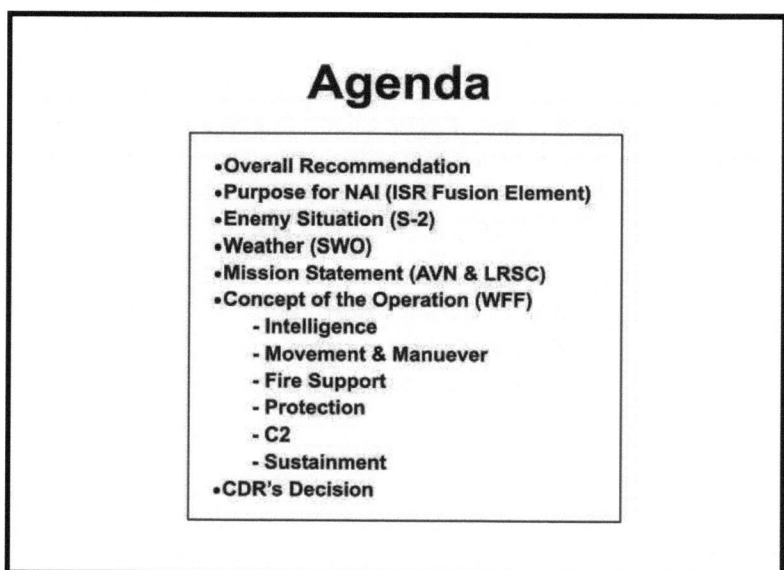

Figure B-13. Slide 3, LRSC insertion conditions check.

Figure B-14. Slide 4, agenda.

Recommendation

(Bottom Line Up Front)

STATUS	WFF SUMMARY
G	Intelligence
G	Movement & Maneuver
G	Fire Support
G	Protection
G	Sustainment
G	C2
G	LRSC
G	Overall Assessment

Figure B-15. Slide 5, recommendation.

ISR Fusion Element

1. Importance of this NAI
    - What we are looking for
    - What could happen if we don't look at this NAI
    - How else can we cover this NAI
    - Why do we need LRS on it

2. How early do we need to be on this NAI?

3. How long do we need to provide coverage on the NAI?

Figure B-16. Slide 6, ISR fusion element.

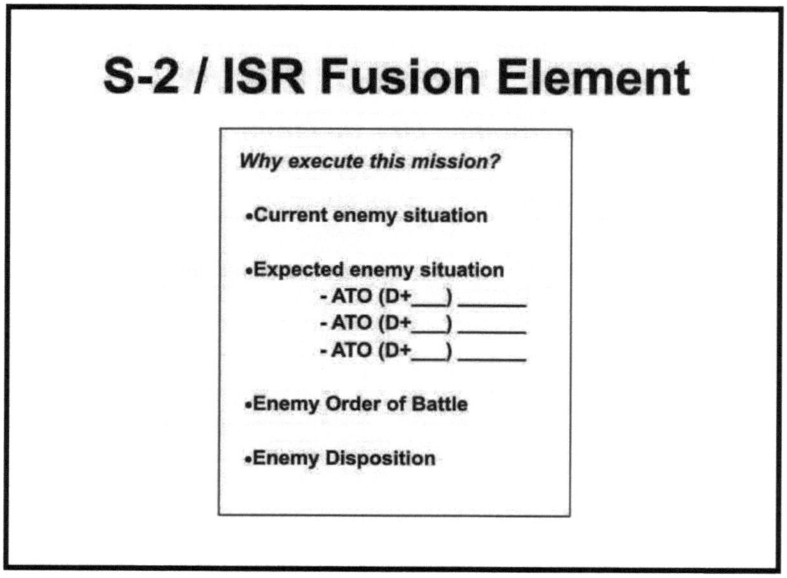

Appendix B

S-2 / ISR Fusion Element

Why execute this mission?

- Current enemy situation

- Expected enemy situation
  - ATO (D+___) _____
  - ATO (D+___) _____
  - ATO (D+___) _____

- Enemy Order of Battle

- Enemy Disposition

Figure B-17. Slide 7, S-2/ISR fusion element.

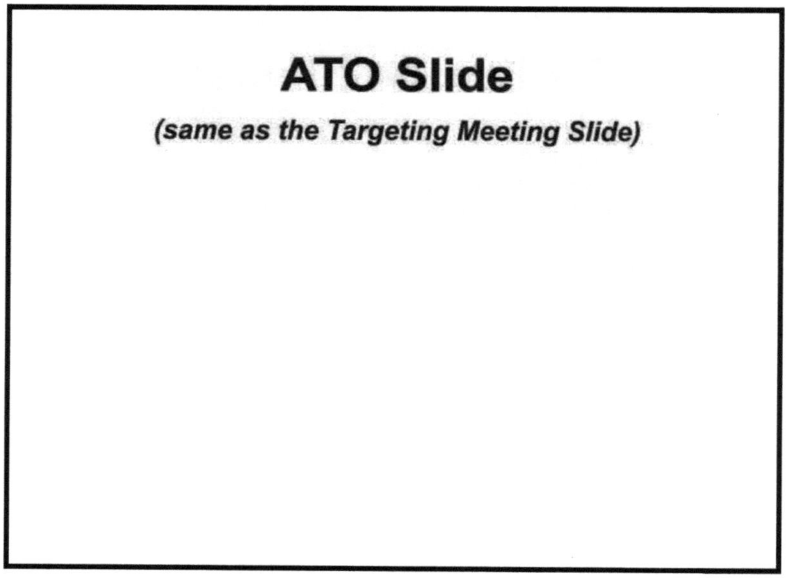

ATO Slide
(same as the Targeting Meeting Slide)

Figure B-18. Slide 8, ATO slide.

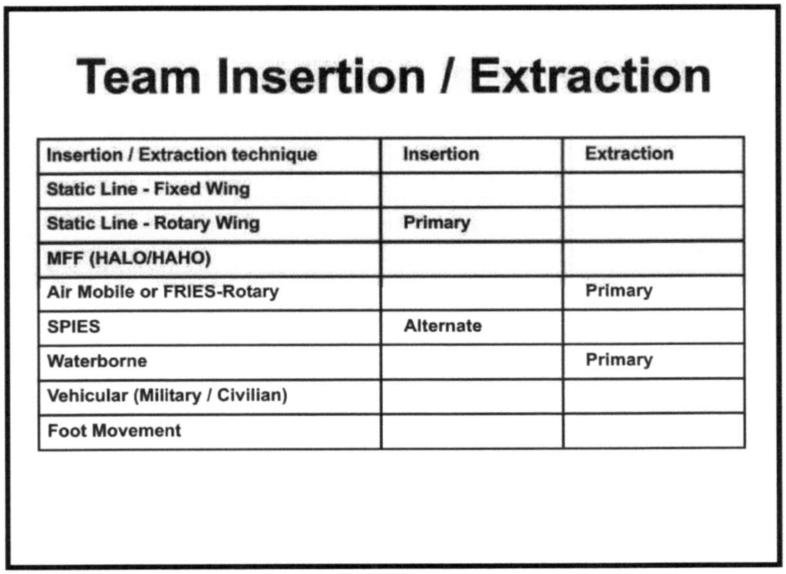

Team Insertion / Extraction

Insertion / Extraction technique	Insertion	Extraction
Static Line - Fixed Wing		
Static Line - Rotary Wing	Primary	
MFF (HALO/HAHO)		
Air Mobile or FRIES-Rotary		Primary
SPIES	Alternate	
Waterborne		Primary
Vehicular (Military / Civilian)		
Foot Movement		

Figure B-19. Slide 9, team insertion and extraction.

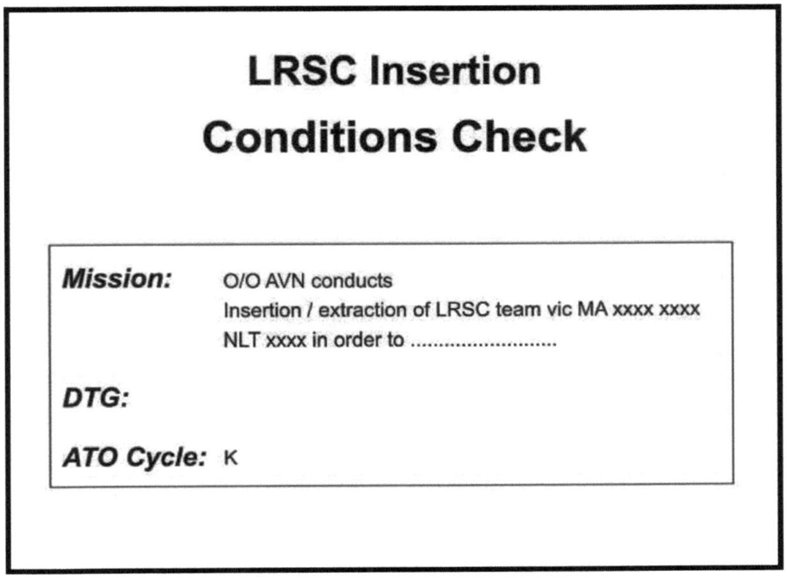

LRSC Insertion Conditions Check

Mission: O/O AVN conducts
Insertion / extraction of LRSC team vic MA xxxx xxxx
NLT xxxx in order to ....

DTG:

ATO Cycle: K

Figure B-20. Slide 10, LRSC insertion conditions check.

Appendix B

Movement & Maneuver

	CONDITIONS
	Aircraft Availability
	Forward Friendly Units Confirmed
	Planning Timeline
	Execution Time
	Higher & Adjacent Units Notified of Plan
	CSAR Plan Coordinated thru CRCC

CONDITIONS	GREEN	AMBER	RED
Aircraft Availability	More than enough	Adequate number	Inadequate number
Forward Friendly Unit Locations Confirmed	Updated <30 min	Updated <2 hrs	Updated <4 hrs
Planning Timeline (Corps FRAGO to F-Hour)	>72 hrs	48 to 72 hrs	>48 hrs
Execution Time (TOT vs EENT/BMNT)	MSN complete before BMNT	MSN in EA complete before BMNT	MSN in EA not complete before BMNT
Higher & Adjacent Units Notified of Plan	Notified verbally and by written order	Notified verbally	Not notified

Figure B-21. Slide 11, movement and maneuver.

Fire Support

	CONDITIONS
	Artillery SEAD Identified (lethal)
	JSEAD Package Identified (lethal/non-lethal)

CONDITION	GREEN	AMBER	RED
Artillery SEAD Identified	Arty SEAD planned, coordinated, and rehearsed	Arty SEAD planned; coordination in progress	Arty SEAD unplanned; coordination incomplete
JSEAD Identified (lethal/nonlethal)	Available and coordinated	Available; coordination in progress	Not coordinated or unavailable

Task Org	C 3/27 18TH FAB NB68719609
JSEAD	1 X EA-6B/2 X F-16CJ 1800 to 1900, 1900 to 2000 EC-130 1900 to 2000
CAS/AI/JAAT	20 Sorties CAS 1900 to 2100 AI 12 X F-15E 1915 to 2015
	As of 031110ZJAN02

Figure B-22. Slide 12, fire support.

Air Protection

	CONDITIONS
	Air Routes/Airspace Control Plan
	Weapons Control Status

CONDITION	GREEN	AMBER	RED
Air Routes and Airspace Control Plan	ACMs disseminated and no conflict	Dissemination in progress	Not disseminated or resolved
Weapon Control Status	Weapons tight		Weapons free

Figure B-23. Slide 13, air protection.

Sustainment

	CONDITIONS
	FARP Availability
	FARP Timeliness

CONDITION	GREEN	AMBER	RED
FARP Availability	Adequate III and V	Marginal III and V	Class V not available
FARP Timeliness	Operational <1 hr	Operational 1 to 2 hrs	Weapons free

Figure B-24. Slide 14, sustainment.

Appendix B

C2

	CONDITIONS	
	C2 Aircraft Available	
	ABCCC/RC-12	
	TACSAT/UHF/HF	
	Relay in Place	
	COMMO Rehearsals Conducted	
	Abort Criteria Established	
	Info Work Space (IWS)	
C2 Aircraft	Available	Unavailable
ABCCC/RC-12	Available	Unavailable
TACSAT/UHF/HF	Available	Unavailable
Retrans in place	Established	Not established
COMMO Rehearsals	Conducted	Not conducted
Abort Criteria	Established	Not established
IWS	Available	Unavailable

ABORT CRITERIA: Loss of more than 2 aircraft
MISSION SUCCESS: Successful Insertion

Figure B-25. Slide 15, command and control.

Intelligence

	Conditions		
	Current Weather and Light Data for EA		
	Current Weather and Light Data Enroute		
	Enemy ADA Capabilities and Locations		
	Target Accuracy		
	Target Timeliness		
CONDITION	GREEN	AMBER	RED
Weather (EA)	Ceiling >3,000' Vis >3mi (4.8km) Winds <10 knots	Ceiling <3,000' Vis <3mi (4.8km) Winds 10-25 knots	Ceiling <500' Vis <2mi (43.2km) Winds <25 knots
Weather (Enroute)	Ceiling >700' Vis >2mi (3.2km) Winds <25 knots	Ceiling <700' Vis 1-2mi (3.2km) Winds 25-30 knots	Ceiling <500' Vis <1mi (1.6km) Winds >30 knots
Enemy ADA	Locations Identified	Locations Templated	Locations Unknown
Target Accuracy	HUMINT	EW	Templated
Target Timeliness	Tgt Updated <30 min	Tgt Updated <2hrs	Tgt Updated <4hrs

Figure B-26. Slide 16, intelligence.

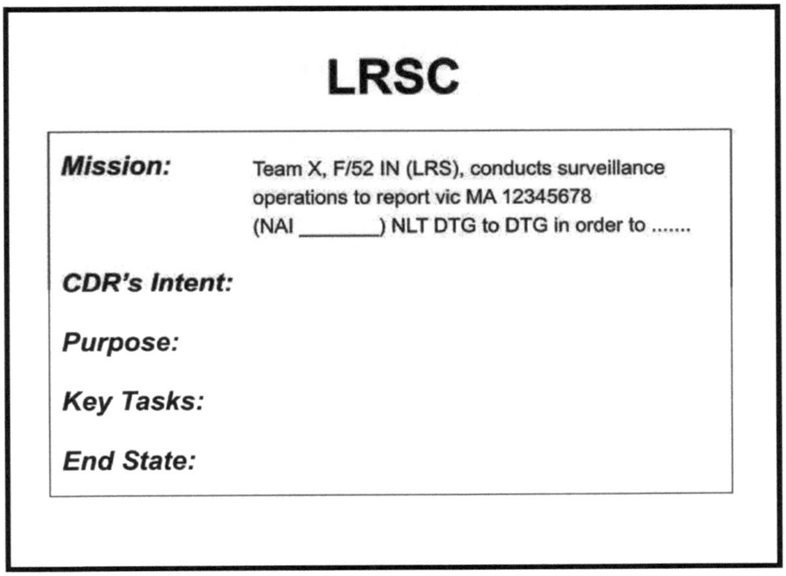

Orders and Briefs

LRSC

Mission: Team X, F/52 IN (LRS), conducts surveillance operations to report vic MA 12345678 (NAI _____) NLT DTG to DTG in order to ....

CDR's Intent:

Purpose:

Key Tasks:

End State:

Figure B-27. Slide 17, LRSC.

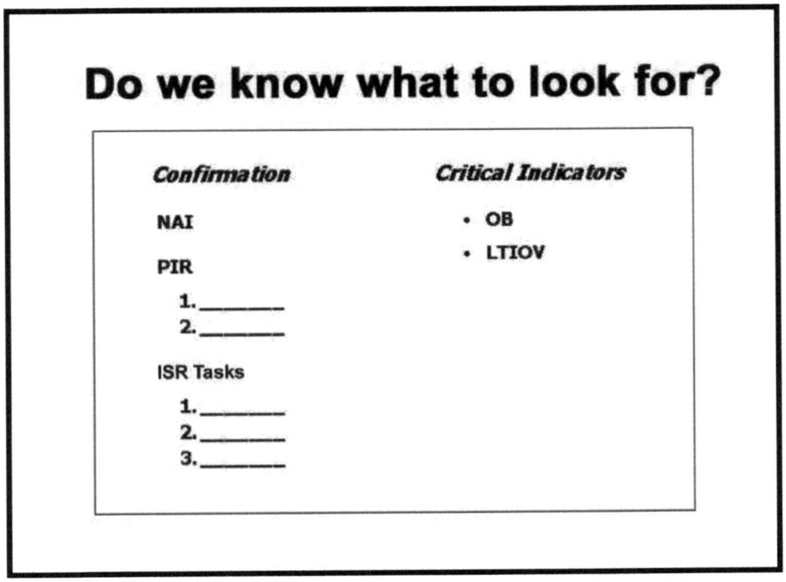

Do we know what to look for?

Confirmation

NAI

PIR
1. _____
2. _____

ISR Tasks
1. _____
2. _____
3. _____

Critical Indicators

- OB
- LTIOV

Figure B-28. Slide 18, "Do we know what to look for?"

LRSC IPB

	CONDITIONS
	Weather and Light
	Enemy situation- Insertion
	Enemy situation- Infiltration
	Enemy situation- Execution
	Enemy situation- Exfiltration
	Enemy situation- Extraction

CONDITIONS	Green	Amber	Red
Weather and Light	Favorable for all phases	Unfav for =/> 1 Phase	100% illum; no terrain; rain = (-30)
Enemy Insertion	Air "GO"; clear LZ; alt LZ	Unclear SEAD result	Enemy confirmed
Enemy Infiltration	LZ – 10 km; no threat	Possible threat	Enemy confirmed
Enemy Execution	Contact not likely	Possible threat	Enemy confirmed
Enemy Exfiltration	PZ – 10km; not likely	Possible contact	Enemy confirmed

Figure B-29. Slide 19, LRSC IPB.

LRSC Movement & Maneuver
(Actions on the Objective)

STATUS	CONDITIONS
	Patrol Base Location
	Forward Friendly Units Confirmed
	Hide Site
	Surveillance Site
	Execution Time

CONDITIONS	GREEN	AMBER	RED
Patrol Base	PB site preplanned; Team consolidated	Passive PB; HS and SS split	No PB; movt during day
Forward Friendly	LNO linked up with FFU	No LNO, but have comm	No LNO and no comm
Hide Site	Pri and Alt planned; subsurface; 1-2km SS	No Alt; -500m surface sites	No Alt; -200m surface; no net
Surveillance Site	Pri and Alt planned; subsurface; + 800m SO	No Alt; -500m surface sites	No Alt; -100m surface; no net
Execution Time	No more than 5 days	6-8 days	8 days or +

Figure B-30. Slide 20, LRSC maneuver.

Orders and Briefs

LRSC Fire Support

STATUS	CONDITIONS
	Insertion--targets planned
	Infiltration--targets planned
	Execution--targets planned on OBJ
	Exfiltration--targets planned
	Extraction--targets planned
	OVERALL

CONDITION	Green	Amber	Red
Insertion	Planned TGTs	Only TRPs	Not available
Infiltration	Planned TGTs	Only TRPs	Not available
Execution	Planned TGTs & RFAs/NFAs	Only TRPs	Not available
Extraction	Planned TGTs	Only TRPs	Not available
Exfiltration	Planned TGTs	Only TRPs	Not available

Figure B-31. Slide 21, LRSC fire support.

LRSC Sustainment

STATUS	CONDITIONS
	CLASS I
	CLASS IV
	CLASS V
	CLASS IX--Batteries
	CLASS VIII--Medical Support
	OVERALL

CONDITIONS	Green	Amber	Red
CLASS I	Adequate	Marginal	Not available
CLASS IV	Adequate	Marginal	Not available
CLASS V	Adequate	Marginal	Not available
CLASS IX	Adequate	Marginal	Not available
CLASS VIII - Med Support	EMT & CLS per TM	Only CLS	No medical skill

Figure B-32. Slide 22, LRSC sustainment.

Appendix B

LRSC C2

STATUS	CONDITIONS
	COB Operational
	AOB Operational
	Link to S2
	Link to S3
	Link to ISR Fusion Element
	Link to FSE

CONDITIONS	GREEN	RED
COB	Operational	No commo
AOB	Operational	No commo
S2	Operational	No commo
S3	Operational	No commo
ISR Fusion Elements	Operational	No commo
FSE	Operational	No commo

Figure B-33. Slide 23, LRSC C2.

LRSC C2
(Communications)

STATUS	CONDITIONS
	Weather/Terrain effects
	Enemy Intercept/Jamming
	Range
	Team Equipment (HF, TACSAT, FM)
	COB/AOB Links (MSE, LAN, FM)

CONDITIONS	GREEN	AMBER	RED
Weather/Terrain	Favorable for all phases	Unfav=/> 1 Phase	Will impact at critical times
Enemy Intercept	Unlikely	Possible	Probable
Range	All supporting units	Some supporting units	No units W/I commo range
Team Equipment	All operational	1 system non-op	No redundant comms
COB/AOB equip	All operational	No Comms w/1 node	No Comms w/TM and/or CP

Figure B-34. Slide 24, LRSC C2, communications.

LRSC Abort Criteria

STATUS	CONDITIONS
	Insertion (MEF– 4 pax and LD commo)
	Infiltration (On order)
	Execution (Compromise on OBJ, no LD commo)
	Exfiltration (Isolated personnel—EPA)
	Extraction

CONDITIONS	GREEN	RED
Insertion	Criteria not met	Criteria met
Infiltration	Criteria not met	Criteria met
Execution	Criteria not met	Criteria met
Exfiltration	Criteria not met	Criteria met
Extraction	Criteria not met	Criteria met

Figure B-35. Slide 25, LRSC abort criteria.

LRSC Risk Mitigation

RISK LEVEL	CATASTROPHIC HAZARDS	CONTROLS IMPLEMENTED	RESIDUAL EFFECT
High	1. Fatality during MFF insertion	1. Individuals at pro-level I; MACO brief prior	Medium
High	2. Fatality while performing surveillance	2. Subsurface surveillance site; claymores set; secure comm link	Medium
	3.	3.	Medium

Figure B-36. Slide 26, LRSC risk mitigation.

Appendix B

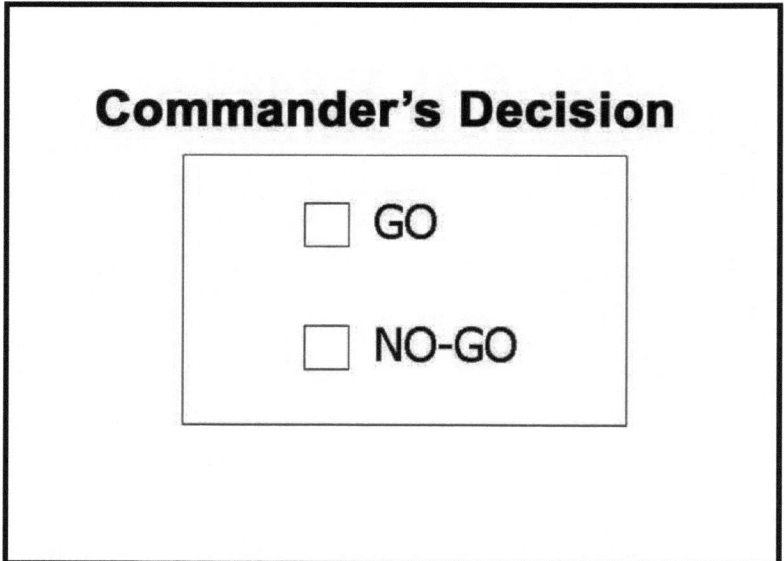

Figure B-37. Slide 27, LRSC recommendation.

Figure B-38. Slide 28, commander's decision.

DEBRIEF

B-11. The commander, his representative, or someone from the S-2 interviews, or debriefs, the team to obtain information about the mission. A debrief is usually conducted just after the mission is over by the commander, his representative, or by someone from the S-2. Figure B-39 shows an example debrief, and Figure B-40 through Figure B-45, pages B-36 through B-48, show example annexes to the same debrief.

Orders and Briefs

```
                (CLASSIFICATION--WHEN FILLED OUT)
DEBRIEF

TEAM: _____

MISSION NUMBER: _____

DTG OF DEPARTURE: _____

DTG OF RETURN: _____

MAPS USED: List maps team actually used on mission.
    1:24,000:
    1:50,000:
    1:250,000:

SPECIAL (PHOTOS):

ENCLOSURES: This includes the Patrol Log, Communications Log,
Surveillance Log, Sketches, and so on.)

1. VISUAL RECONNAISSANCE.
    a. Primary loading date, alternate landing zone data, or both
(size, MOG, obstacles, slope, vegetation, immediate surroundings).
    b. Significant enemy sightings during the mission (SALUTER).
    c. Indications of enemy activities such as tracks, trash,
and sounds.

2. ORGANIZATION.
    a. General Patrol Personnel.
    b. Special Patrol Personnel.

3. EQUIPMENT.
   a. Individual Primary Small Arms.
   b. Alternate or crew served weapons.
   c. Mines.
   d. Grenades.
   e. Booby-traps.
   f. Demolitions.
   g. Special weapons.
   h. FM radios (VHF).
   i. HF radios.
   j. Smoke grenades.
   k. VS-17 panels.
   l. Strobe lights.

                (CLASSIFICATION--WHEN FILLED OUT)
```

Figure B-39. Example debrief.

Appendix B

```
                    (CLASSIFICATION--WHEN FILLED OUT)
   m. Pen flares.
   n. Compasses, PSN-11, and so on.
   o. Flashlights.
   p. Individual items.
   q. Optical equipment.
   r. Maps.
   s. Photographic equipment.

4. MISSION. This is an exact duplicate of the mission assigned to
the team.
5. SIR to PIR or ISR tasks to PIR. This also includes SIR to
   intelligence requirements or ISR tasks to intelligence requirements.
6. TERRAIN.
   a. Land Forms.
   b. Vegetation.
      (1) Lowland.
      (2) Ridge and mountain sides.
      (3) High ground, ridge tops and hill tops.
   c. Rivers and Streams.
      (1) Location.
      (2) Width.
      (3) Depth.
      (4) Current (speed and direction).
      (5) Slopes of the bank.
      (6) Composition of the soil on bottom of banks.
      (7) Dimensions of the dry bed.
      (8) Navigability of large streams.
      (9) Potability of water.
   d. Observations of Civilian Activity.
      (1) Where were people seen?
      (2) When were they observed?
      (3) How many people were seen?
      (4) Were they civilian or military?
      (5) What was their ethnic group(s), language(s), and so on?
      (6) What clothing (color, condition), footgear, and headgear
were they wearing?
      (7) What kind of equipment did they have (color, size, shape,
condition)?
      (8) What were the people doing?
      (9) What was the physical condition of the people observed
(morale and discipline)?
   e. Observations of Structures and Other Manmade Objects.
      (1) Location(s).
      (2) Quantity.

                    (CLASSIFICATION--WHEN FILLED OUT)
```

Figure B-39. Example debrief (continued).

(CLASSIFICATION--WHEN FILLED OUT)

 (3) Shape(s), size(s), purpose(s).
 (4) Construction materials.
 (5) Markings.
 (6) Contents of structure.
 (7) Estimate of last use.
 (8) Indications of occupancy.
 (9) Animals or animal pens in vicinity.
 (10) Crops in vicinity.
  (a) Type.
  (b) Field or paddy size.
  (c) Care of the area.
  (d) Stage of development.
  (e) Food and water storage area(s).
 (11) Trenches, fighting holes, bunkers, or advanced warning positions (location, size, number, description of each).
  (a) Stage of development.
  (b) Efforts to hide from aerial view.
f. **Observations of Animals.**
 (1) What type, where, and when?
 (2) Wild or tame?
 (3) Condition of health?
 (4) Presence of vectors?
g. **Enemy Tactics.**
 (1) What was the enemy's reaction time, if contact was made?
 (2) How does the enemy employ counterreconnaissance assets?
 (3) Was the patrol tracked at any time?
 (4) Did the enemy try to break contact?
 (5) What action did the enemy take when aircraft flew over?
 (6) What reaction did the enemy have when it was attacked?
 (7) Any signals used (visual, sound, voice)?
 (8) Unit discipline.
 (9) Are enemy personnel well trained?
 (10) What type of equipment is the enemy using?
h. **Trails.** Identify by number as located during the mission.
 (1) Direction and location.
 (2) Width.
 (3) Estimate of use (animals or man).
 (4) Overhead canopy.
 (5) Undergrowth along sides of trails.
 (6) Direction signs, symbols, target indicators on trail.
 (7) Surface characteristics (hard packed or soft earth, dead vegetation).
 (8) Description of vehicle tracks.
i. **Roads.**
 (1) Direction (in degrees) and location.
 (2) Width (in meters).

(CLASSIFICATION--WHEN FILLED OUT)

Figure B-39. Example debrief (continued).

(CLASSIFICATION--WHEN FILLED OUT)

   (3) Surface material. This includes sand, packed earth, gravel surfaced, asphalt, concrete, and so on.
   (4) Indications of movement on the road.
   (5) Maintenance of road (craters repaired, pot holes, and so on).
   (6) Road drainage characteristics (crown and camber, drainage ditches, and so on).
   (7) Obstructions.
   (8) Road signs, markers, and so on.
   (9) Description of vehicle tracks (if applicable):
  j. **Soil.**
   (1) Appearance (color):
   (2) Composition (clay, sand, rocky, roots, loam, and so on):
   (3) Hardness (dry, wet, muddy, very muddy):
   (4) Ease of digging (fast, medium, slow):
   (5) Standing water.
  k. **Map Correction.** This includes landforms, tree lines, waterways, roads, trails, built up areas, and so on.

7. **WEATHER.**
  a. **Visibility.**
  b. **Cloud Cover.**
  c. **Rainfall.**
  d. **Ground Fog.**
  e. **Winds.**
  f. **Temperatures and Humidity.**
  g. **Illumination at Night.**
  h. **Effects of Weather on Team Personnel.**

8. **COMMUNICATIONS.** State whether or not you have or had--
  a. Jamming?
  b. Problems at relay site?
  c. Difficulties in radio sets?
  d. Signs of enemy direction-finding capabilities?
  e. Most effective frequency?
  f. Most effective time?
  g. Most effective antennae?

9. **NARRATIVE.**
This narrative is a chronological detailed statement emphasizing time, movement activities, and observations within the area of observation. Provide an overlay.
  a. **Date-Time Group and Place.**

(CLASSIFICATION--WHEN FILLED OUT)

Figure B-39. Example debrief (continued).

```
            (CLASSIFICATION--WHEN FILLED OUT)
    b. Movement.
        (1) Movement by foot (direction and distance).
        (2) Danger areas (locations).
        (3) Actions on objective (direction, distance, and so on).
        (4) Extraction (direction, distance, and so on).
10. ADDED INFORMATION.
    a. Anything not already covered.
    b. General estimate of military activity in the area.
    c. Signals.
11. RECOMMENDATIONS.
    a. Items of equipment or material that can or should be improved
to further enhance operational ability.
    b. Operational techniques that could have improved the mission.
    c. Lost or unrecovered equipment.
12. CONDITION OF PATROL.
    a. Water intake.
    b. Food intake.
    c. Minor injuries and illness.
    d. Remarks (include when team will be ready to execute another
mission):

Team Leader
    Print   _____
    Sign    _____
    Date    _____

Debriefer(s)
    Print   _____
    Sign    _____
    Date    _____

DTG of Debrief  _____

            (CLASSIFICATION--WHEN FILLED OUT)
```

Figure B-39. Example debrief (continued).

```
                    (CLASSIFICATION--WHEN FILLED OUT)
INTELLIGENCE ESTIMATE
1.  MISSION
2.  AREA OF OPERATION
      a. Weather.
          (1) Existing situation.
              (a) 72-hour forecast.
              (b) Light data (includes BMNT, EENT, MR, MS, and
illumination).
              (c) Trends (includes temperature, precipitation
and wind).
          (2) Effects of enemy.
          (3) Effects on friendly.
      b. Terrain.
          (1) Existing situation.
              (a) Observation and fields of fire.
              (b) Avenues of approach.
              (c) Cover and concealment.
              (d) Obstacles.
              (e) Key terrain.
          (2) Effects on enemy courses of action.
          (3) Effects on friendly courses of action.
      c. Other Characteristics. This includes information about
population, culture, and religion.
          (1) Existing situation.
          (2) Effects on enemy courses of action.
          (3) Effects on friendly courses of action.
3.  ENEMY SITUATION
      a. Disposition.
      b. Composition.
          (1) Identified units.
          (2) Unidentified units.
      c. Strength.
          (1) Committed forces.
          (2) Reinforcements.
          (3) Air.
          (4) Air defense.
          (5) CBRN.
      d. Recent and Present Significant Activities.
      e. Peculiarities and Weaknesses.
          (1) Personnel.
          (2) Intelligence.
          (3) Operations.
          (4) Logistics.
          (5) Personalities.
                    (CLASSIFICATION--WHEN FILLED OUT)
```

Figure B-40. Example intelligence estimate annex.

(CLASSIFICATION--WHEN FILLED OUT)

4. ENEMY CAPABILITIES
   a. Composition.
   b. Analysis and discussion.
5. CONCLUSIONS.
   a. Intelligence.
   b. Weather and Terrain.
   c. Most Probable Enemy Course of Action.
   d. Most Dangerous Course of Action.
   e. Enemy Vulnerabilities.

(CLASSIFICATION--WHEN FILLED OUT)

Figure B-40. Example intelligence estimate annex (continued).

Appendix B

```
(CLASSIFICATION--WHEN FILLED OUT)
```
COMMUNICATION ANNEX
1. **SITUATION**
    a. **Enemy Forces.**
       (1) Weather and terrain that may affect communications.
       (2) Electronic warfare capabilities.
       (3) Direction finding capabilities.
    b. **Friendly Forces.**
2. **MISSION.** This does not apply.
3. **EXECUTION.**
    a. Plan and location for initial entry report (with a diagram).
    b. Team no-communications with COB (or MSS) and AOB.
    c. Team no-communications plan for communications.
    d. Security plan for communication windows.
    e. Type of antennas to use.
    f. Team no-communications drill.
       (1) HF.
       (2) FM.
    g. Specific instructions to key individuals during HF transmissions and window times.
4. **SERVICE AND SUPPORT**
    a. **Radios.**
    b. **Batteries.**
    c. **Antenna Kits.**
    d. **Related Equipment.**
    e. **Sterilization Plan.**
5. **COMMAND AND SIGNAL**
    a. **Locations.** Location of team leader and assistant team leader during communications window.
    b. **Long-Range Communications.**
       (1) Windows.
       (2) Frequencies and times.
           (a) Primary.
           (b) Alternate.
           (c) Guard.
       (3) Duress code word.
           (a) COB (or MSS) or AOB.
           (b) Team.
       (4) Azimuths.
       (5) Addresses.
           (a) COB (or MSS) or AOB.
           (b) Team.
       (6) Other code word.
       (7) Reports.
    c. **FM Communications.**
       (1) Call signs (internal and external).
       (2) Frequencies (internal and external).
       (3) Brevity plan.
    d. **Reports Other Than SOP.**

Figure B-41. Example communications annex.

```
            (CLASSIFICATION--WHEN FILLED OUT)
FIRE-SUPPORT ANNEX
1. SITUATION.
   a. Enemy Forces.
      (1) Enemy AD assets and fire support:
      (2) Enemy equipment.
          (a) Rockets:
          (b) Cannon:
          (c) Missiles:
          (d) CBRN:
   b. Friendly Forces.
      (1) Higher headquarters' concept of fires:
      (2) Adjacent units' concept of fires:
      (3) Supporting air and naval forces:
   c. Attachments and Detachments.
  2. MISSION. Team Leader gives information about the fire plan:
3. EXECUTION.
   a. Concept of Fires:
   b. Air Support.
      (1) General (commanders' intent):
      (2) Air interdiction:
      (3) Close air support:
      (4) Electronic combat:
      (5) Reconnaissance and surveillance operations:
      (6) Miscellaneous:
   c. Field Artillery Support.
      (1) General (concept of fires):
      (2) Artillery organization for combat:
      (3) Allocation of ammunition:
      (4) Miscellaneous:
   d. Naval Gunfire Support.
      (1) General (concept of naval gunfire):
      (2) Naval gunfire organization:
      (3) Miscellaneous:
   e. Nuclear Operations:
   f. Smoke Operations:
   g. Target Acquisition:
   h. Coordinating Instructions.
      (1) Target list:
      (2) Fire support coordination measures:
      (3) Targeting products and attack-guidance matrix:
      (4) Rules of Engagement:
4. SERVICE AND SUPPORT.
   a. Command and Signal:

            (CLASSIFICATION--WHEN FILLED OUT)
```

Figure B-42. Example fire support annex.

Appendix B

(CLASSIFICATION--WHEN FILLED OUT)

LINKUP ANNEX
1. **SITUATION.**
   a. **Enemy Forces.**
      (1) Disposition, composition, strength, and identification:
      (2) Terrain at link-up site:
      (3) Obstacles near or around the linkup site:
   b. **Friendly Forces.**
      (1) Linkup unit:
      (2) Designated liaison team:
2. **MISSION.** Team Leader gives information about the linkup operation.
3. **EXECUTION.**
   a. Commander's intent.
   b. Concept of the operation.
      (1) Maneuver.
      (2) Fires.
      (3) Intelligence.
      (4) Electronic warfare.
      (5) Engineers.
      (6) Other.
   c. **Task to Maneuver Units.**
      (1) Elements.
      (2) Individuals.
   d. **Task to Combat Support Units.**
   e. **Coordinating Instructions.**
      (1) Time of linkup.
      (2) Location of linkup site.
      (3) Rally points.
      (4) Actions on enemy contact.
      (5) Actions at linkup site.
      (6) Rehearsals.
4. **SERVICE AND SUPPORT.**
5. **COMMAND AND SIGNAL.**
   a. **Chain of Command.**
      (1) Locations of key individuals:
      (2) Location of liaison team:
   b. **Signal.**
      (1) Frequencies and call signs:
      (2) Far-recognition signal and identification.
         (a) Day--primary and alternate:
         (b) Night--primary and alternate:
      (3) Near-recognition signal and identification.
         (a) Day--primary and alternate:
         (b) Night--primary and alternate:

(CLASSIFICATION--WHEN FILLED OUT)

Figure B-43. Example linkup annex.

```
            (CLASSIFICATION--WHEN FILLED OUT)
VEHICLE-MOVEMENT ANNEX
1. SITUATION.
    a. Enemy Forces.
        (1) Disposition, composition, strength, and identification.
Find in enemy situation template overlay.
        (2) Weather. Obtain from company OPORD weather forecast.
        (3) Terrain along and adjacent to route.  Use map
reconnaissance and intelligence portion of company OPORD, and the
terrain analysis.
        (4) Vegetation. Get this from the map, safe area intelligence
description, terrain analysis.
        (5) Obstacles and potential ambush sites.  information
received from S-2; current or prior intel on enemy location and
activities, previous patrol logs or from enemy sit-temp from
company OPORD.
    b. Friendly Forces.
        (1) Units along route.  found in friendly situation from
company OPORD.
        (2) Unit providing transportation.  found in Co OPORD, before
or during coordinations.
2. MISSION. This lists information about the vehicle-movement
operation).
3. EXECUTION.
    a. Commander's Intent.
    b. Concept of the Operation.  Describe in general terms from SP
to detrucking point.
        (1) Maneuver.  in detail, the plan for completing the
insertion from start to finish.
        (2) Fires.  QRF, on call artillery support.
        (3) Intelligence.
        (4) Electronic warfare.
        (5) Engineering. Any assets required to precede insertion
vehicle such as mine clearing.
        (6) Route.  Use overlay.
        (7) Other.
    c. Task to Maneuver Units.
        (1) Teams. Who is responsible for preparing the vehicle(s)?
        (2) Individuals. Brief vehicle drivers on the routes, action
on enemy contact, and vehicle interval and speed.
    d. Coordinating Instructions.
        (1) Time of departure or return.  SP time or time of return.
        (2) Loading and order of movement.  Team SOP; vehicle
configuration; and position of support vehicles, location of
insertion vehicle, or both.

            (CLASSIFICATION--WHEN FILLED OUT)
```

Figure B-44. Example vehicle movement annex.

Appendix B

(CLASSIFICATION--WHEN FILLED OUT)

(3) Actions on enemy contact. Action of support and insertion vehicle or, if vehicles stop completely, action of team, or both.

(4) Actions at the dismount point. Is this a rolling insertion, or will it stop with the lead and chase vehicle setting up a roadblock? How will the team offload? How will it pull security?

(5) Rehearsals. Practice--
- The load plan
- Loading onto vehicles.
- Off-loading from vehicles
- Reacting to contact.
- Conducting communications checks.

(6) Inspections. Inspect vehicles for the following:
- Serviceability,
- Fuel.
- Seats.
- Air guard.
- Sand bags on the floor.
- Fire extinguishers.
- Tools
- Tow bar
- Spare tire
- Canvas cover on or off IAW SOP
- Physical condition of the driver.

4. **SERVICE AND SUPPORT.** Team Leader lists any special equipment needed.

5. **COMMAND AND SIGNAL.**
  a. **Command.**
    (1) Chain of command. This includes convoy commander and drivers and QRF personnel.
    (2) Location of key leaders. Team leader, ATL, RATELO, convoy commander.
  b. **Signal.**
    (1) Special signals for movement:
    (2) Communication in and between vehicles.
- Type of radios.
- Frequencies.
- Call signs.

(CLASSIFICATION--WHEN FILLED OUT)

Figure B-44. Example vehicle movement annex (continued).

(CLASSIFICATION--WHEN FILLED OUT)

AIR INFILTRATION/EXFILTRATION ANNEX

1. SITUATION.
   a. **Enemy Forces**. This lists the enemy unit's composition and disposition.
      (1) ADA. Brief type, range (with overlay), means of target acquisition, and minimum altitude for engagement.

      (2) LZ and DZ. Brief enemy units on LZ and DZ (detailed) (if briefed in intelligence annex, then IAW intelligence annex).
      (3) Confirmed positions along flight route. Brief detailed enemy situation (if not covered in intel annex).
   b. **Weather**. This follows the intelligence annex.
   c. **Friendly Forces**.
      (1) Unit providing lift.
         (a) Insertion. Brief friendly unit supporting.
         (b) Extraction. Brief friendly unit supporting.
      (2) Unit providing armed escort support. Brief friendly supporting unit, type of armed escort, all weather capabilities.

2. MISSION. This lists information about the air movement.

3. EXECUTION.
   a. **Concept of the operation**. Cover the whole operation in general terms using a map, sand table or overlay.
   b. **Task Organization**. Lists the task organization for the operation.
   c. **Maneuver**.
      (1) Flight routes. Must use overlay.
         (a) Primary routes.
         (b) Check points. (time, distance, and description). (*Use map.*)
         (c) False insertions. (*Use map.*) The team will conduct a false insertion at GL.
         (d) Alternate route. (*Use overlay.*)
         (e) Checkpoints. (*Use map.*)
         (f) False insertions. (*Use map.*)
      (2) Actions at LZ or DZ.
         (a) Actions upon landing.
         (b) Assembly plan.
         (c) Cache of equipment. Brief method and location of cache, camouflage, security, type of report to be sent, any taskings for the cache
         (d) Actions on enemy contact. To include en route, while landing, while loading/unloading, when aircraft is in the loiter area, after aircraft has departed-must include method of marking team for the aircraft if the team is in contact.

(CLASSIFICATION--WHEN FILLED OUT)

Figure B-45. Example air infiltration/exfiltration annex.

Appendix B

(CLASSIFICATION--WHEN FILLED OUT)

   (e) Emergency extraction. Brief emergency PZ or the use of SPIES.

   (f) Loiter times and areas. Brief using map-include time and specific directions for the aircraft, such as the height of hover or any other actions at the loiter area and actions if no communications with team.

  d. **Coordinating Instructions**.

   (1) Insertion.

    (a) Departure (NLT) time.

    (b) PZ or airfield location. (Use map.)

    (c) Flight time. Brief total time of flight.

    (d) Insertion time. Brief TOT on DZ or LZ.

    (e) Transportation. Method of transportation to PZ or airfield.

    (f) Decision point. Decision point and actions taken for downed aircraft before and after crossing. Brief decision point (use map) and team actions if aircraft down before or after the decision point, to include a plan for the aircrew.

    (g) Abort criteria. Get this information from the company order.

    (h) Actions. On enemy contact en route.

    (i) Evasion plan. See EPA or explain.

    (j) Time schedule.

    (k) Load plan. Brief from team SOP.

    (l) Rehearsals.

   (2) Extraction.

    (a) Date-time-group. Time of extraction as specified in order or on call

    (b) Type. Type of extraction.

    (c) Communications. Communication checkpoint (CCP) location (use map), and distance and direction to PZ from CCP.

    (d) PZ location. Brief location (use map).

    (e) LZ location. Brief location (use map).

    (f) Load plan. Brief from team SOP.

    (g) Actions on enemy contact. While aircraft is en route, after aircraft lands, and while loading.

    (h) Lack of aircraft plan. Brief plan if the aircraft fails to arrive, or how long to wait at PZ?

4. **SERVICE AND SUPPORT.** List equipment for FRIES/SPIES, additional headsets for the aircraft, aircraft armament, equipment carried by aircrew and so forth.

5. **COMMAND AND SIGNAL.**

 a. **Command**.

  (1) Chain of command. For the team as well as for the aircrew (in flight and on the ground if the aircraft goes down).

  (2) Locations of key leaders. During flight (TL, ATL, LNO).

(CLASSIFICATION--WHEN FILLED OUT)

Figure B-45. Example air infiltration/exfiltration annex (continued).

(CLASSIFICATION--WHEN FILLED OUT)

  b. **Signal**.
    (1) Insertion.
      (a) Primary frequency. Brief primary frequency for aircraft or vehicle
      (b) Alternate frequency. Brief alternate frequency for aircraft or vehicle
      (c) Call signs. Brief call signs for aircraft or vehicle
      (d) Code words. Brief any appropriate code words for insertion
    (2) Extraction.
      (a) Primary frequency. Brief primary frequency for aircraft or vehicle
      (b) Alternate frequency. Brief primary frequency for aircraft or vehicle
      (c) Call signs. Brief call signs for aircraft or vehicle
      (d) Code words. Brief any appropriate code words for extraction
      (e) Far recognition. Brief far recognition signal (usually FM)
      (f) Near recognition. Brief near recognition for day or night (IR, chemical, VS 17, flashlight, IR flashes, and so forth.)
      (g) LZ markings. Describe how the LZ will be marked (inverted Y or other method).
      (h) Signaling devices. Describe what types of signaling devices each member of the team carries.
      (i) Authentication. Method of authentication with the aircraft or vehicle when CCP is inbound.
      (j) Actions in absence of communications plan. Brief plan in case of no communications with the aircraft or vehicle.

(CLASSIFICATION--WHEN FILLED OUT)

Figure B-45. Example air infiltration/exfiltration annex (continued).

Appendix C
Planning Area Facilities and Sites

The goal of planning with limited access is to allow LRS teams an environment without distractions and a reduced risk of mission compromise. Facilities include fixed sites (best) or field sites.

FACILITIES

C-1. Each LRSU benefits by having access to a secure planning facility. Ideally the facility should enough space available to allow planning by the required number of teams to support operations. The location and type of facility used depends on availability, security, and deployment requirements--detailed planning is the same, regardless. Once the planning facility is established, the teams conduct detailed mission planning. Access to the team is limited to such individuals as the team LNO, company commander, BFSB and R&S squadron S-2 and others on the access roster. If available, the team rehearses in a secured area nearby ideally on terrain similar to the area where they will operate.

FIXED SITE

C-2. When available, the unit conducts detailed planning at a fixed site on its home installation. The next best option is to use a remote fixed site. The site should offer separate sleeping and planning areas (Figure C-1).

SLEEPING AREAS

C-3. Each sleeping area has electricity, heating, air conditioning (if possible), cots and showers.

PLANNING AREAS

C-4. Each planning area should be large enough for a LRS team to plan the mission (Figure C-2). It should have a place where the team can post all mission-essential information, a sand table, separate bins for classified and unclassified trash, RFI logs, a time schedule, and an access roster.

COMBINED PLANNING AND SLEEPING AREAS

C-5. Due to any number of constraints, the team must be able to adapt to limited space, and be ready to live in the planning area.

FIELD SITE

C-6. If no fixed site is available, or if the deployment plan dictates, the unit can conduct detailed planning at a *field* site such as an aircraft hanger, a boat(s), or a tent(s) (Figure C-3). Everyone at the planning site may be subject to the same restrictions and limitations--leaders, teams, and supporting personnel. Each team needs--

- Separate field tents for planning and sleeping.
- Electrical source.
- Latrines.
- Passive and active security measures such as wire obstacles or guards.
- Site maintenance support.
- Communications with the LRSC operations section and with the rehearsal area.
- Planning aids.

Appendix C

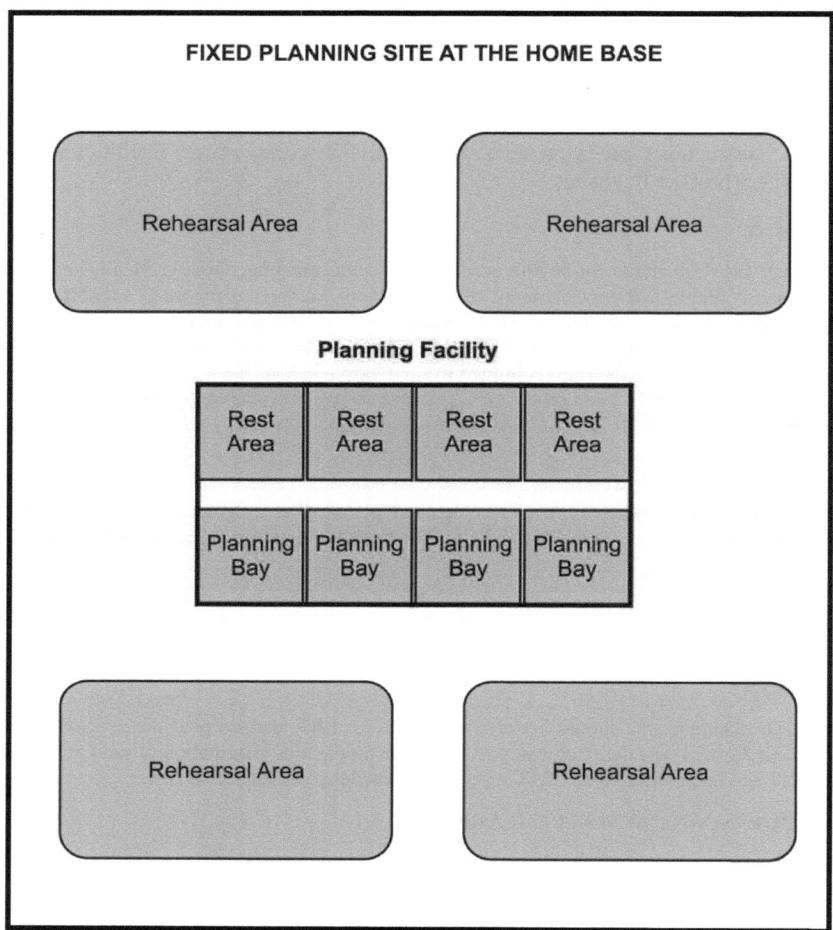

Figure C-1. Example fixed site for planning.

Planning Area Facilities and Sites

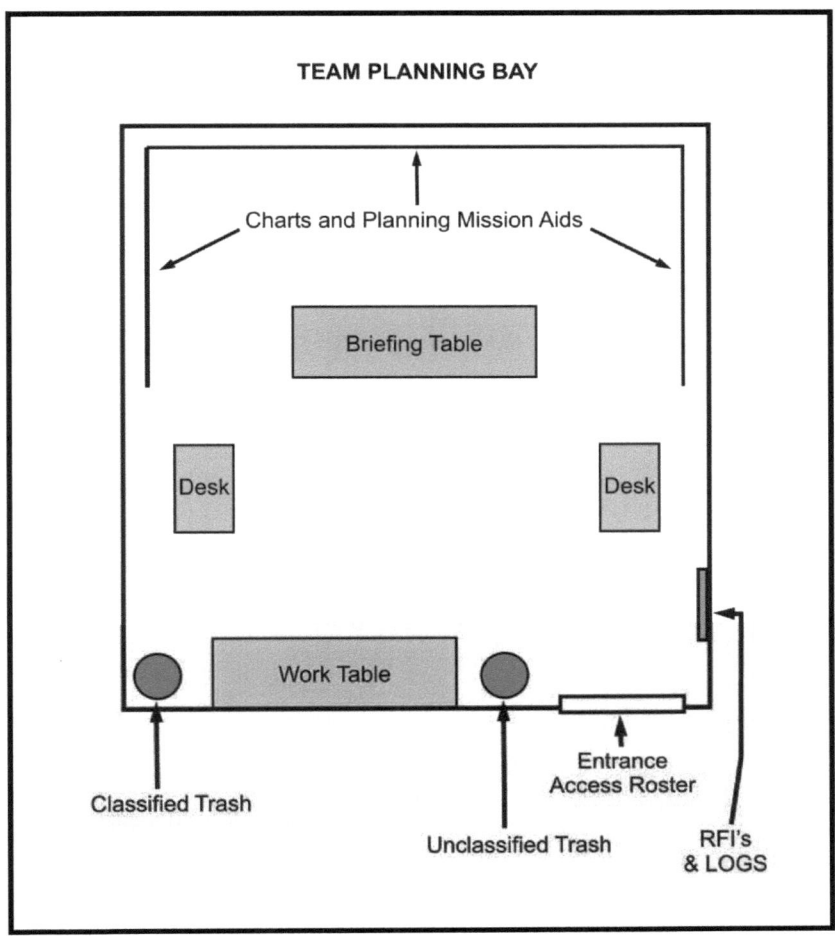

Figure C-2. Example planning area.

Appendix C

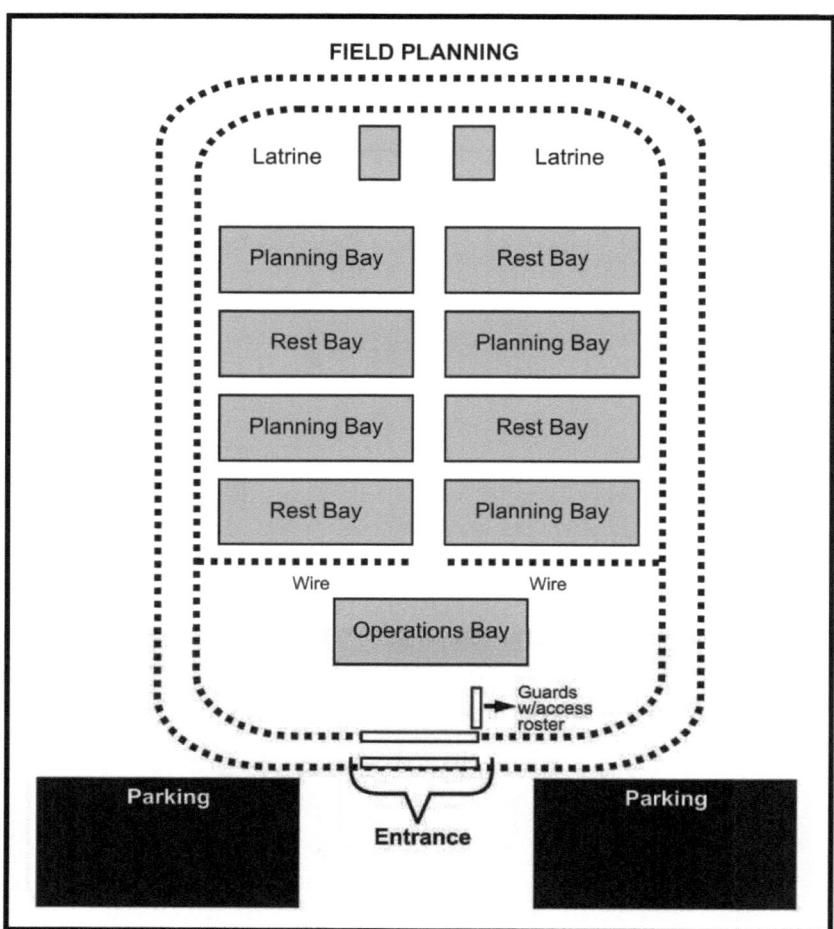

Figure C-3. Use of intermediate staging area for planning.

Appendix D
Geographic Environments

Teams are greatly affected by extreme climates and terrain such as jungles, deserts, mountains, extremely cold areas, and urban areas. Field manuals describe conditions and operational techniques for each. Extreme temperatures, humidity, and elevations also affect the lift capability of transporting aircraft.

JUNGLE OPERATIONS

D-1. Dense jungle restricts ground and air observation as well as electronic surveillance. None of these observation methods work well for collecting information on a dismounted surveillance target. Such a target leaves little evidence that it has passed. LRS teams have the dismounted skills, stealth, and communications needed to collect battlefield information under such conditions. (FM 90-5 provides more information on jungle operations.)

INFILTRATION

D-2. Dismounted, helicopter, and small boat insertion work well in jungle terrain. The limited availability of LZs could require the team to insert by rappelling or FRIES. Careful coordination with adjacent or friendly forward units is necessary for foot or boat movements to prevent fratricide.

EXFILTRATION

D-3. Teams may be recovered by all available means, but communication and coordination are key when operating in a jungle environment. Dismounted exfiltration routes must be coordinated before the teams use them. Linkup operations with friendly forces require careful and deliberate coordination to the lowest element possible (company or platoon). The SPIES is ideally suited for extraction of a team from dense vegetation.

COMMUNICATIONS

D-4. Dense vegetation, high humidity, and frequent rainfall make HF communications difficult. The vegetation affects radio ranges and makes setting up antennas more difficult. Also, radio components experience higher failure rates in wet environments. UHF TACSAT is best when there are holes in the canopy.

DESERT OPERATIONS

D-5. To survive in the desert, LRS teams must approach each task systematically so that it becomes a habit. Weather and terrain are the main enemies in any military operation. However, in the desert, these threats greatly increase (FM 90-3).

OPERATIONAL CONSIDERATIONS

D-6. Leaders must consider the following in planning LRS operations in a desert environment:

Batteries

D-7. Teams must test all batteries with a battery tester. The heat reduces battery life by one-third. Teams must not leave batteries on the ground. The sand and other elements will cause a rapid drain of power.

Water

D-8. Teams can conduct operations for only a few days without a resupply or cache of water. During the 24 hours before insertion, each Soldier must ensure he is fully hydrated. Each Soldier should carry a minimum of 11 quarts of water, which is how much water he must drink each day when moving. This is three 2-quart canteens plus one 5-quart bladder. Even when not moving, a Soldier in a desert environment requires 7 quarts of water a day.

Appendix D

Insertion

D-9. When inserting by helicopter, the team should consider using LZs that may provide a lower dust signature such as a salt marsh or other hard-packed area. The distance from the LZ to the hide or surveillance site must be carefully considered. Water consumption is a prime consideration.

Movement

D-10. Teams average 1 kmph/day and 3 kmph/night dismounted. They should use GPSs, because terrain features may be few or shifting, and maps inaccurate. Walking on rocks and shale can help prevent the enemy from tracking the team. Movement on wet or dark sand is generally quicker. Loose or dune sand leaves clearer tracks and reduces movement speed.

Hide or Surveillance Site

D-11. Teams should consider locating hide and surveillance sites together due to potential extended observation distances. Hide and surveillance sites can be camouflaged with a variety of materials including the diamond desert net, burlap or bed sheets (color matched) pulled tight over a frame.

- Surveil from a point higher than the NAI; afternoon heat (between about 1100 and 1600 hours) obscures optics at ground level due to the mirage effect, among other things. This makes vehicles difficult to identify beyond 2 to 3 kms.

- To identify vehicles at night, consider moving the surveillance site to within 1 to 2 km of the objective or NAI.

- When constructing subsurface hides, dig them in sandy soil. To prevent sides from caving in, shore them up.

MOUNTAIN OPERATIONS

D-12. Irregular mountain topography normally provides good cover and concealment. Observation varies depending on trees and scrub growth. Surveillance sites near ridges and peaks often provide broad areas of observation. Helicopter movement of teams is often limited by altitude capability, erratic wind conditions, and the lack of landing sites. Communications are generally difficult; relay stations might be needed for communication between the teams and base stations (FM 3-97.6).

- During mountain operations, UHF TACSAT is best for primary communications, with HF as secondary (FM 3-97.6).

- Use of mountaineering equipment is a must. Teams should also plan for scaling equipment and other specialized gear.

- Teams must also be prepared to infiltrate by FRIES or SPIES due to rock formations. HALO/HAHO insertion may also be considered if suitable DZs can be located.

COLD WEATHER OPERATIONS

D-13. In extreme cold, teams are hampered by the need to maintain body warmth. In deep snow, the teams must operate on skis or snowshoes; dogsleds or skimobiles might also be required. Long-range weather forecasts are an important planning consideration.

- Deep snow can conceal stationary surveillance sites, but increase the difficulty of orientation and the concealment of moving teams.

- Magnetic storms, aurora effects, and ionosphere disturbances can seriously degrade radio communications.

- Trafficability and load-bearing qualities of ice and snow crust are important planning considerations.

- Survival is difficult in extreme winter conditions. To operate for extended periods at maximum efficiency, the team must establish a warming area. They can use the heat from candles while in a hide or surveillance site. Teams may require the use of extreme cold weather (ECW) sleeping bags and tents. Goggles or dark glasses are required during operations due to the threat of snow blindness.

- Northern summer conditions are characterized by long periods of daylight, numerous water obstacles, and marshy areas. When aircraft or ground operations are restricted, the teams can use boats designed to navigate northern waterways.

Appendix E
Contingency Plans

Operations seldom proceed as planned. LRSU operations are high risk conducted by skilled and experienced Soldiers. Key to successful operations is the anticipation of problems and opportunities that may arise. Visualizing and planning for contingencies allow leaders to gain and maintain flexibility.

BRANCHES AND SEQUELS

E-1. A contingency plan provides for major contingencies that can reasonably be anticipated during the course of an operation in a particular geographic area. Contingency plans normally take the form of branches and sequels.

BRANCHES

E-2. A branch is a contingency plan (an option built into the basic plan) for changing the mission, disposition, orientation, or direction of movement of the force to aid success of the current operation, based on anticipated events, opportunities, or dispositions caused by enemy action. Branches are developed by the LRSC, LRSD and the LRS team. For example: all three organization develop branch contingency plans for loss of communications, the initiation of a LRS team E&R plan and the need to conduct resupply. The LRS team will plan a branch in the case of compromise on the objective. The LRS team may also anticipate a change of mission from surveillance to target acquisition if particular circumstances arise during the conduct of a mission. LRS leaders must anticipate changes and plan accordingly.

SEQUELS

E-3. A sequel is an operation that follows the current operation. They are future operations that anticipate the possible outcomes--success, failure, or stalemate--of the current operation. Sequels are normally not planned for below the LRSC level. The LRSC is normally alerted to the need to plan sequels for employment of LRSDs or LRS teams based on orders from the BFSB or R&S squadron.

CONTINGENCY PLAN MATRIX

E-4. A contingency matrix can help LRS leaders plan, brief, and track contingencies. Leaders can identify events by phase, schedule planning and briefing logically or chronologically, and effectively execute and monitor contingencies. Figure E-1, shows an example completed contingency matrix.

Appendix E

INFILTRATION PHASE				
	TEAM	**CP**	**COB and AOB**	**LNO**
Aircraft down before FLOT	Inform COB and DOB Follow air crew EPA.	Initiate recovery Ready medical Ready reinsert	Monitor EPA or E&R nets Confirm EPA or E&R frequency Confirm signals with LNO, CP	LNO with aviation unit for recovery or reinsertion
Aircraft down cross FLOT	Inform COB and DOB Follow EPA	Initiate recovery Ready medical Ready QRF	Monitor EPA or E&R nets Confirm EPA or E&R frequency Confirm signals with LNO, CP	LNO CSAR: Recovery platform Gunship escort SEAD, FA, TACAIR
Enemy contact	Break contact Consolidate and reorganize Inform COB and DOB	Initiate recovery Ready medical Ready reinsert	Monitor team's recovery Confirm team's frequency Confirm signals with LNO, CP	LNO with aviation unit for recovery or reinsertion
Failure of ANGUS	Troubleshoot Monitor frequency Initiate no-commo plan	Initiate no-commo plan for recovery Ready reinsert	Troubleshoot Monitor no-commo plan for recovery	LNO with aviation unit for recovery
EXECUTION PHASE				
	TEAM	**CP**	**COB and AOB**	**LNO**
Compromise without CX	Break contact Inform COB and DOB BPT CM or extract	Inform higher Ready QRF Initiate recovery BPT Execute CSAR	Monitor CSAR, E&R nets Confirm CSAR and E&R frequency Confirm signals with LNO, CP	BPT LNO CSAR (platforms, escort, fires)
Compromise with CX	Break contact Inform COB and DOB BPT extract	Inform higher Ready QRF Initiate recovery	Monitor CSAR, E&R nets Confirm CSAR and E&R frequency Confirm signals with LNO, CP	LNO CSAR (platforms, escort, fires)

Figure E-1. Example completed contingency matrix.

Contingency Plans

EXFILTRATION PHASE				
	TEAM	**CP**	**COB and AOB**	**LNO**
Enemy contact	Break contact Consolidate and reorganize Inform COB and DOB	Initiate recovery Ready medical Ready reinsert	Monitor team's recovery Confirm team's frequency Confirm signals with LNO, CP	LNO with aviation unit for recovery and reinsertion
Failure of aircraft to arrive	Inform COB and DOB BPT move to alternate PZ BPT initiate EPA	Inform higher Ready new extraction time or location	Monitor team's recovery Confirm team's frequency Confirm signals with LNO, CP	LNO with aviation unit for recovery
RECOVERY PHASE				
	TEAM	**CP**	**COB and AOB**	**LNO**

Figure E-1. Example completed contingency matrix (continued).

Appendix F
Coordination for Army Aviation

This appendix provides an example OPORD (Figure F-1) and annexes for coordinating the following:

- Army aviation fire support (Figure F-2).
- Intelligence (Figure F-3).
- Rehearsal areas (Figure F-4).
- Vehicular movement (Figure F-5).

Appendix F

```
OPERATION ORDER

1. Situation.
    a. Enemy Situation. Location, activity, course of action of
enemy air defense assets that could influence operation.
    b. Friendly Situation. Location, activity, course of action of
friendly assets that could influence operation.
    c. Weather.
        (1) Weather forecast, 72-hour period. This covers the day
prior to, the day of, and the day after insertion.
        (2) Weather decision time.
        (3) Point of contact for weather issues.
        (4) Course of action for weather delay.
        (5) Course of action for cancellation of insertion due to
weather.

2. Mission. Complete mission statement pertaining to the insertion
phase only. (Discussion of ground tactical plan is unauthorized.)

3. Execution.
    a. Concept of the operation. Brief overview of what the team
wants to accomplish during the helicopter insertion.
        (1) Coordinating instructions.
            (a) Type or number of aircraft to use.
            (b) Type and number of escort aircraft available.
            (c) Type and number of escort aircraft to use.
            (d) Locations or times when escorts will be available.
            (e) Types and locations of organic weaponry and special
equipment on each aircraft.
            (f) Locations of primary and alternate PZs.
            (g) Markings that indicate primary and alternate PZs.
            (h) Desired landing direction.
            (i) Aircraft's ETA at PZ.
            (j) Load plan or procedures to follow.
            (k) Number of personnel.
            (l) Approximate total weight of cargo (equipment and
personnel)--varies by team.
            (m) Desired aircraft configuration:
                - Seats in or out of aircraft.
                - Doors open or closed.
                - Special insertion equipment.
            (n) Locations and types of aircraft weaponry and special
equipment organic to the aircraft.
            (o) Locations of key leaders during flight--team leader,
RATELO, ATL.
            (p) Flight routes to RP based on overlay and map.
            (q) Flight routes to primary and alternate LZs based on
overlay and map.
            (r) Identification and location of prominent air
reference points.
```

Figure F-1. Example OPORD.

```
            (s) False insertions.
                - Type (touch down or hover).
                - Time (before or after insertion of team).
                - Location and order of false insertions to conduct.
            (t) Deception planning.
                - Use of SEAD.
                - Altitude of flight.
                - Additional deception for flight crew to use.
            (u) Action in the event that the aircraft crash lands.
                - Before the decision point.
                - After the decision point.
        (2) Landing zone operations.
            (a) Location of primary and alternate LZs based on
overlays and map.
            (b) Time warnings to use.
            (c) Desired landing direction.
            (d) Off-load plan.
            (e) Criteria for aborting.
            (f) Loiter area location and time.
            (g) Actions on contact.
                - During flight.
                - During landing.
                - After takeoff.
                - Emergency pickup location.
                - Additional contingencies.
        (3) Extraction.
            (a) Locations of primary and alternate PZs.
            (b) Location of the no-communications PZ.
            (c) Location of emergency PZ.
            (d) Date-time group of extraction.
            (e) Location of communications check point.
            (f) Method of marking PZ.
                - Day or night.
                - Near or far.
                - With or without communications.
            (g) Landing heading.
            (h) Load plan.
            (i) No-communications plan.
            (j) Special extraction equipment needed.
            (k) Actions on enemy contact during extraction.
            (l) Additional contingencies.

4. Service and Support.
    a. Equipment carried by air crew.
    b. Priority of destruction for aircraft.
    c. Bump plan.

5. Command and Signal.
    a. Ground tactical commander who assumes command if aircraft
crashes.
    b. Frequencies and call signs.
    c. Code words.
```

Figure F-1. Example OPORD (continued).

```
Coordination for Fire Support

1.  Identification of unit.
2.  Time of departure or return.
3.  Fire-support overlay.
4.  Target list.
5.  Type of fire support available.
6.  Ammunition available.
7.  Priority of fires.
8.  Method of requesting fires.
9.  Fire-control measures (RFA or NFA).
10. Communications.
    a. Call signs.
    b. Frequencies.
    c. Emergency signals or markings.
    d. Method of correction.
    e. Time to process request.
11. Special instructions.
```

Figure F-2. Example fire support annex.

```
Coordination for Intelligence

1. Enemy forces.
    a. Unit identification.
    b. Strength.
    c. Composition.
    d. Disposition.
    e. Probable course of action.
    f. Equipment assets.
    g. Most recent activity.
    h. Known or suspected location.
    i. Anticipated activities in next 24 hours.
    j. Photographs available.

2. Friendly forces.
    a. Units operating in area.
    b. Locations if known.
    c. Host nation.
        (1) Any changes to area study information.
        (2) Population support.
        (3) Attitude toward US forces.
        (4) Times civilian population is active.
        (5) Curfews or checkpoints, and so on.
            (a) Times.
            (b) Locations.
    d. Imagery.
        (1) Line-of-sight surveys available.
        (2) Imagery of target area.
        (3) Imagery of infiltration or exfiltration locations.
        (4) Availability of updated maps and other
reference material.
        (5) Video assets available.
```

Figure F-3. Example intelligence annex.

```
Coordination for Rehearsal Area

1.  Identity of the unit that must rehearse.

2.  Terrain in rehearsal area must resemble terrain in the mission
area.

3.  Availability of area security.

4.  Availability of aggressors.

5.  Availability of blanks, pyrotechniques, and live ammunition.

6.  Availability of mock-ups.

7.  Time the area is available.

8.  Boundary for training area.

9.  Transportation to area.

10. Coordination with other units using rehearsal area.
```

Figure F-4. Example rehearsal area annex.

```
Coordination for Vehicular Movement

1.  Identification of supporting unit.

2.  Number and type of vehicles and tactical preparation.

3.  Entrucking point.

4.  Load time.

5.  Departure time.

6.  Load plan.

7.  Preparation of vehicle for movement.
    a. Driver responsibilities (equipment, weapon, and so on).
    b. Team responsibilities.

8.  Time vehicle is available for inspection or rehearsals or
preparations.

9.  Routes.
    a. Primary and alternate routes (use overlays and map).
    b. Checkpoints.
    c. Primary and alternate detrucking points.

10. Actions on enemy contact.
    a. Ambush.
    b. Checkpoints/roadblocks.

11. Actions on breakdown.
    a. Before decision point.
    b. After decision point.

12. Call signs or frequencies.

13. Special instructions.
```

Figure F-5. Example vehicular movement coordination annex.

Appendix G
Hide and Surveillance Sites

During surveillance, which is the LRS team's primary mission, the team leader reconnoiters, then selects positions for the surveillance and hide sites. Where to construct the positions depends on his METT-TC analysis conducted during the planning phase and his continued analysis once in the vicinity of the objective. The two sites communicate by wire, VHF, or messenger.

Selected team members observe or surveil the objective from the surveillance site. Some members rotate between the hide and surveillance sites. Others run the team's HF or TACSAT directly from the hide site or from a separate location chosen specifically for conducting communications.

SURFACE SITES

G-1. The enemy situation may prohibit moving to a subsurface site, so camouflage must be done correctly during occupation of both sites and improved when circumstances allow (Figure G-1).

Figure G-1. Two-man surface site using ghillie suits.

ADVANTAGES OF SURFACE SITES

- Simple construction.
- Few materials.
- Quick setup.
- Little soil removal if any.
- Optical standoff.
- Quick escape.

DISADVANTAGES OF SURFACE SITES

- Little protection from small-arms fire.
- No protection from indirect fires or CBRN hazards.
- Risk of compromise by dogs, civilians, and enemy patrols.

Appendix G

CONSTRUCTION MATERIALS FOR SURFACE SITES

- Natural vegetation.
- Ghille suits.
- Ground blinds.
- Poncho(s), waterproof.
- Yeti or camouflage net to prevent reflection.
- One 550-pound cord or bungee cords.
- Chicken wire (optional).
- Burlap or canvas cloth (optional).

CONSIDERATIONS FOR SURFACE SITES

G-2. Team members--

- Avoid cutting vegetation, use man-made or natural camouflage.
- Keep equipment packed when not in use.
- Remain in uniform and keep on load-carrying equipment.
- Maintain security around the clock.
- Construct small, easy to conceal two-man site or,
- Construct three-man site for longer stays (one rests while others surveil).
- Rotate surveillance teams just after dark and just before daylight.
- Set up communications between the hide and surveillance sites.
- Take their rucksacks to the surveillance site.
- Remain in hide site during day when conditions only allow limited visibility surveillance.
- Use nets or natural camouflage to construct all-round concealment for the surveillance site. Ensure that site is hidden from every angle, including overhead.
- Determine the location of the hide, surveillance, and communication (if used) sites, based on METT-TC, but especially based on terrain.
- Try to change direction when moving from the hide site to the surveillance site. For example, move in a dogleg or fishhook, or take an indirect route.
- Never wear a ghillie suit during movement, because pieces can rip off in vegetation and leave a trail. Instead, put the suit on just before occupying the surveillance site.

HASTY SUBSURFACE SITES

G-3. The team constructs a hasty subsurface site when they have too little time to construct a complete subsurface site. A hasty subsurface site is especially useful in the absence of natural cover and concealment. They plan the site so they can improve it to a full subsurface site as time and the situation allow (Figure G-2 and Figure G-3).

Hide and Surveillance Sites

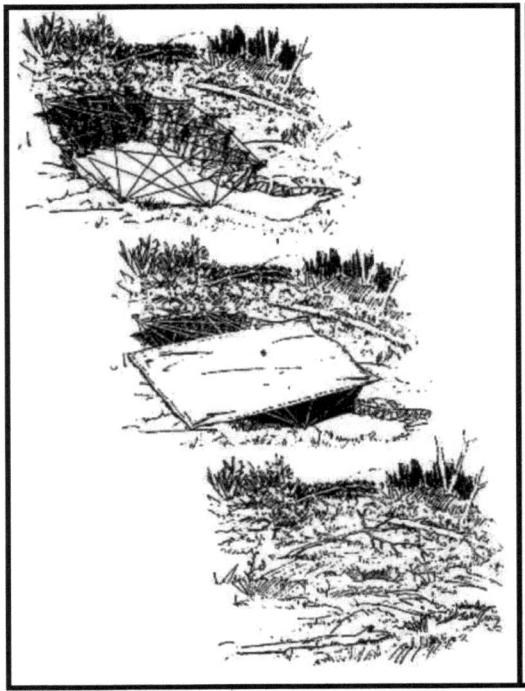

Figure G-2. Suspension line-weave site.

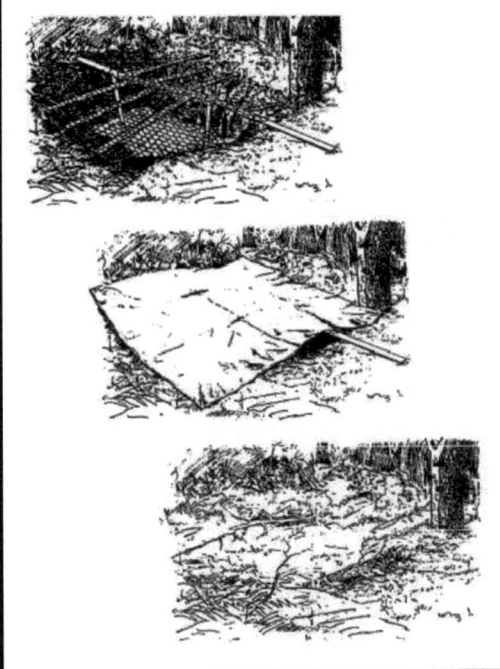

Figure G-3. Polyvinyl chloride site.

ADVANTAGES COMPARED TO SURFACE SITE

- Lower profile.
- Better protection against small-arms and indirect fires.
- Excellent camouflage.

DISADVANTAGES COMPARED TO SURFACE SITE

- Requires more construction tools.
- Challenge of concealing removed soil.
- Greater construction time.
- More construction noise.

CONSTRUCTION MATERIALS NEEDED

- Ponchos or other waterproofing material.
- Yeti or small camouflage net.
- Entrenching tool or D-handled shovels.
- One 550-pound or bungee cord.
- Sandbags.
- Polyvinyl chloride pipe with connectors.
- Fiberglass rod.
- Aluminum conduit.

Appendix G

- Plywood.
- Chicken wire (optional).
- Burlap or canvas (optional).
- Small saplings, stripped and lashed together in place of pipe or fiberglass rods.

FINISHED SUBSURFACE SITE

G-4. The team generally uses finished subsurface sites for stay-behind missions and when they anticipate having to remain underground for extended periods of time.

LOCATION

G-5. Dig the site in a well-concealed area, away from enemy observation, and well away from any populated areas.

CONSTRUCTION

G-6. Use any available containers, such as rucksacks, sandbags, or socks, to remove the dirt. Some of the removed soil will later be placed on top of the site. Leftover dirt must also be camouflaged.

Overhead Cover

G-7. Overhead cover is constructed strong enough that it can be walked on.

Ingress and Egress

G-8. Construct two sets of entrances and exits, primary and secondary (emergency), and cover and conceal each.

Size of Site

G-9. The site must accommodate the whole team. It must allow freedom of movement and have room for separate and comfortable sleeping positions.

Materials

G-10. Materials needed depends on the design.

- Fifty 2 x 4 x 12's (2 inches by 4 inches by 12 feet boards).
- Six 4 x 4 x 6's (4 inches by 4 inches by 6 feet boards).
- Sufficient gravel to cover the floor.
- Eighteen inches of cover over entire site.
- Backhoe or Soldiers with shovels.
- Sandbags, 100 each.
- One large general-purpose tent to cover construction until complete.

ADVANTAGES OF SUBSURFACE SITES

- Little risk of compromise.
- Protection from artillery and small-arms weapons fire.
- Protection from nuclear attack.
- Excellent camouflage.

DISADVANTAGES OF SUBSURFACE SITES

- Concealment of leftover soil away from the site.
- Construction noise.
- Construction resources (time, manpower, materials, and equipment).

USE OF EXISTING SITES

G-11. Look for and use depressions or predug holes, such as former fighting positions, and improve them as the situation allows. During heavy rains, do not use streams and waterways as they could flood.

CONCEALMENT

G-12. To conceal the site, use yeti nets, man-made and natural camouflage, or chicken wire.

COVER

G-13. Build a barricade to provide shelter.

STOCKAGE

G-14. Stock it with rations, water, ammunitions, batteries, and so on, and arrange equipment, such as rucksacks and communications equipment, so it can be grabbed in an emergency.

STANDING OPERATING PROCEDURE

G-15. Have an SOP for exiting the site. If the team conducts surveillance from the site, then how the team leaves the site depends on the location of the enemy objective (Figure G-4). The team should also prepare a deception plan to cover exiting by the secondary (emergency) exit, in case the enemy finds the primary entrance.

Appendix G

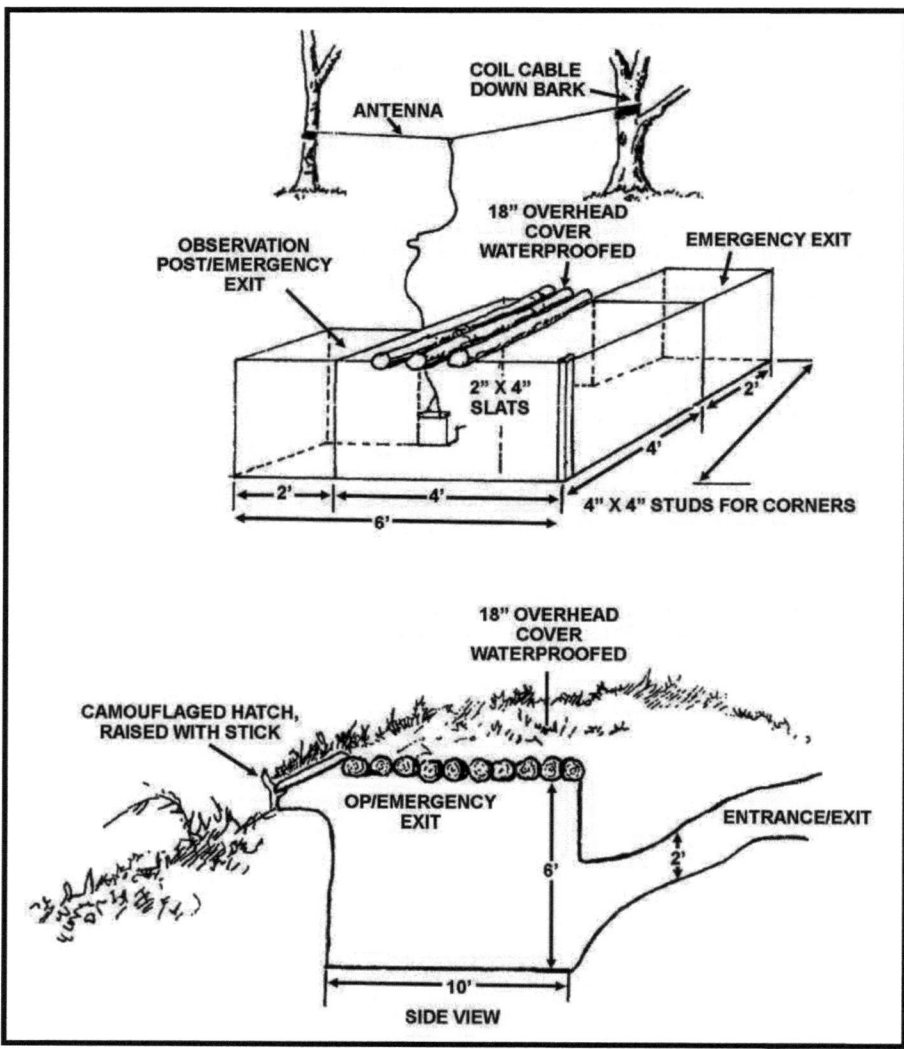

Figure G-4. Example subsurface site.

WASTE MANAGEMENT

G-16. Use a portable camping toilet or line an MRE box with a trash bag. Cover waste odor with lime (best) or baking soda. Remove the waste in zipper storage bags, leftover MRE bags, or any other type of sealable container.

PERSONAL EQUIPMENT

G-17. Disassemble shovels and carry them in rucksacks, and leave on load-carrying equipment.

SITE SELECTION

G-18. When selecting a site, the leader should--

- Consider LOS to target.
- Ensure the site is in range of available observation equipment so that it can meet the reporting requirements.
- Ensure the site has adequate overhead cover and concealment.
- Ensure the site is located away from natural lines of drift, roads, trails, railroad tracks, and major waterways.
- Ensure the site is defendable for at least a short time.
- Ensure the site has primary and secondary (emergency or alternate) hasty exits.
- Ensure the site has a concealed and serviceable entrance, and that Soldiers make little noise getting in and out of the hide site.
- Ensure the site works within the factors of METT-TC relative to other site positions (hide, surveillance, and communication).
- Ensure the site is located well away from any man-made objects.
- Ensure the site is located downwind of inhabited areas.
- Ensure the site uses, but is not dominated by, high ground.

LEADER RECONNAISSANCE

G-19. The team leader selects tentative sites during the planning phase. He physically reconnoiters (in a stay-behind); observes from aircraft; studies photographs, line-of-site data, soil and drainage data, or conducts a map reconnaissance. At a minimum, he selects primary and alternate hide and surveillance sites (thus, four sites in all). Before the team occupies the sites, the team leader physically reconnoiters the tentative sites chosen during planning. If necessary, he moves the site to a better location.

OCCUPATION OF HIDE SITE

G-20. The leader can occupy the hide site by any of several methods:

FISHHOOK OR DOG-LEG METHOD

G-21. The team occupies the hide site off the direction of march (Figure G-5).

Appendix G

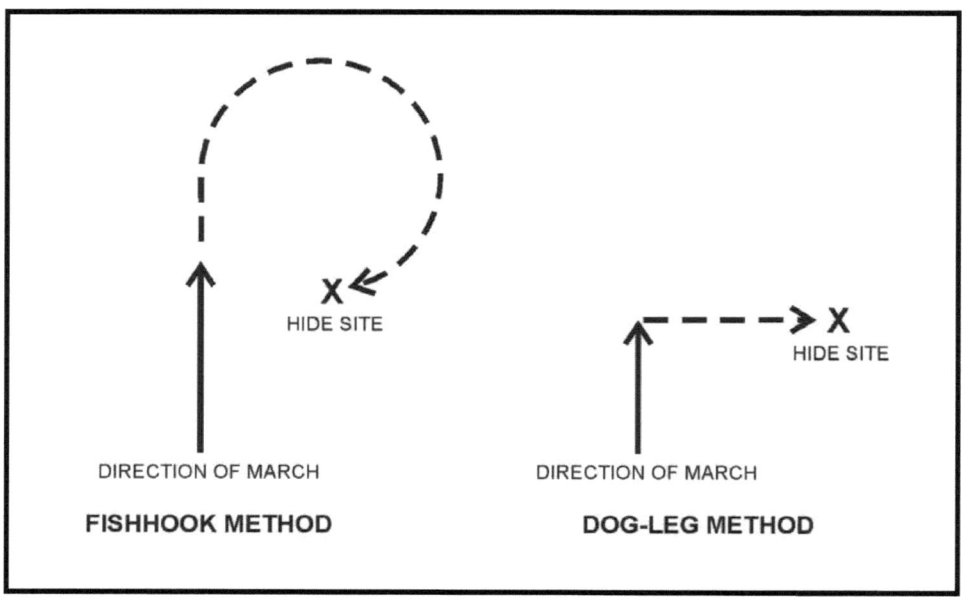

Figure G-5. Fishhook and dog-leg methods.

OCCUPATION BY FORCE

G-22. The team only occupies a hide site by force if it must, such as when time is a major limiting factor. In such a case, the team leader reconnoiters and the team moves directly into the tentative site (Figure G-6).

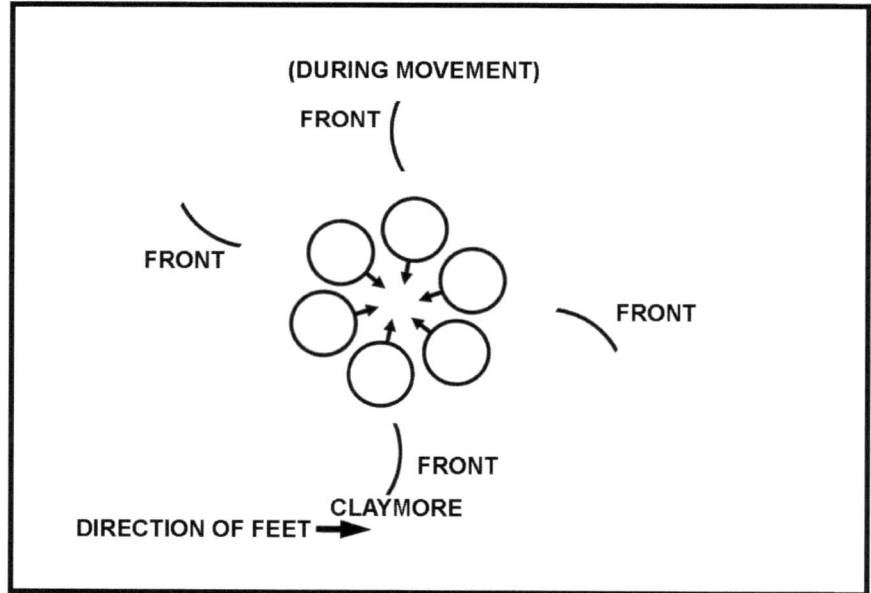

Figure G-6. Forcible occupation of site.

ACTIONS IN HIDE SITE

G-23. The team maintains security at all times. Soldiers are positioned either back-to-back or feet-to-feet, using all-round security.

- The team waits 15 minutes before moving or unpacking equipment--a listening halt. They do not lean against small trees or vegetation. They place Claymores at least in the four cardinal directions.
- If communication is to be conducted from the hide site, they construct the antenna before daylight. They only raise it off the ground once they are ready to establish communication.
- Wear their load-carrying equipment at all times and camouflage all around the position.
- The best time to rotate teams is at dusk and dawn. The surveillance team takes their rucksacks or assault packs. The team rests during the day.

PRIORITY OF WORK

G-24. Except for security, work priorities vary based on METT-TC. The team has security, alert, guard and rest, evacuation, rendezvous, scheduling, maintenance, hygiene, and meal plans.

- Conduct stand-to starting before first light and continue until after full light. Conduct stand-to again starting before dark and continue until after dark. Vary start times to avoid patterns. Conduct stand-to at anytime a heightened level of enemy activity is expected.
- Maintain equipment, radios, weapons, and camouflage.
- Perform personal hygiene and preventive medicine. Conduct isometric exercises.

SITE STERILIZATION

G-25. Before departing hide and surveillance locations, team members must ensure sites and routes have been sterilized.

- Carry out all foreign debris.
  If possible, avoid burying waste and trash. Animals will uncover it and expose it to enemy patrols. If it must be buried, enclose it in sealed containers or cover the scent with CS or lime, then bury it at least 18 inches deep.
- Sterilize the sites with displaced earth. Bury the overhead materials in the site itself.
- Camouflages the area by blending the site with local surroundings.
- As team members withdraw from the site, ensure routes are camouflaged to prevent detection.

Appendix H
Battle Drills

A battle drill is a collective action executed by a platoon or smaller element without the application of a deliberate decision making process. Well-rehearsed battle drills are critical to the success of a LRS team. LRS teams are lightly armed, with limited amounts of ammunition, and normally have no immediately available fire support. Teams only have resources for basic life-saving first aid in the event of casualties. A LRS team might only get one chance to defeat or disengage from an enemy force. Therefore, it is critical all team members respond instantly and instinctively when in contact. The team must rehearse battle drills thoroughly before actual enemy contact is made. LRS team battle drills are a supplement to ARTEP 7-1-DRILL.

BREAK CONTACT

H-1. A team should break contact as soon it can, since it lacks the capability to stay and fight. The team fires and maneuvers in two- or three-Soldier groups. Team members can use fragmentation or smoke grenades to cover their withdrawal, continuing until they successfully break contact. Doing so could require repeated bounds. After breaking contact, the team consolidates at a rendezvous or rally point and reorganizes.

PRINCIPLES

H-2. General principles for breaking contact include--

- Maintaining a high initial volume of fire to kill or suppress the enemy.
- Using smoke to screen movement. Hexachloroethane (HC)smoke is best, but white phosphorous (if available) kills, wounds, and screens.
- Dually priming Claymores (one command-detonated fuse or one timed fuse from 45 seconds to 2 minutes) and preparing them for immediate deployment. This can be integrated into a rucksack destruction plan.
- Rehearsing "man down" (wounded) actions.
- Rehearsing "RTO down" vital communication retrieval actions.

H-3. Following the successful execution of the break contact battle drill, the team leader uses METT-TC to determine the next action of the team.

INITIAL CONTACT

H-4. When the team makes initial contact, all team members seek cover and concealment. Each Soldier makes every attempt to avoid masking the fires of other team members. The lead element (first, second, and third Soldiers), deploy and take cover within a few steps of their original locations, and lay down a base of fire. The trail element (fourth, fifth, and sixth Soldiers) deploy at the assistant team leader's command. He calls left or right, depending on the direction of enemy contact and the location of the lead element. The team deploys roughly on line, and starts firing. This lead element then bounds back (Figure H-1 and Figure H-2).

Appendix H

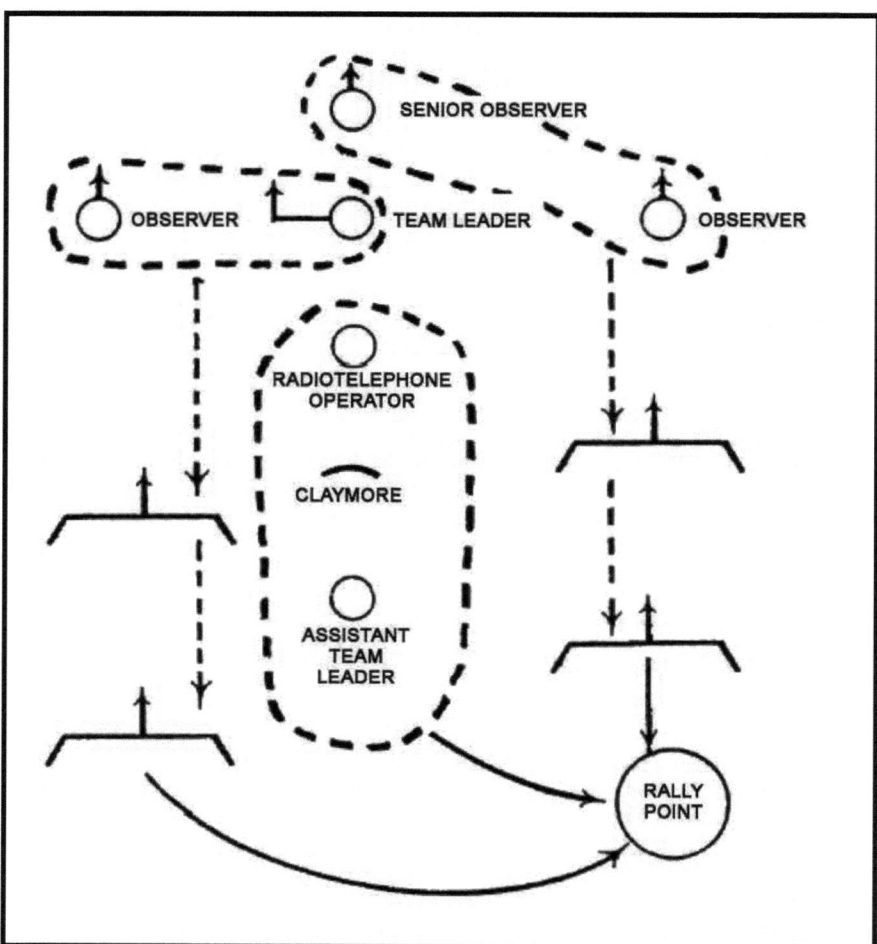

Figure H-1. Break contact front (diamond or file).

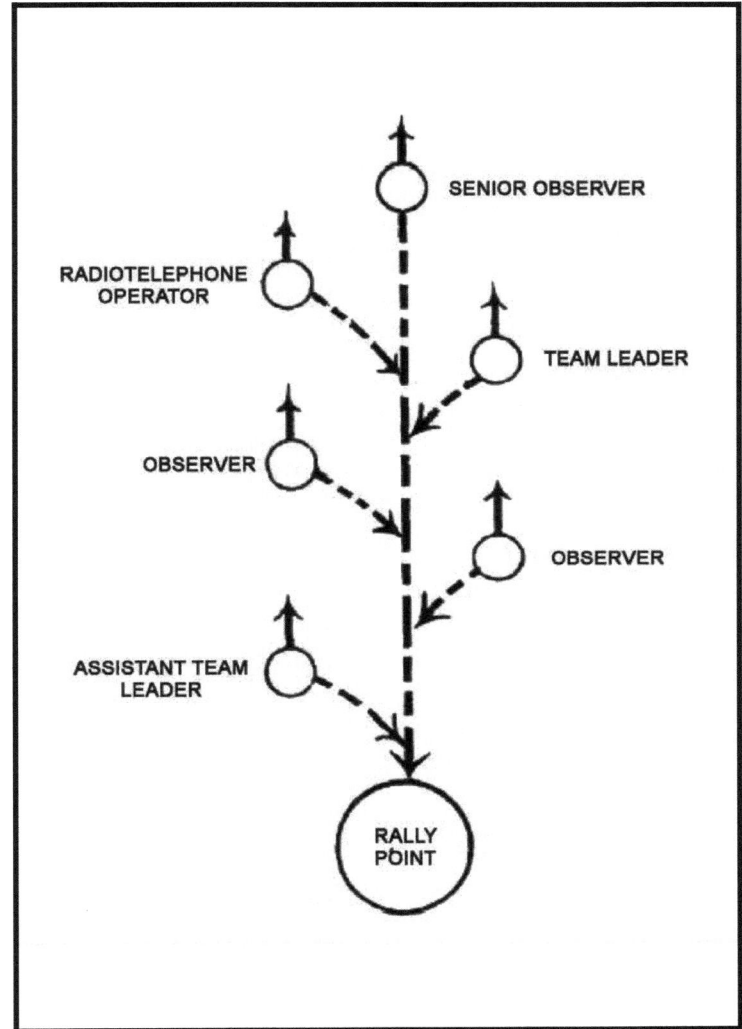

Figure H-2. Break contact front, left and right (Australian peel).

FIRST BOUND

H-5. The lead element (first, second, and third Soldiers) bound back first, followed by the trail element. The elements can alternate bounds to keep the enemy guessing.

SMOKE

H-6. The elements throw smoke on all bounds.

RELOADING OF MAGAZINES

H-7. Everyone changes their magazines on the move.

Appendix H

MALFUNCTIONS

H-8. If a team member's weapon malfunctions, he immediately moves and conducts corrective action on the bound.

SAFETY

H-9. All team members keep weapons on "Safe" during bounds.

FORMATION

H-10. Team members bound on line so as to not cross into lines of fire.

ASSEMBLY

H-11. Once the team has broken contact, they reassemble on line and move out of the contact area.

CONTACT LEFT OR RIGHT

H-12. The team turns toward the contact, takes a knee, and returns fire. One element bounds back while the other team suppresses the enemy. The second team bounds back and continues until contact is broken. Both teams assemble on line and move out of the area (Figure H-3).

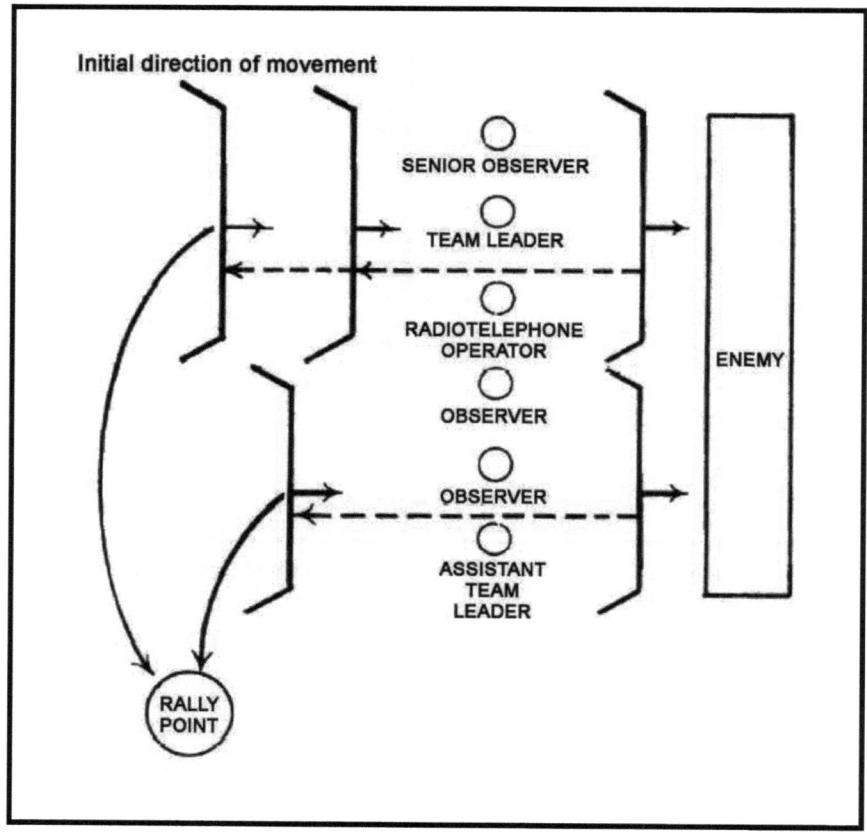

Figure H-3. Break contact left, right (diamond or file).

REACT TO AIR ATTACK

H-13. The first Soldier who hears or sees an aircraft gives the "freeze" signal. The first Soldier who sees an attacking aircraft alerts, "Aircraft, front (left, right, or rear)." The team moves quickly into a line formation, well spread out, perpendicular to the aircraft's direction of flight. As each Soldier comes on line, he goes prone, using available cover. Between attacks, the team should seek better cover and concealment. If the team leader wants the team to move out of the area, he gives the clock direction and distance (Figure H-4).

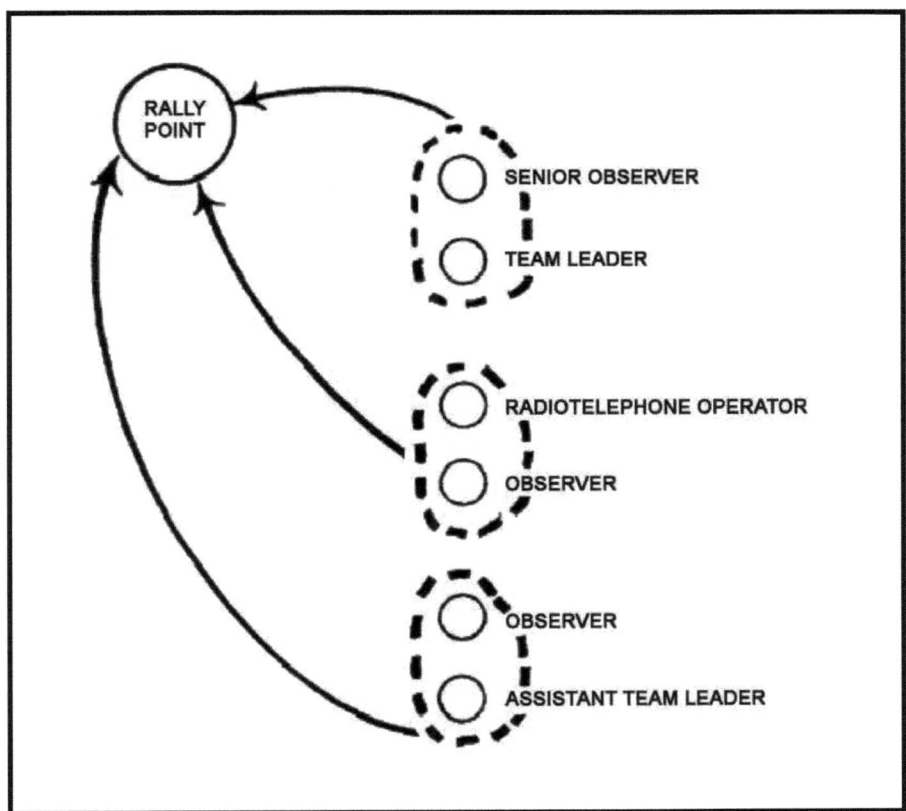

Figure H-4. React to enemy air attack.

H-14. After the team consolidates and reorganizes, it moves to the last rally point. The team should engage the aircraft only as a last resort. Using the head-on method, they mass fires to engage attacking aircraft. They engage slow-moving aircraft at 50 meters and fast-moving aircraft at 200 meters. If the team receives fire, the team leader decides whether to continue the mission, move out of the area, or return fire on the aircraft.

H-15. Another technique is to disperse into two groups of three Soldiers each or three groups of two Soldiers each. On sight of the aircraft, the team leader designates a rally point and gives the command to disperse. On linkup, the team leader assesses the situation and either calls for extraction or continues the mission.

Appendix H

REACT TO INDIRECT FIRE

H-16.   On receiving indirect fire, the team deploys and takes cover. If more rounds impact, the team leader gives the clock position and the direction and distance to move. The team consolidates while moving or at a distance given by team leader. Once the team consolidates and reorganizes, it moves out quickly. The enemy might adjust fires as the team moves. The team should remain oriented on the 12 o'clock position. They may elect to move to the last rally point or as directed by the team leader. He must also decide whether to continue the mission or move out of the AO (Figure H-5).

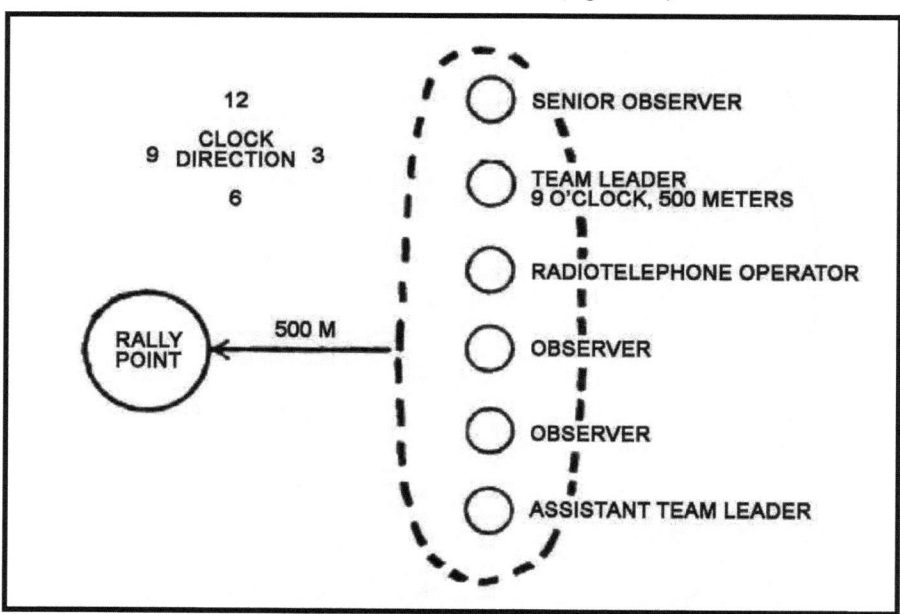

Figure H-5. React to indirect fire or air attack.

REACT TO FLARES

H-17.   If the team encounters flares, it should execute the following actions:

GROUND FLARES

H-18.   The team moves out of the illuminated area and takes cover. Each Soldier closes his firing eye to protect his night vision. The team leader decides the next direction to move.

OVERHEAD FLARE WITH WARNING

H-19.   The team assumes a prone position--behind concealment, when available--before the flare bursts. Each Soldier closes his firing eye to protect his night vision.

OVERHEAD FLARE WITHOUT WARNING

H-20.   The team gets into a prone position, making the most use of nearby cover, concealment, and shadows until the flare burns out. Each Soldier closes his firing eye to protect his night vision. The team leader gives the direction of movement.

BREAK FROM HIDE OR SURVEILLANCE SITE

H-21.   When a hide or surveillance site is compromised, the element involved might have to execute a breakout drill. The drill must be well rehearsed to ensure speed, a base of fire, screening, and survivability.

The type of breakout drill used depends on the type of site and the number of personnel in the site. Proper rehearsals determine the best methods of site construction and the materials needed for a breakout. After constructing the site, the team employs M18 Claymores and smoke to assist execution of the drill.

- Emplace smoke grenades in the ground around the site so they can be activated with a pull cord instead of throwing them after clearing overhead cover.
- Rig Claymores in tandem for one-step activation.
- Keep equipment packed at all times. Prepare an assault pack with all of the site's mission-essential equipment.

H-22. Upon site completion, the leader visually recons the immediate vicinity and designates covered and concealed positions for each team member. The leader designates responsibilities and order of movement from the time claymores are detonated. This helps eliminate confusion and increase battle drill speed and effectiveness.

H-23. When compromise is inevitable, the element notifies other team members not in the position of the situation. Claymores are rigged to all detonate at one time. Smoke grenades are thrown and each Soldier moves to his designated covered and concealed position, and then he lays down a base of fire to cover bounding team members. After the breakout, the element links up with the rest of the team or continues the mission (Figures H-6A through H-6D, this page through B-11).

Appendix H

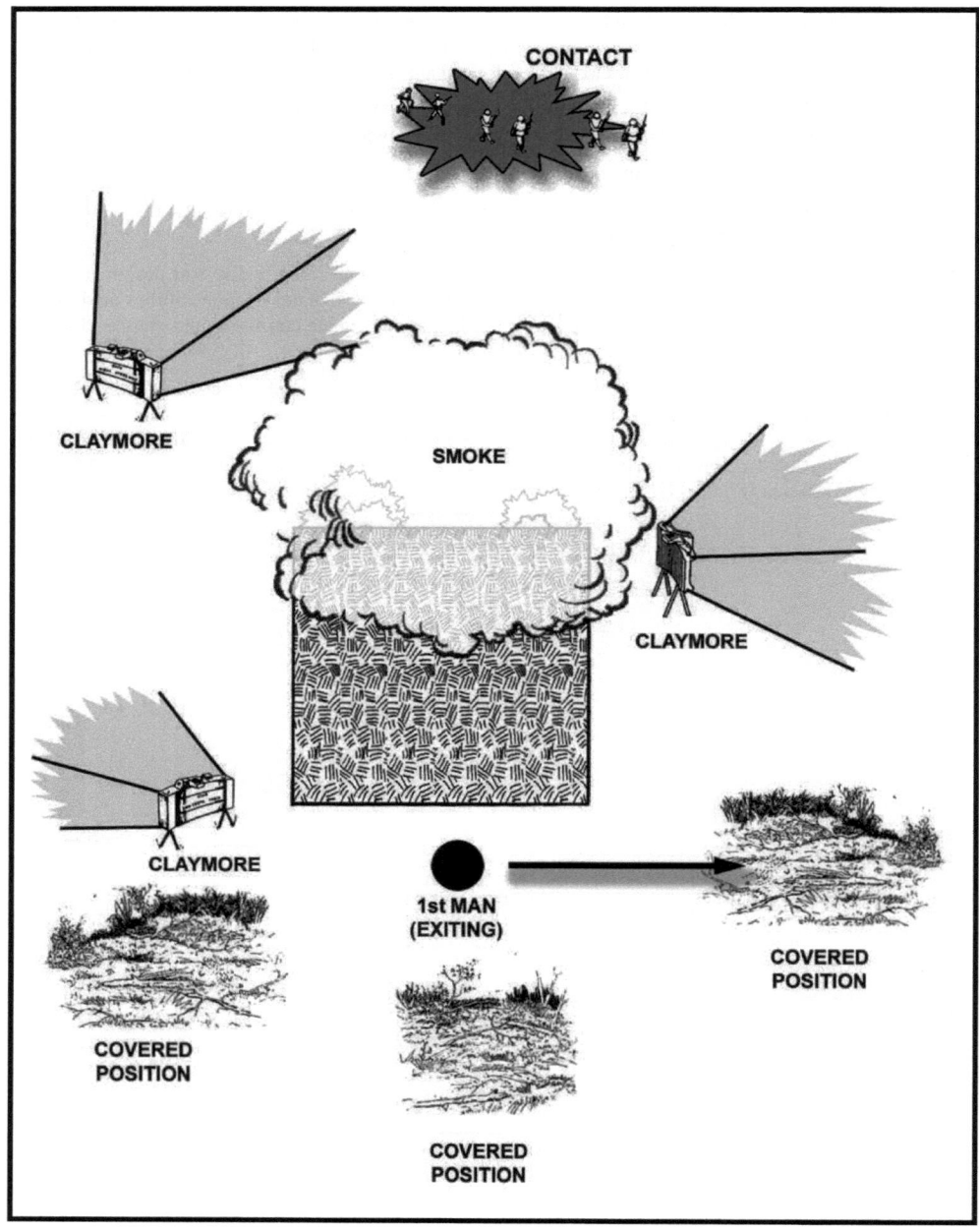

Figure H-6A. Break contact from hide or surveillance site.

Battle Drills

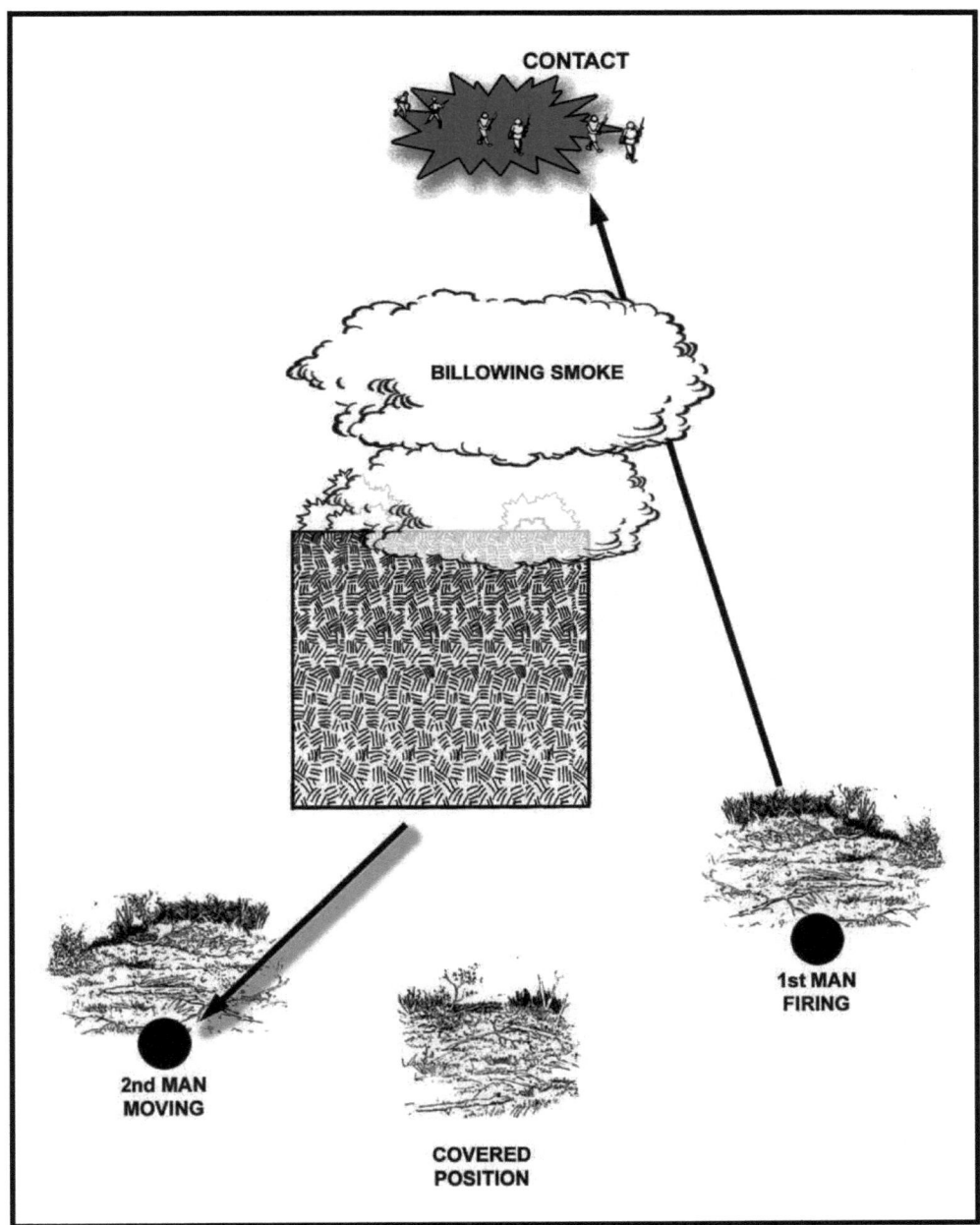

Figure H-6B. Break contact from hide or surveillance site (continued).

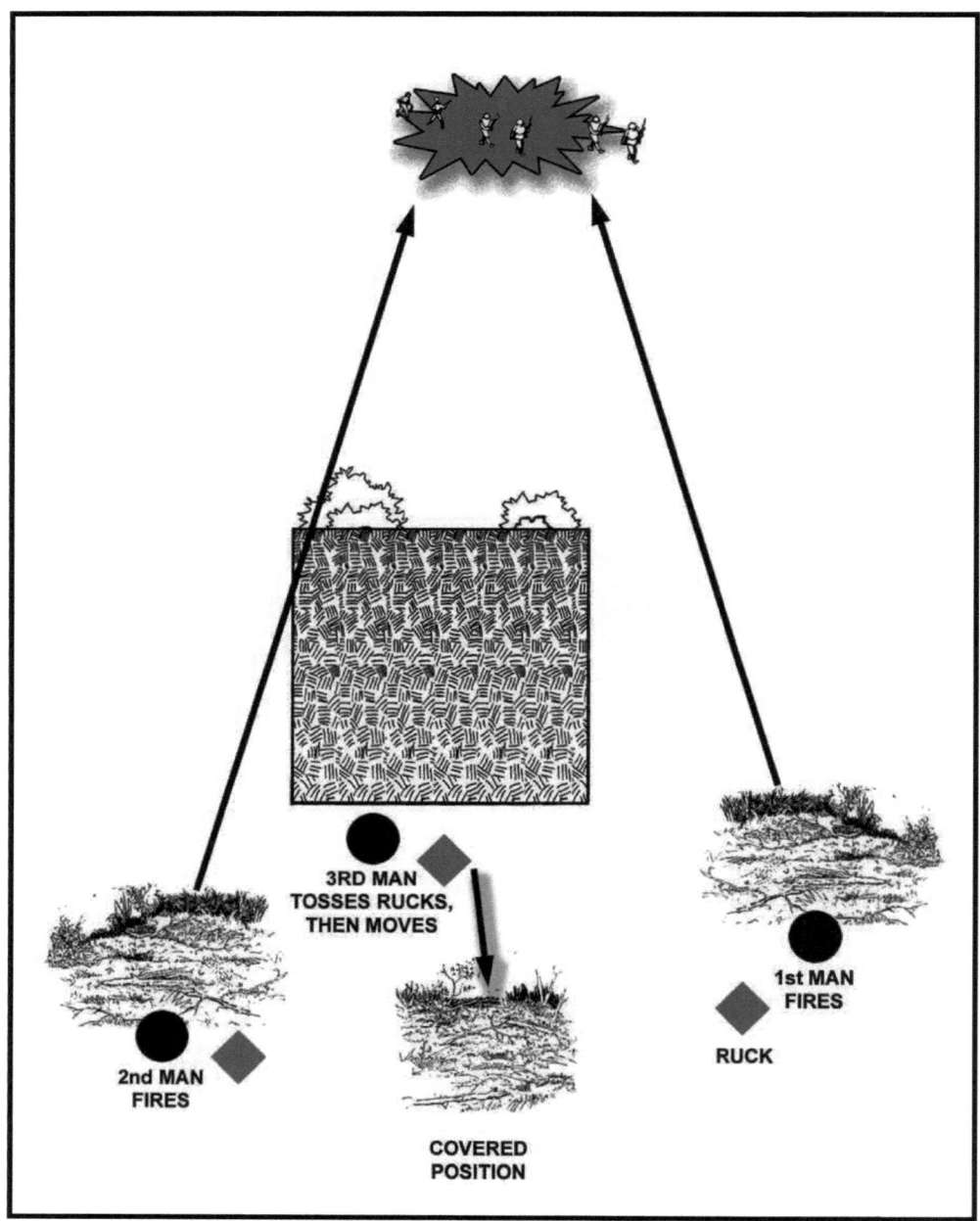

Figure H-6C. Break contact from hide or surveillance site (continued).

Battle Drills

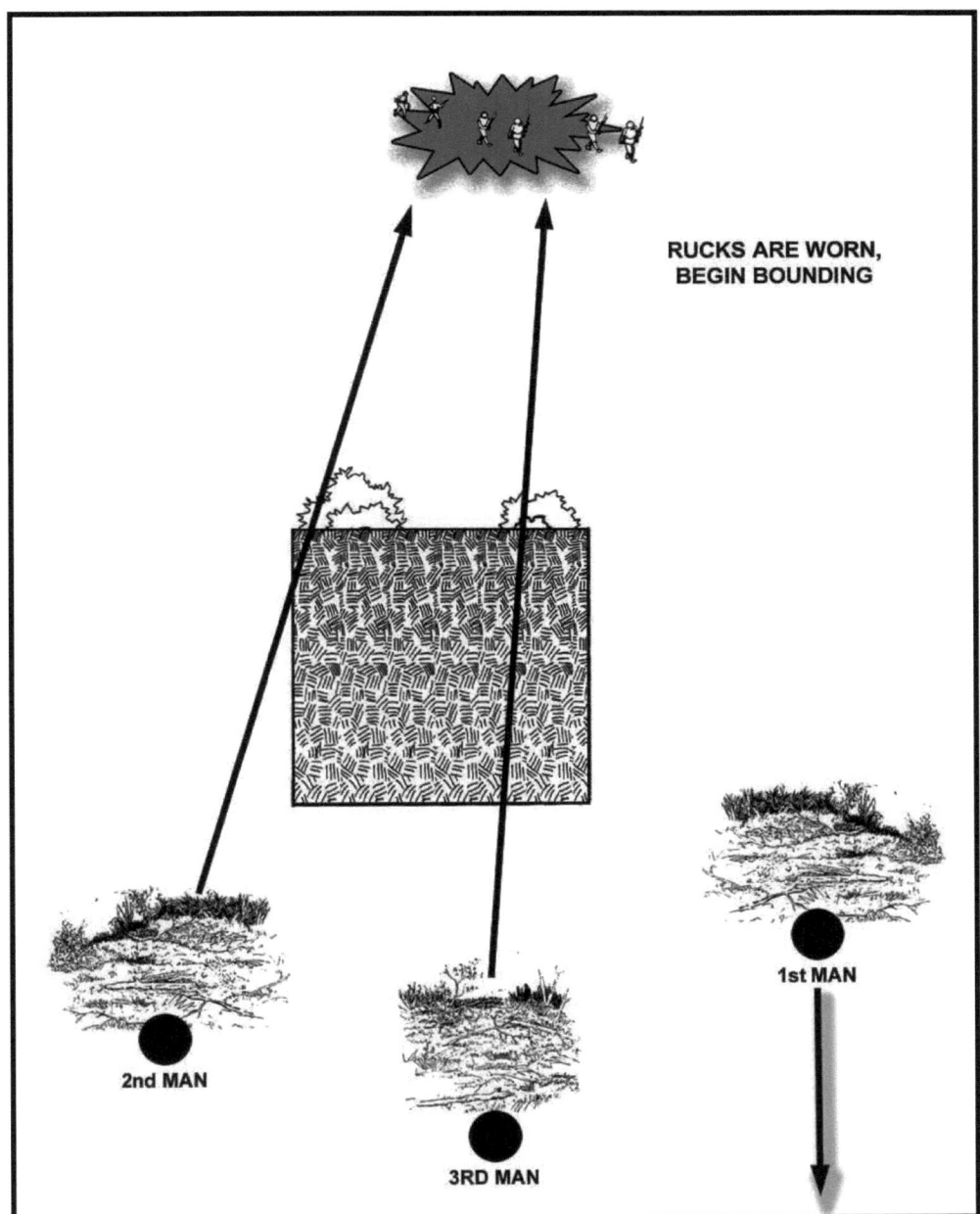

Figure H-6D. Break contact from hide or surveillance site (continued).

Appendix I
Tracking and Countertracking

The LRS teams must learn crucial tracking skills and techniques for use while in enemy territory. This allows them to provide immediate intelligence on the frequency and flow of enemy traffic on a trail. Tracking is also useful when a LRS team conducts a PR mission to retrieve a downed pilot.

Good tracking skills enhance countertracking skills, and good countertracking skills assist in the success or failure of a mission by allowing the team to effectively evade anyone tracking them while in enemy territory.

Both skill sets increase the Soldier's general awareness and reduces the chance of being caught off guard.

CONCEPTS OF TRACKING

I-1. To become a tracker, the LRS Soldier must develop and refine traits such as patience, persistence, acute observation, good memory, and attention to detail. These all help when tracking signs weaken and the tracker must rely on intuition. As he evaluates sign, he forms an opinion about the enemy's training, equipment, and morale. Six factors help the tracker form a picture of the enemy.

DISPLACEMENT-TYPE SIGN

I-2. "To displace" means "to move something from its original position." Thus, "displacement" means "the act of displacing" or "the signs or evidence that something has moved."

Survey Area

I-3. The tracker looks for displacement signs in a full 15-meter deep, 180-degree arc and up to the height of a tall man (Figure I-1). He evaluates all of the signs for trends and patterns.

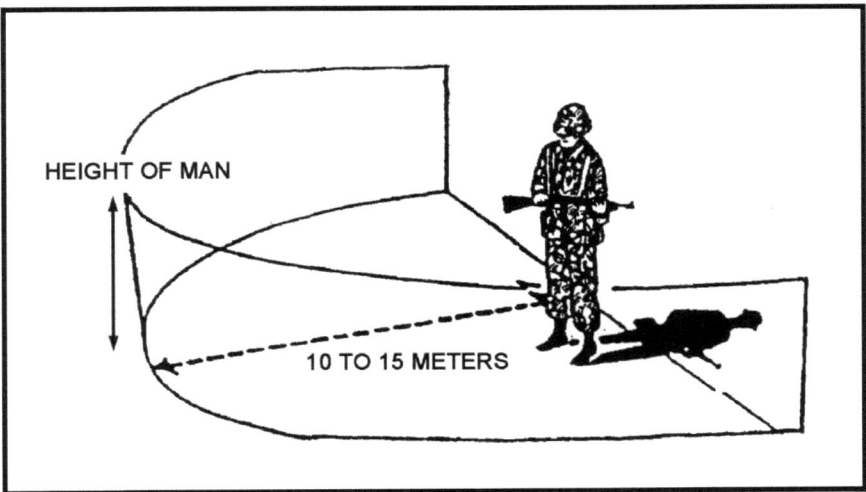

Figure I-1. Areas surveyed for indicators by tracker.

Human Passage

I-4. Some displacement type signs that indicate the likely presence of other humans include footprints; scraped and broken vegetation; bits of thread or clothing on the ground or on vegetation; limbs freshly

Appendix I

broken or grass bent over on a windless day; the sudden excited cries of animals or sounds of sudden movement; disturbed insect life; and upturned rocks.

Armed Soldier

I-5. A footprint, found near a waist-high scuffmark on a tree, can indicate the passage of an armed Soldier (Figure I-2).

Figure I-2. Examples of displacement.

Footgear

I-6. A footprint can reveal what footgear the enemy is wearing, if any; the lack of proper equipment; the direction of movement; the number and gender of people moving and their rate of movement; the weight of their loads; the amount of time passed since they made the track; and whether they know they are being tracked (Figure I-3).

Rapid Movement

I-7. If the footprints are deep and the pace long, the party is moving fast. Long strides and deep prints, with toe prints deeper than heel prints, indicate that they are running.

Heavy Load

I-8. If the prints are deep, short, and widely spaced, with signs of scuffing or shuffling, the party is probably carrying a heavy load.

Gender

I-9. To determine the gender of the party, study the sizes and positions of the footprints. Women tend to walk with their toes pointed inward, whereas men usually walk with their feet pointed straight ahead or outward. Women's prints and their strides are usually shorter than those of men.

Backstep

I-10. If a party knows or suspects that someone is following them, they might try to hide their tracks. Parties walking backwards have a short, irregular stride. The prints will have an unusually deep toe, and the soil will be kicked backward, confirming the direction of movement.

Tracking and Countertracking

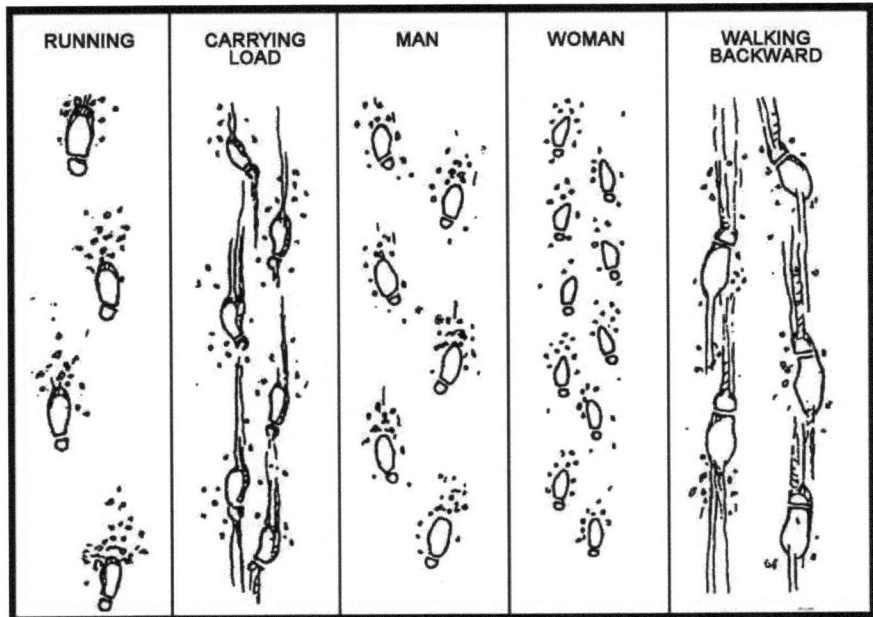

Figure I-3. Types of footprints.

Box Method

I-11. To use the 36-inch-box method, mark off a 30- to 36-inch cross-section of the trail, count the prints in the box, then divide by two to determine the number of parties who used the trail. The M16 rifle is 39 inches long. You can use it to measure (Figure I-4).

Identify a Key Print

I-12. The figure shows the use of a left boot print as the key print. In this situation, you draw a line from the heel across the trail.

Move Forward

I-13. Move to the key print for the opposite foot and draw a line through the instep. Your two lines form two opposite sides of a box. The edges of the trail form the other two sides. Together, the four lines form a box.

Count

I-14. Count every whole or partial print that falls inside of the box. Because any person who was walking normally would have stepped in the box at least once, the number of footprints in the box will help you to determine the total number of people in the party.

Appendix I

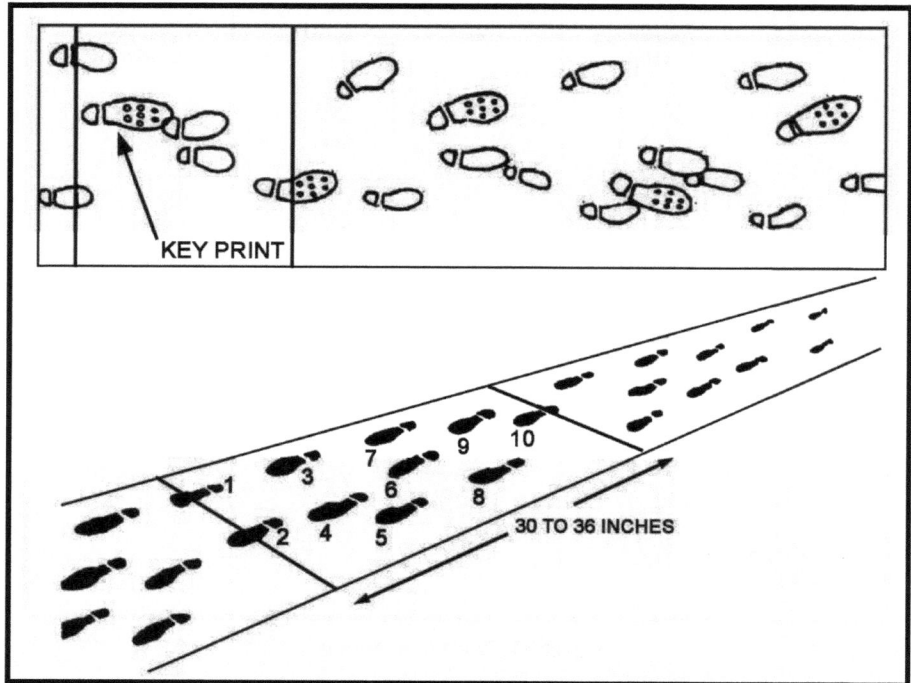

Figure I-4. Box method for determination of number of footprints.

STAINS

I-15. One obvious example of staining is blood on the ground or foliage. Other examples include mud dragged by footgear and crushed vegetation on a hard object. Crushed berries also cause stains, and, finally, the movement of water makes it cloudy.

EFFECTS OF WEATHER

I-16. The weather may help or hinder the tracker to determine the age of signs. Wind, snow, rain, and sunlight all affect tracking. Assess recent weather aids in determining the age of a track. A tracker can use the last rain or strong wind as a measure to show the amount of time it has been there.

LITTER

I-17. A poorly disciplined unit leaves a trail of litter.

DECEPTION TECHNIQUES

I-18. Deception applies when the followed party tries to slow the tracker by, for example, leaving false trails, walking backwards, brushing out trails, splitting up, fading out, fanning out, and walking over rocky ground or through streams.

INTERPRETATION OF COMBAT INFORMATION

I-19. The tracker makes a mental image of his quarry based on his learned concepts. When reporting to the commander, he reports what he believes, but should not state it as a fact. The commander considers the facts alongside the tracker's beliefs and any other information he has.

ORGANIZATION OF TRACKING TEAM

I-20. Tracking units can be any size as long as they have these three elements: a leader, a tracker, and security. Often tracking teams consist of two types:

TRACKER AND COVER PERSON

I-21. Each team member is equally skilled. They can move fast because they know each other's abilities and weaknesses and can compensate for each other.

TRACKING TEAM LEADER, TRACKER, RTO, AND TWO SECURITY PERSONNEL

I-22. The advantages of a tracking team with this many members are increased observation and security. The disadvantage is the size of the team.

TRACKER AND DOG TEAM

I-23. Trackers work more effectively with dog teams than without.

TRAITS OF A TRACKING DOG

I-24. The dog(s) follows a trail faster and can continue to track at night. Despite years of domestication, dogs retain most of the traits of their wild ancestors. If put to controlled use, these traits are effective when tracking.

Endurance

I-25. A dog can hold a steady pace and effectively track for up to eight hours. The speed can be up to 10 miles per hour, only limited by the speed of the handler. The speed and endurance can be further increased by the use of vehicles and extra teams.

Mental Characteristics

I-26. Dogs are curious by nature. They can also be aggressive or lazy, cowardly or brave. Dogs' sensory traits are what make them seem intelligent.

Aggressiveness

I-27. Tracking dogs are screened and trained to function as aggressive trackers, eager to please their handlers.

SENSES OF A TRACKING DOG

I-28. Knowledge of the dog's senses and how he uses them helps the evader to think ahead of the dog.

Sight

I-29. A dog's vision is the lesser of the sensing abilities. He can see in black and white and has trouble spotting static objects at more than 50 yards. He can spot moving objects at considerable distances, but he does not look up unless he is trained. His night vision is no better than that of a human.

Sound

I-30. Dogs can hear quieter and higher frequencies than humans. A dangerous problem for the evader is the dog's ability to hear. Even more dangerous is their ability to locate the *source* of a sound. Dogs can hear 40 times better than a human.

Smell

I-31. The dog's sense of smell is about 900 times better than a human. It is by far the greatest asset to the tracker and largest threat to the evader. Dogs can detect minute substances or disturbances on the ground, or even in the air. Using distracting or irritating odors such as CS powder or pepper only bothers a dog for 3 to 5 minutes. After the dog discharges the odor, he can pick up a cold trail even quicker. The dog

Appendix I

smells odors from the ground and air and forms scent pictures. He puts together these scent pictures from several sources.

Individual Scent

I-32. This is the most important scent when it comes to tracking. Vapors from body secretions work their way through the evader's shoes onto the ground. Sweat from other parts of the body rubs off onto vegetation and other objects. Scent is even left in the air.

Reinforcing Scent

I-33. Objects that reinforce the scent as it relates to the evader are introduced to the dog. Some reinforcing scents could be on the evader's clothing or boots, or they could be made of the same material as is used in his clothing. Even the smell of boot polish can help a dog find a person.

Ecological Scent

I-34. For the dog, the most important scent comes from the earth itself. The strongest smell comes from disturbances in ecology such as crushed insects, bruised vegetation, and broken ground. Over varied terrain, dogs can smell particles and vapors carried by the evader wherever he walks.

UNFAVORABLE TRACKING CONDITIONS

I-35. Few conditions are ideal for dog tracker teams. During training, the teams learn the difficulties that they will face and develop skills to cope with them:

Unverified Start Point

I-36. The dogs may follow the wrong route or scent.

Heat, Low Humidity, and Dry Ground

I-37. These all cause rapid evaporation of scent.

Wind

I-38. Wind disperses scent, causing the dog to track downwind.

Heavy Rain

I-39. This washes scent away.

Distracting Scents

I-40. These divert the dog's attention from the trail. Some distracting scents are blood, meat, manure, farmland, and populated areas.

Elements that Cover Scent

I-41. Some elements in nature cover the scent picture partially or completely. For example, sand can blow over the tracks and help to disguise it; snow and ice can form over the track and make it nearly impossible to follow; and water can completely obscure a trail.

FAVORABLE TRACKING CONDITIONS

I-42. Some conditions favor the teams:

Fresh Scent

I-43. Scent is probably the most important factor for dog tracker teams. Fresh scent increases the chance of success.

Verified Starting Point

I-44. Introducing a definite scent to the dogs early on increases the chance that the dogs will follow the correct trail.

Unclean Evader

I-45. An unclean evader leaves a more distinctive scent.

Fast Moving Evader

I-46. A fast moving evader causes more ground disturbances than a slower moving evader and, because he is sweating more than the slower mover, he also leaves a stronger scent trail.

Night and Early Morning

I-47. At these times, the air is thicker, so scent lasts longer.

Cool, Cloudy Weather

I-48. This limits evaporation of scent.

Lack of Wind

I-49. This keeps the scent close to the ground. It also keeps it from spreading around, allowing the dog to follow the correct route.

Thick Vegetation

I-50. This restricts the dissemination of scent and holds the smell.

COUNTERTRACKING

I-51. To avoid or evade the enemy, the LRS team must constantly use countertracking techniques. Knowledge of tracking is probably the best way to successfully evade trackers. Knowledge of trackers and dog teams greatly assists the survivor when evading the enemy. The two main types of trackers and methods of evading each follow:

Visual Trackers

I-52. Visual trackers cue, obviously, on visible signs. Evading them requires that you reduce visual signs and confuse the tracker(s):

Outdistance the Tracker

I-53. Put time and distance between you and the tracker. This increases the chance that the track will disappear or the tracker team will track too slowly to keep up with you (the team).

Change Direction Frequently

I-54. Change direction often--at least every 1,000 meters--and cover your changes. This can confuse less skillful trackers and buy you some time. If a tracker picks up sign, he can send a party ahead as far as he feels confident that the tracks may be found on a track trap such as road or muddy bank. Track traps allow trackers to gain time and distance on the evaders. Methodical tracking is slow and arduous. By changing direction, you can prevent these cutting parties from finding sign where they expect it and can enlarge the area that the tracker must cover. If you must cross a track trap such as a road, approach at an off-angle to your azimuth and cross the road or track trap. After crossing, go a short distance, and then change direction sharply, carefully covering the signs.

Use Streams for Deception

I-55. Approach a stream at an off-angle. Move downstream for 100 to 200 meters, leaving false trails every so often. At the end of the false trail, leave a set of tracks and countertrack them just enough that the tracks can be seen just barely. This will make the tracker think that he has found the right set of prints. Find a place to enter the stream such as a log and move in the opposite direction from where you entered. Go to a place where you can leave the stream without leaving sign and do so. This will buy you time, because the trackers will have to check out false trails and scour the banks for tracks.

Appendix I

Move on Hard Surfaces

I-56. Any surface that is too hard to retain the indentations of feet makes visual tracking almost impossible. Ensure that the soles of your footwear are as clean as possible to avoid leaving deposits on the surface. Enter the hard-surfaced area from an off-angle, and leave as little sign as possible. Even as little as 100 meters of hard surface can gain you a great amount of time over a tracker team.

Camouflage Sign

I-57. Any deliberate attempt at camouflage will slow any following parties by making them look harder for sign. It will also make the aging of sign more difficult. The best technique is to leave as little sign as possible. This is an individual responsibility. One person at the rear of the formation cannot cover all the sign left by a team of six. Each person must take special care at track traps and things such as ant mounds that leave definite signs for trackers. When crossing roads or track traps, try to step in the footprints of the Soldiers before you. This will make the job of countertracking for the last Soldiers easier. If you cannot completely camouflage the tracks, try to age them by brushing them or sprinkling debris in them.

Walk Backwards

I-58. Walking backwards produces a different type of print than walking forward, but only the most skillful trackers will pick up the difference. Used with other deception methods, this one can create some confusion and buy some time.

Split Up the Team

I-59. Multiple sets of tracks slow the tracker team, because following six sets of prints takes much longer than following one set. The team can then link up at a rally or rendezvous site. This is a great technique for use in tall grass or other areas where leaving tracks is unavoidable. Animal tracks increase the confusion.

Use Animal Trails

I-60. Wildlife that shares trails with the team will soon obscure any signs of human passage.

Set Booby Traps

I-61. Even if the tracker team fails to detonate the device, they will become more cautious. A few hasty devices such as a grenade and trip wire will slow them down considerably. The more they encounter, the slower they will move.

Dog Trackers

I-62. Dogs track by scent, so countertracking against them is more difficult than countertracking a visual tracker. One of the main ways to defeat a dog is to defeat his handler. A dog is only as effective as the handler allows:

Change Direction Frequently

I-63. One way to defeat the handler is to change direction frequently. A relatively inexperienced handler might perceive the dog's resulting frequent direction changes as a sign of indecisiveness. He might then think that the dog has lost the scent. Changing direction frequently in difficult terrain such as brush has the added advantage of entangling dog and handler, thus fatiguing and stressing them.

Use Hard Surfaces

I-64. Dogs track best on loose textured surfaces. Hard surfaces seldom retain scent as well. Use these as is tactically feasible. The effect of this technique depends on the experience of the dog.

Travel in More Frequented Areas

I-65. While not a preferred technique for a LRS team, traveling in more frequented areas might be necessary to throw dogs off the scent. Having to differentiate between scents can cause them to lose the scent of the team.

Travel in Streams

I-66. While traveling in streams, avoid contact with anything that could catch and reinforce your scent such as a branch or rock.

METHODS OF AVOIDING DETECTION

I-67. To avoid detection--

Wear Common Footwear

I-68. Some types of soles, such as the ripple type, are uncommon, which allows for easier tracking. Flat soles leave less sign than treaded soles, but are impractical for traveling over rough terrain. Wearing sandbags over boots reduces sign. However, burlap rips off and leaves telltale sign, so watch out for this should you choose this method.

Leave No Scent

I-69. Avoid leaving scent. While in a position or on the move, leave as little scent picture as possible. Food, blood, urine, and feces all leave strong, reinforcing scents for a search dog. Be especially careful in areas where you must remain for some time and in areas that you know are patrolled by dog teams.

Behave Unpredictably

I-70. Go against everything that human nature tells you to do. Go into difficult terrain and behave as most people moving through such an area would not do.

Practice Tracking

I-71. Know your enemy. Practice tracking and you will notice things that you should avoid doing when you are the one being tracked. For the team to successfully countertrack as a unit, each team member must have a basic working knowledge of tracking.

Appendix J
Night Operations

Night-fighting skills are necessary and are combat multipliers. Infantry forces use these skills to gain tactical and psychological advantages. Night operations do not rely on technology for success--commanders can plan and execute them with or without the use of night vision devices. In fact, for LRSU, night operations are the norm.

This appendix lists the psychological, physiological, and physical effects of night-fighting. It also discusses the techniques used to maintain direction, control, and surprise. Some of these apply in other limited visibility conditions such as fog, rain, snow, and sandstorms.

NIGHT VISION

J-1. At night, the eye uses spiral eye cells called rods. These cells cannot differentiate color, and are easily blinded by light. This creates a central blind spot, which causes the viewer to miss larger objects as distances increase.

PROTECTION OF NIGHT VISION

J-2. Soldiers who work and perform tasks in daylight experience a reduction in night vision. Exposure to intense sunlight for two to five hours significantly decreases visual sensitivity for up to five additional hours. Sunlight magnified by reflective surfaces such as sand or snow reduces the rate of adaptability to the dark and it reduces general night vision even more. These effects are cumulative and may persist for several days. Consequently, Soldiers scheduled for night operations should wear military, neutral-density (N-15) sunglasses (or the equivalent) in bright sunlight.

NIGHT-VISION SCANNING

J-3. Soldiers use the night-vision-scanning technique to overcome physiological limitations and reduce illusions. It also protects their night vision and dark adaptation capabilities. The night-vision-scanning technique involves scanning slowly and regularly from right to left or from left to right (Figure J-1). Granted, this is the same as scanning in daylight. The difference is that, at night, Soldiers must also avoid looking directly at faintly visible objects.

Appendix J

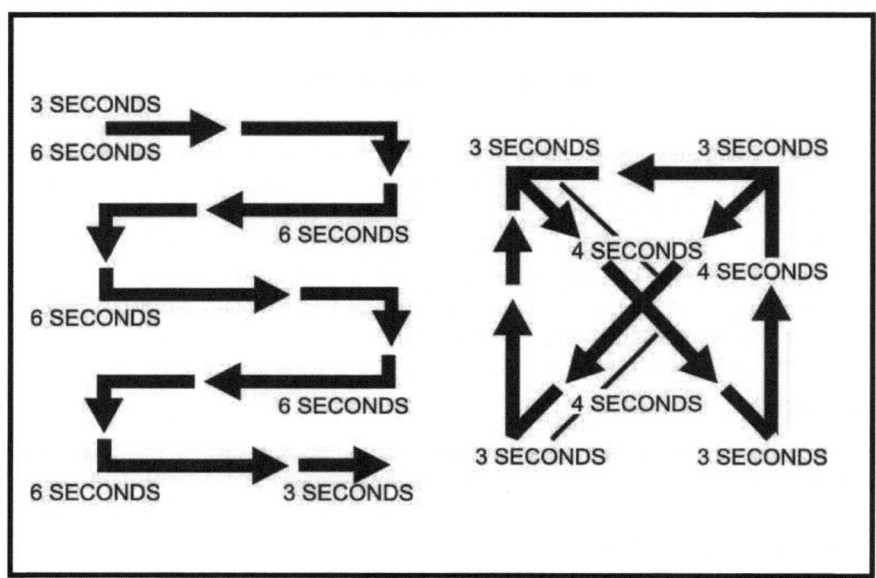

Figure J-1. Typical scanning patterns.

OFF-CENTER VISION

J-4.   Looking straight at an object (using *central vision*) works in daylight, but not at night. This is due to the aforementioned central blind spot. To compensate for this blind spot, Soldiers use *off-center (peripheral) vision*. In other words, instead of looking directly at an object, they look 10 degrees above, below, or to either side of it (Figure J-2).

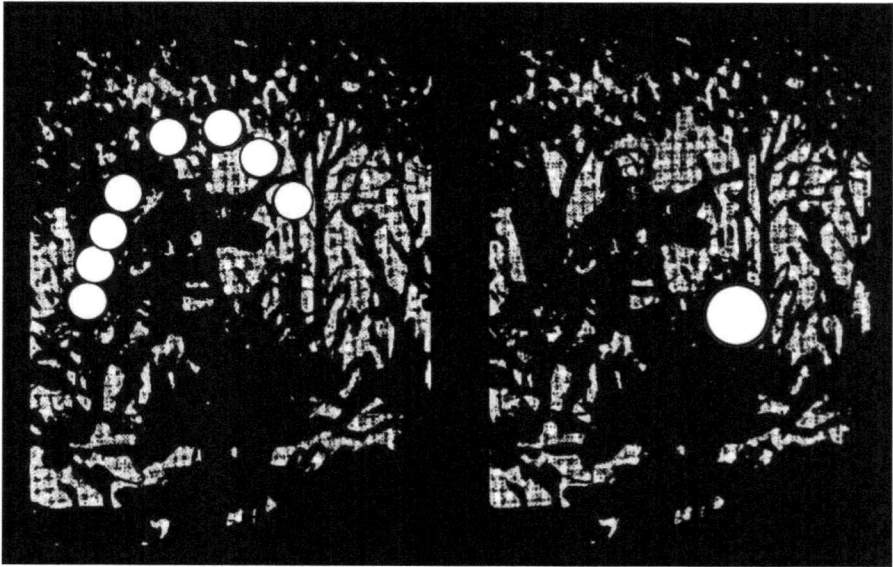

Figure J-2. Off-center viewing technique.

Adaptation to the Dark

J-5. Adaptation to the dark means that as darkness increases, so does visual sensitivity. Just like the aperture in a camera, the pupil opens to let in more light. Soldiers are unique, so they adapt to darkness at slightly different degrees and rates. After the first 30 minutes in the dark, visual sensitivity maxes out (to about 10,000 times that of normal), and it increases little after that time.

- Visual sensitivity in the dark is affected by exposure to bright lights such as matches, flashlights, flares, and vehicle headlights. Full recovery from this exposure might take up to 45 minutes.
- Use of night vision goggles impedes dark adaptation. However, if a Soldier adapts to the dark *before* he dons the goggles, then he will adapt to the dark fully about two minutes after removing the goggles.
- Color perception decreases as light decreases. Soldiers can distinguish light and dark colors only to the degree of reflected light intensity.
- Darkness reduces visual acuity to one-seventh of daylight levels, so Soldiers can see only large objects.

Bleached-Out Effect

J-6. Even when the Soldier practices off-center viewing, the image of an object that he views for more than two to three seconds tends to bleach out into one solid color. As a result, he can no longer see the object, which can increase operational risk. To overcome this effect, the Soldier must know about this phenomenon and avoid looking at any object for more than two or three seconds. By shifting his eyes from one off-center point to another, he can continue to see the object in his peripheral vision.

Shape or Silhouette

J-7. At night, Soldiers must identify objects by their shapes or silhouettes. Knowing the architectural designs of structures common to the AO determines the success of this technique. For example, in the US, a church is often characterized by a high roof and steeple, but churches elsewhere have different architectures.

Light Sources and Distances

J-8. Table J-1 shows how far the naked eye can detect light sources at night.

Table J-1. Light sources and distances.

Sources	Distances		
Vehicle headlight	4.0	to	8.0 kilometers
Muzzle flash from single cannon	4.0	to	5.0 kilometers
Muzzle flash from small-arms weapon	1.5	to	2.0 kilometers
Bonfire	6.0	to	8.0 kilometers
Flashlight	0.0	to	2.0 kilometers
Lighted match	0.0	to	1.5 kilometers
Lighted cigarette	0.5	to	0.8 kilometers

Note: From the air, these distances can increase two to three times.

Appendix J

HEARING

J-9.   Hearing is more acute in the dark for several reasons: mental concentration increases; background noises tend to diminish; and, lower temperatures and higher humidity carry sound farther. Practice and training help the Soldier overcome fear of night sounds. Training helps him to discriminate (distinguish) multiple sounds, faint sounds, and the directions from which sounds originate. Table J-2 shows how far away the Soldier can hear particular sounds at night.

Table J-2. Sounds and distances.

Source	Distance		
Cannon shot	0.0	to	15.0 kilometers
Single shot from a rifle	2.0	to	3.0 kilometers
Automatic weapons fire	3.0	to	4.0 kilometers
Tank movement			
On a dirt road	0.0	to	2.0 kilometers
On a highway	3.0	to	4.0 kilometers
Motor vehicle movement			
On a dirt road	0.0	to	500.0 kilometers
On a highway	0.0	to	1.0 kilometers
Movement of troops on foot			
On a dirt road	0.0	to	300.0 meters
On a highway	0.0	to	600.0 meters
Small-arms weapon loading	0.0	to	500.0 meters
Metal on metal	0.0	to	300.0 meters
Conversation between a few men	0.0	to	300.0 meters
Steps of a single Soldier	0.0	to	40.0 meters
Axe blow, sound of saw	0.0	to	500.0 meters
Blows of shovels and pickaxes	0.0	to	1,000.0 meters
Screams	0.0	to	1,500.0 meters
Oars on water	0.0	to	2,000.0 meters

SMELL

J-10.   Smell is the Soldier's least used sense. Typically, he only uses two percent of its potential. Different diets produce different human odors. For example, habitual meat eaters smell different from habitual vegetarians. Most enemy have different diets from those of US Soldiers. Once US Soldiers learn the enemy's characteristic odor, they can easily detect him at night. Practice improves skill and confidence.

J-11.   Facing into the wind at a 45-degree angle makes sensing odors easier. The Soldier relaxes, breathes normally, sniffs sharply, thinks about specific odors, and concentrates. Table J-3 shows the distances at which the human nose can typically sense particular odors.

Table J-3. Odor sources and distances.

SOURCE	DISTANCE
Diesel fuel	0.0 to 500.0 meters
Cigarette smoke	0.0 to 150.0 meters
Heat tab	0.0 to 300.0 meters

FATIGUE

J-12.   Too much work and too little sleep make for a tired Soldier, especially in conditions of great stress. Tired Soldiers affect the unit's capabilities. Following a work-rest schedule can help prevent collective fatigue. It builds-in recovery time to maintain unit effectiveness. Leaders--

- Ensure that each Soldier sleeps or rests during part of each of his off-shift periods.
- Rotates cross-trained Soldiers through various duties to reduce errors.
- Should assign two Soldiers to each job that requires discrimination between factors, such as OP procedures or writing and encrypting messages.
- Can experiment until he finds the best schedules. A four-hours-on, four-hours-off schedule works in good weather; a two-hours-on, four-hours-off schedule works better for bad weather. Other schedules can also help. No schedule suits everyone, but a particular schedule might work best for a particular team.

J-13.   Sleep order depends on task seriousness, complexity, and tedium level. For example, team leaders and RTOs might rate Priority 1 or 2 in this system. So, if someone has to miss sleep to check the OP, the team leader might check it once, his assistant twice, and an observer thrice. The team leader must get the most sleep, since he makes the most serious decisions and processes the most complex information.

J-14.   Some Soldiers operate at their greatest efficiency early in their awake cycle, and vice versa. Leaders try to have decision makers perform their most critical tasks when they tend to think most clearly. Leaders evaluate and plan this ahead of time.

> **CAUTION**
>
> The intense concentration required to use night vision devices can degrade the other senses. Leaders should prepare Soldiers for night operations by having them use all of their senses. On some operations, this might require that some of them avoid using night vision devices.

SELECTION OF ROUTE

J-15.   The leader determines the route for night movement based on METT-TC. Since more than one route might satisfy METT-TC, leaders should select the easiest one to navigate. Night travel is strenuous, and often done when Soldiers are tired. This adds to physical and psychological stress. Simple navigation is easier to direct and control.

J-16.   He analyzes the selected route farther using the factors of OAKOC. METT-TC might weight some of these factors, such as terrain, cover, or avenues of approach, over other factors.

J-17.   Before analyzing the route, the leader divides it into segments or legs. Each leg starts with a checkpoint and ends with a change in direction or prominent terrain feature. The leader orients and controls the team's movement on the checkpoints along the route. He uses OAKOC to analyze each leg, and to

Appendix J

determine probable hasty ambush sites, likely areas for enemy movement, and locations with improved observation.

J-18. The leader also identifies a contiguous feature, or *catchpoint*, such as a river, road, or ridge, on the far side of each checkpoint. If the team misses the checkpoint, they head for the nearest catchpoint. This is a quick, easy way to reorient movement.

J-19. The leader tries to reconnoiter the route before he moves the unit. Ideally, he reconnoiters day and night. During his reconnaissance, he adds, confirms, or adjusts orientation aids. These can include any of the following:

- Terrain features (hills, cliffs, rivers, ridges, draws).
- Man-made features (towers, buildings, bridges, and roads).
- Ground surveillance radar (GSR).
- Wire.
- Illumination rounds.
- Night vision devices.
- Machine gun tracer fire. When mortar illumination rounds or tracer fire is used to locate positions, the leader plans the fire patterns so that the team can see them.

J-20. A reorientation plan is one of the final ingredients in route selection. The leader plans for reorientation throughout the movement using checkpoints, catchpoints, and position locators. Nevertheless, units do get lost. Therefore, the leader must plan how to recover and reorient his team and complete the mission. He plans this during the reconnaissance and adds checkpoints as needed. He uses distant terrain features to resection off indirect fire. Planning how to react should the unit become lost reduces the negative effects should it occur.

NIGHT WALKING

J-21. Leaders must train their units to move silently. Night movement requires different muscles than day movement, and so requires practice.

J-22. Whereas daylight travel stresses the calf muscles, walking at night places more strain on the muscles of the thighs and buttocks. Soldiers must get used to taking short, careful steps, stressing the use of the larger muscle groups in their thighs. This method of balanced, smooth walking at night reduces the chance of tripping over roots and rocks, and it reduces noise. Sufficient practice helps make crossing terrain at night seem as natural and easy as walking on a sidewalk in the daytime. Soldiers conditioned to move this way can travel far with little fatigue.

J-23. To walk at night, the Soldier looks ahead, shifts his weight completely to his left foot, and then slowly lifts his right foot about knee high. He balances on his left foot while he eases his right toes down and out, feeling lightly for twigs or trip wires. Still keeping his weight on his left foot, he gently touches his right toes to the ground about 6 inches to the front of his left foot. He lightly feels the ground under the outside of the toes of his right boot. Then, he feels with his right boot for any twigs, loose rocks, or holes. If he finds none, he finally settles his foot on the ground. Once he is confident of solid, quiet footing, he slowly shifts his weight forward onto the right foot, hesitates until he has his balance, then repeats the sensing process with his left foot.

J-24. Crossing fords and streams requires extensive team-level training. The team must establish and maintain security when crossing these obstacles. To cross a ford, the Soldier slips silently into the water, gains and maintains his footing, and remains alert. He starts crossing by sliding his leading foot forward and dragging his rear foot, as if shuffling. This helps him keep his balance in the current. After everyone crosses, the leader counts heads, and the team moves out.

SIGNALS

J-25. The team uses simple, familiar signals to pass information, identify locations, control formations, or initiate activity. Each basic signal has alternate signals as backups, and everyone must know these as well. Signaling at night helps the leader control what is happening. It supports security and surprise. It requires different methods than daylight signaling, for example, arm-and-hand signals might not be visible at night. The most common signals use sound, touch, and sight. The leader chooses signals based on the unit's activity and desired results, then he briefs the Soldiers and has them practice.

AUDIO SIGNALS

J-26. These include radio, wire, telephones, messengers, and grating or clicking of objects.

Radio and Telephone Signals

J-27. When using the radio and telephone at night, operators take precautions. They know that noise travels farther at night than during the day. They lower the volume as much as practical. They use headphones or earphones to reduce unnecessary noise. They know the possibility of loud static. They use signals such as breaking squelch a specified number of times.

Messages

J-28. Messengers carry written messages to avoid confusion and misinterpretation. When this is not possible, leaders ensure the messenger understands the message by having him repeat it word for word.

Oral Signals

K-1. Oral communication at night should be whispered. To do this, the Soldier takes a normal breath, exhales half of it, and then whispers into the other person's ear using the remainder of his breath.

VISUAL SIGNALS

J-29. These can be active or passive and include many options. For visual signals to work, everyone must see and recognize them.

Passive Visual Signals

- Sticks indicating direction.
- Light-colored paint.
- Tape.
- Rock formations.
- Markings on the ground.
- Powder.

Active Visual Signals

- Flares.
- Flashlights.
- Illumination rounds (M203, mortar, artillery).
- Chemical lights.
- Infrared or incandescent strobe lights.
- Either AN/PVS-5 or -7 night vision devices (infrared), or both.
- Burning fuel (saturated sand in a can).
- Luminous tape or compass dial.

Appendix J

Uses of Visual Signals

J-30. These signals can be used to identify a critical trail junction, mark a rally or rendezvous point, mark caches, or report that a danger area is clear. White powder can be used to indicate direction at a confusing trail intersection. A flashlight with a blue filter (with an "X" cut out of the filter) can signal "All clear" to a unit crossing a danger area. The possibilities are endless. However, the leader ensures that each Soldier in the team understands every signal used.

TOUCH SIGNALS

J-31. These consist of wire, string, or rope used in the hide or surveillance position to communicate without disclosing the position. They usually consist of wire loosely secured to an arm or leg. By prearrangement, two pulls on the wire might mean that a ground-mounted force is approaching, while three pulls might indicate a convoy.

TARGET DETECTION

J-32. Successful night movement and target engagement depend on knowing the enemy--how he attacks, defends, and uses terrain. Studying the enemy's techniques and patterns aids in target detection. Nature provides an endless array of patterns. Man invariably disturbs or alters these patterns so they are detectable. Sensing the enemy at night requires leaders and Soldiers to exercise patience, pay attention to detail, and practice. Patience and confidence are critical to effectively sense a target at night. While moving through an area, Soldiers look calmly and methodically for patterns rather than on details. For example, they look for straight lines, light variations, and any other odd-looking, obviously man-made, or disturbed patterns. The team looks for sentries or positions at the entrances to draws, on hills overlooking bridges or obstacles, and on the military crests of prominent terrain features. This is where the enemy will go for observation. The team also looks for supporting positions. They must know the ranges of enemy weapons, including supporting weapons, as well as the ranges of the enemy's night vision and line-of-sight observation devices. They must search thoroughly for enemy positions and for any other signs of enemy activity. Typical indicators of enemy activity include sounds, odors, displacement, weathering, littering, and camouflage:

Sounds

J-33. A Soldier places an ear to the ground or drives a stick 6 inches into it. Ground is denser than air, so sounds can travel farther through it, although determining direction is difficult. Rain and winds mask sounds. Rain causes Soldiers to seek shelter in static positions or, if moving, to put down their earflaps. Both actions degrade the ability to hear someone stalking them.

Odors

J-34. When the sun sets, the air cools and odors float downhill. When the sun and temperatures rise, so do odors. The more odors rise, the more likely they are to be picked up and carried by the wind.

Displacement

J-35. The team checks for displaced stones, leaves, or logs. The undersides of these objects are usually darker in color than their top sides, and they are usually also damp. Crumbled rocks have lighter colored faces and chips. At night, Soldiers need flashlights to detect these signs, so the team must place security well out. When viewed through infrared devices, broken and crushed vegetation look very different from undisturbed vegetation.

Weathering

J-36. Weathering indicates a recent disturbance. At night, the team usually needs experience--and light--to detect weathering.

Littering

J-37. Littering proves humans were present and can give clues about their discipline (or lack of it), distraction, supplies, and morale. However, it might be deliberate, that is, litter and other items left behind might contain booby traps.

Camouflage

J-38. Natures few straight lines tend to stand out. So do contrasting or unnatural colors, tones, and textures such as green leaves among dead branches. For example, an infrared source will show newly cut foliage.

MOVEMENT

J-39. Team leaders determine the best formation and movement techniques based on METT-TC. The file works well at night--it is easy to control and allows rapid movement through dense terrain. The only problem is that it does not allow the team to mass fire to the front. However, its advantages usually outweigh its disadvantages. To aid in movement control and security—

- Each Soldier must be within reach of the Soldier to his front.
- No Soldier moves unless told to do so.
- The leader(s) does the talking.
- The leader(s) positions himself far enough forward to make timely decisions to eliminate confusion.

Appendix K
Example Evasion and Recovery Plan

Figure K-1, page K-2, shows an example plan of action for an evasion.

Figure K-2 and Figure K-3, pages K-5 and K-6, show the front and back of an example of DD Form 1833 TEST (V2).

Appendix K

```
CLASSIFIED FOR TRAINING
(WHEN FILLED OUT)

EVASION PLAN OF ACTION
```

1. **GENERAL.**

   TEAM CALL SIGN: _____ DTG PREPARED: _____

   MISSION/TARGET NUMBER: _____ UNIT: _____

DUTY POSITION	RANK AND NAME	SSN	BLOOD TYPE
TL			
ATL			
RTO			
SSO			
Asst RTO/ S-6			
S-6			

2. **PREPARATION.** Before and during final inspections, the TL or ATL—

   a. ___ : Reviews ISOPREP cards.

   b. ___ : Reviews safe area for evasion intelligence description.

   c. ___ : Retains identification cards and tags.

   d. ___ : Sterilizes uniform (retain name tape and US army tape).

   e. ___ : Ensures that no one carries personnel effects.

   f. ___ : Performs PCI on all survival kits and equipment.

Figure K-1. Example plan of action for an evasion.

3. IMMEDIATE ACTIONS AFTER BEGINNING EVASION.

   a. INITIAL EVASION GOALS, FIRST 48 HOURS, INJURED AND UNINJURED. Think about RV locations and procedures during infiltration, actions on the objective (AOO), and exfiltration; team internal linkup locations and procedures, that is, where to go and what to do, for example, hide near crash or separation area, or evade alone or as a team; prepare movement plans, intended actions, and length of stay at initial rendezvous or linkup point:

   b. EVASION MOVEMENT PROCEDURES (GENERAL). Consider noise, movement time of day, camouflage, enemy sightings, danger areas, borders, and so on:

4. EXTENDED EVASION GOALS AFTER 48 HOURS (SPECIFIC).

   a. DETRIMENTS TO TRAVEL. These include difficult terrain, major rivers, heavily populated areas, curfews, minefields, concealment, weather factors, time of day or night, and times and means of movement:

   b. BORDER SITUATION (IF APPLICABLE). State the present condition of border areas, including neutral, friendly, and enemy border and coastline areas:

   c. RECOMMENDED TRAVEL ROUTES – Consider water, concealment, attitude of population, density, terrain, and advantages and disadvantages of the chosen routes:

   d. CBRN CONSIDERATIONS (IF APPLICABLE). Consider prevailing wind conditions, terrain, enemy capabilities, and so on:

   e. LIKELY SOURCES OF FOOD AND WATER. Locations and methods of obtaining food and water such as locations of caches, cache report information, and any other planned resupply points:

   f. SITUATION.

      1) ENEMY. Describe enemy force deployment, counterreconnaissance capabilities, search and interrogation tactics, expected treatment of POWS, and recommended resistance techniques.

      2) FRIENDLY. State which units are operating in or near the mission area of operations (AO); list CSAR and recovery assets available, partisan and foreign friendly forces in the AO and their recovery capabilities, and any established conventional or unconventional, assisted-evasion networks.

5. SAFE AREA FOR EVASION. Complete required items for infiltration, NAI and target area, and exfiltration as required.

   a. LOCATION AND CATEGORIES. Describe general location and categories of safe area for evasion, which area offers the best chance of survival and escape and rescue.

      1) Describe planned method for making contact with recovery assets or nets:

      2) Cover infiltration, NAI area and exfiltration:

   b. DISTINCTIVE FEATURES (AIR). Describe distinctive features of areas to enable visual identification from the air. Cover infiltration, NAI area, and exfiltration:

   c. SIGNIFICANT TERRAIN FEATURES. Describe significant terrain features to aid in visual identification on the ground. Cover infiltration, NAI area, and exfiltration:

   d. CONTACT POINT AND PROCEDURES. Describe designated contact points, contact procedures, intentions at selected areas, and criteria for moving to the next contact point. Cover infiltration, NAI area, and exfiltration:

Figure K-1. Example plan of action for an evasion (continued).

Appendix K

```
6. SAR CONTACT PROCEDURES:

The primary method and procedure for contacting the recovery element:

    a. ORG FREQUENCY CALL SIGN

    b. RADIO TYPE AND LISTENING PERIOD:

    c. RADIO TYPE AND TRANSMITTING PERIOD:

    d. GROUND-TO-AIR SIGNALS:

    e. AUTHENTICATION CODES (FROM THE ATO SPINS):

        1)  COLOR OF THE DAY:

        2)  LETTER OF THE DAY:

        3)  NUMBER OF THE DAY:

        4)  WORD OF THE DAY:

        5)  BULL'S EYE:

    f. CODE WORDS:

    g. RELAY CODEWORD: 0 1 2 3 4 5 6 7 8 9

    h. BONA FIDES:

    i. AIR-TASKING ORDER SPINS CHANGE OVER TIMES:

    j. RECOVERY ACTIVATION SIGNAL. Describe the signal and procedures for its use in detail.

    k. LOAD SIGNAL. Describe the signal and procedures for its use in detail.

    l. SIGNAL PLACEMENT AND TIMES:

    m. ENROUTE EVASION PLAN PRIOR TO THE DECISION POINT: Include any actions the aircrew and team will take:

7. ADDITIONAL COMMENTS ABOUT THE TEAM ESCAPE AND RESCUE PLAN. Address attitudes, intentions, and local intelligence information:

8. ACTIONS REQUIRING YOU TO CONDUCT ESCAPE AND RESCUE:

9. SURVIVAL EQUIPMENT CARRIED BY TEAM AND INDIVIDUAL, INCLUDING SPECIAL EQUIPMENT:

10. EMERGENCY RESUPPLY PLAN: Describe contents and how emergency resupply will be delivered.)
```

Figure K-1. Example plan of action for an evasion (continued).

CONFIDENTIAL (When Blocks 50 - 54 filled in)

ISOLATED PERSONNEL REPORT (ISOPREP)

PRIVACY ACT STATEMENT

AUTHORITY: 10 U.S.C. Sections 133, 3012, 3051 and 8012; E.O. 9397.
PRINCIPAL PURPOSE(S): To protect recovery forces from enemy entrapment and facilitate the recovery of isolated persons.
ROUTINE USE(S): To be completed by designated personnel subject to isolation due to hostile activity. Contains personal information that may be used to ensure positive identification. The form will be unclassified/official use only. Blocks 50, 51, 52, 53, 54, 55 and 56 are optional blocks and are only utilized when directed by unit SOP or Service doctrine. When blocks 50, 51, 52, 53, 54, 55, and 56 are completed this form becomes classified CONFIDENTIAL and must be handled appropriately. This form may only be transmitted via .mil to .mil email accounts.
DISCLOSURE IS MANDATORY. The information is necessary since it affects the entire personnel recovery process. Exceptions on disclosure are made for government contractors.

SECTION 1 - PERSONAL INFORMATION

1.a. LAST NAME*	1.b. FIRST NAME*	1.c. M.I.	2. GO BY NAME	3. GENDER*	4. GRADE*	5. SSN*	6. COALITION ID	7. DOB *(YYYYMMDD)**

8. BRANCH OF SERVICE/AGENCY/DEPT*	9. CURRENT UNIT*	10. BLOOD TYPE*	11. HEIGHT *(in.)**	12. WEIGHT *(lbs.)*	13. HAIR COLOR*	14. EYE COLOR*	15. ETHNIC GROUP
16.a. CITIZENSHIP*	b. IF OTHER, SPECIFY:		17. ACCENT		18. RELIGIOUS PREFERENCE		19. BLOOD CHIT NUMBER

20. IDENTIFYING SCARS/MARKS/TATTOOS*	21. KNOWN MEDICAL CONDITIONS AND PRESCRIPTIONS

SECTION 2 - UNIFORM DATA

22. SHIRT SIZE:	23. PANT SIZE:	24. HAT SIZE:	25. BOOT TYPE:	26. BOOT SIZE:	27. BOOT WIDTH:

SECTION 3 - TRAINING/CAPABILITY DATA

SERE TRAINING	a. TYPE	b. YEAR *(YYYY)*	c. MONTH	d. LOCATION /OTHER *(Specify)*	e. COMMENTS
28. TRAINING 1				/	
29. TRAINING 2				/	
30. TRAINING 3				/	

31. PRIMARY LANGUAGE CAPABILITY	a. LANGUAGE *(If Other, specify)*	b. READING	c. WRITING	d. SPEAKING	e. COMMENTS

32. OTHER LANGUAGE CAPABILITIES:

SECTION 4 - REINTEGRATION INFORMATION

33. PRIMARY NEXT OF KIN		34. PARENT NO. 1 *(Contractors Optional)*		35. PARENT NO. 2 *(Contractors Optional)*	
a. NAME:		a. NAME:		a. NAME:	
b. ADDRESS:		b. ADDRESS:		b. ADDRESS:	
c. CITY:		c. CITY:		c. CITY:	
d. STATE	e. ZIP CODE:	d. STATE	e. ZIP CODE:	d. STATE	e. ZIP CODE:
f. TELEPHONE:		f. TELEPHONE:		f. TELEPHONE:	
g. SAME AS: ☐ PARENT NO. 1 ☐ PARENT NO. 2				37. HOME OF RECORD *(Contractors Optional)*	

36. CHILDREN AT HOME *(Contractors Optional)*				
a. NAME(S)	b. DOB(S)	a. NAME(S)	b. DOB(S)	a. ADDRESS:
				b. CITY:
				c. STATE / d. ZIP CODE:

38. SPECIAL FAMILY SITUATIONS:

SECTION 5 - JPRC/PRCC USE ONLY

39. DATE MISSING *(YYYYMMDD)*	40. DATE RECOVERED *(YYYYMMDD)*	41. DATE ENTERED REINTEGRATION *(YYYYMMDD)*	42. DATE RELEASED TO UNIT CONTROL *(YYYYMMDD)*

43. NOTES

DD FORM 1833 TEST (V2), MAY 2008 CONFIDENTIAL (When Blocks 50 - 54 filled in) PREVIOUS EDITION IS OBSOLETE. Adobe Designer 7.0

Figure K-2. Example of DD Form 1833 TEST (V2) (front).

Appendix K

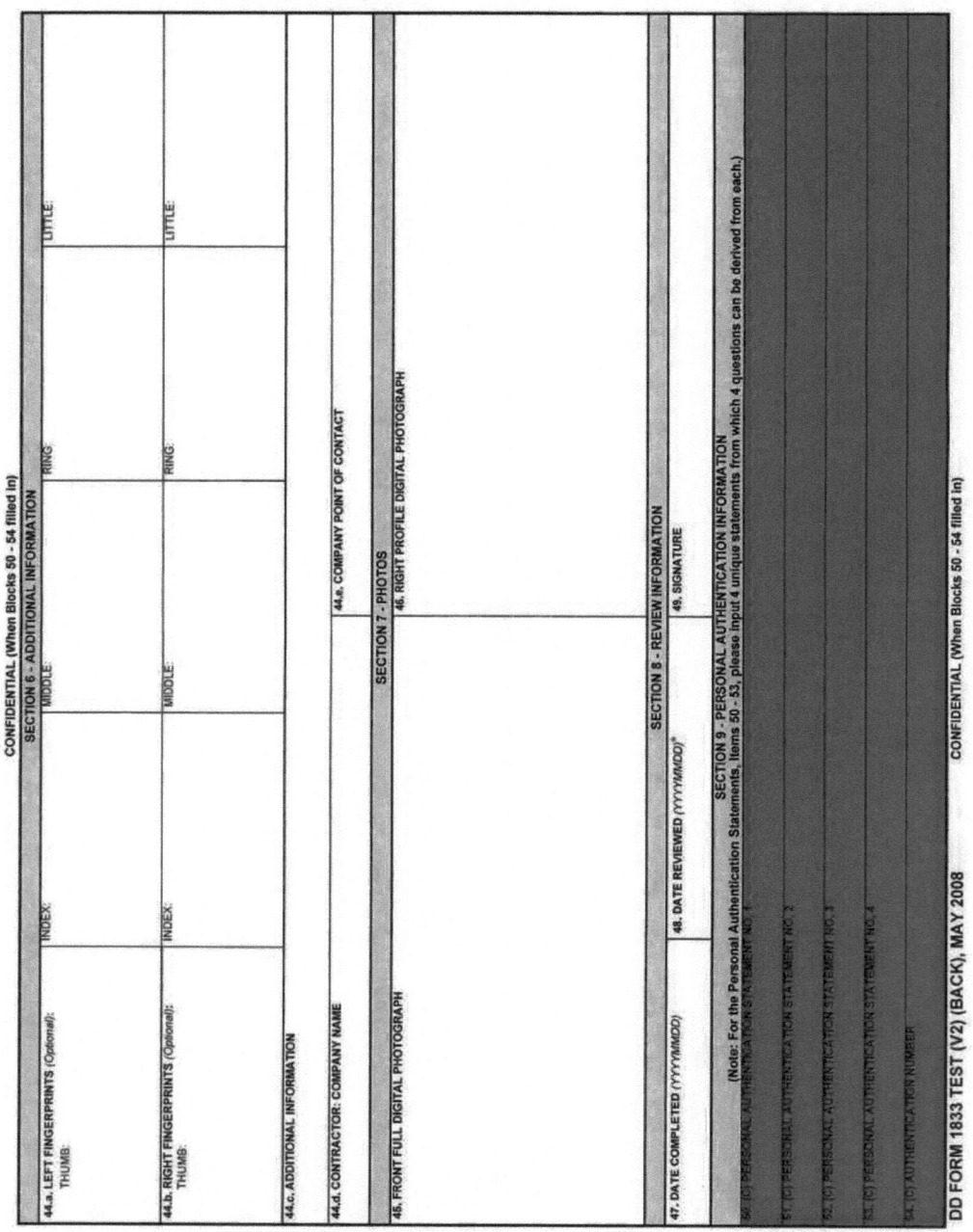

Figure K-3. Example of DD Form 1833 TEST (V2) (back).

Glossary

Section I. ACRONYMS AND ABBREVIATIONS

1SG	First Sergeant		CAS	close air support
	A		CATS	combined arms training strategy
AAA	antiaircraft artillery		CBRN	chemical, biological, radiological, and nuclear
AC2	Army airspace command and control		CCA	close combat attack
AES	advanced encryption standard		CCIR	commander's critical information requirements
ALE	automatic link establishment		CLS	combat lifesaver
ANDVT	advanced narrowband digital voice terminal		COA	course of action
AO	area of operations		COB	company operations base
AOB	alternate operations base		COMSEC	communications security
AOI	area of interest		COO	combined obstacle overlay
ARNG	Army National Guard		CYRIL	proword used to describe a situation report
ARNGUS	Army National Guard of the United States			**D**
ASCOPE	areas, structures, capabilities, organizations, people, events		DAG	division artillery group (OPFOR)
			DAMA	demand-assigned, multiple access
ASI	additional skill identifier		DAR	designated area of recovery
ASIP	Advanced System Improvement Program		DF	direction finding
			DOD	Department of Defense
ATOSPIN	air tasking order special instructions		DP	decision point
			DSVT	digital secure-voice terminal
ATV	all-terrain vehicle		DTG	date-time group
	B		DV	digital voice
BCT	brigade combat team		DZ	drop zone
BDA	battle damage assessment			**E**
BFSB	Battlefield Surveillance Brigade		E&E	escape and evasion
BLOS	beyond line-of-sight		E&R	evasion and recovery
BMNT	begin morning nautical twilight		ECCM	electronic counter-countermeasures
BRS	base radio station		ECW	extreme cold weather
BSC	brigade support company		EMT	emergency medical technician
	C		EOM	end of message
CAB(M)	combat aviation brigade (medium)			
CAF	combat air force			

23 June 2009   FM 3-55.93   Glossary-1

Glossary

EPA	evasion plan of action		HC	hexachloroethane
EW	electronic warfare		HF	high frequency
F			HHT	headquarters and headquarters troop
FAC	forward air controller		HMMWV	high-mobility, multipurpose wheeled vehicle
fax	facsimile (graphics only)			
FC	Fires Cell		HPT	high payoff target
FH	frequency hopping		hq	headquarters (graphics only)
FID	foreign internal defense		hr	hour (graphics only)
F-Kill	firepower kill		HUMINT	human (collected) intelligence
FLIR	forward-looking infrared		HVT	high-value target
FLOT	forward line of own troops		**I**	
FM	frequency modulation; field manual		IBCT	Infantry brigade combat team
FO	forward observer		IFF	identification friend or foe
FOT	frequency of transmission		IMETS	Integrated Meteorological System
FRAGO	fragmentary order		IMINT	imagery intelligence
FRIES	fast-rope insertion and extraction system		IND	individual
			INTREP	intelligence report
FRP	final reference point		INTSUM	intelligence summary
FSO	fire support officer		IPB	intelligence preparation of the battlefield
G				
G-2	Assistant Chief of Staff, Intelligence		IR	information requirement
			IRP	initial reference point
G-3	Assistant Chief of Staff, Operations and Plans		ISOPREP	isolated personnel report
			ISR	intelligence, surveillance, and reconnaissance
G-6	Assistant Chief of Staff, Communications			
			IWEDA	Integrated Weather Effects Decision Aid
GI&S	geospatial information and services			
			J	
GMV	ground mobility vehicle		J-2	joint staff intelligence directorate
GPS	global positioning system		J-3	joint staff operations and plans directorate
GSO	ground safety officer			
GSR	ground-surveillance radar		J-6	joint staff communications-electronics directorate
H				
HAHO	high altitude, high opening		jm	jumpmaster (graphics only)
HALO	high altitude, low opening		JOG	joint operations graphic
HCA	humanitarian and civic assistance		JPRA	Joint Personnel Recovery Agency

J-SEAD	joint suppression of enemy air defense	**M**	
JSRC	joint search and rescue center	**M**	meter
JTF	joint task force	**MA**	mission analysis
K		**MASINT**	measurement and signals intelligence
Kbps	kilobits per second	**MBITR**	multiband intra-team radio
KIA	killed in action	**MCL**	minimum clear (takeoff) length
K-Kill	catastrophic damage	**MCOO**	modified combined obstacle overlay
km	kilometers	**MCP**	main command post
kmph	kilometers per hour	**MCW**	minimum clear (takeoff) width
L		**MDMP**	military decision-making process
LAN	local area network	**METT-TC**	mission, enemy, terrain and weather, troops and support available, time, and civil considerations
lat	latitude (graphics only)		
LBE	load-bearing equipment		
LCE	load-carrying equipment		
LEA	law enforcement agency	**MFF**	military free fall
LNO	liaison officer	**MHz**	megahertz
LOC	lines of communication	**MIA**	missing in action
long	longitude (graphics only)	**MIJI**	meaconing, intrusion, jamming, and interference
LORAN-C	Long-Range Navigation-- Revision C	**mm**	millimeter
LOS	line of sight	**MOU**	Memorandum of Understanding
LQA	link quality analysis	**MPF**	mission planning folder
LRAS3	Long-Range Advanced Scout Surveillance System	**MRE**	meal ready to eat
		MSE	mobile subscriber equipment
LRS	long-range surveillance	**MSIP**	multispectral image processor
LRSC	long-range surveillance company	**MSS**	mission support site
LRSD	long-range surveillance detachment	**N**	
		NAI	named area of interest
LRSLC	long-range surveillance leaders course	**NCO**	noncommissioned officer
		NCOIC	noncommissioned officer in charge
LRSU	long-range surveillance unit	**NEO**	noncombatant evacuation operations
LTIOV	latest time information of value		
LWCSS	Lightweight Camouflage Screening System	**NFA**	no-fire area
		NIMA	National Imagery Mapping Agency
LZ	landing zone		

Glossary

NLOS	non-line of sight	POL	petroleum, oil, and lubricants
NLT	not later than	POW	prisoner of war
NM	nautical mile (1,852 meters or 6,076 feet)	PR	personnel recovery
		PSG	platoon sergeant
NSA	National Security Agency	PZ	pickup zone
NSN	national stock number		**R**
NSTV	nonstandard tactical vehicle	R&S	reconnaissance and surveillance
NVG	night vision goggles	RAG	regimental artillery group (OPFOR)
NVIS	near-vertical incidence sky wave		
	O	RASP	recruitment, assessment, and selection program
OAKOC	observation and fields of fire, avenues of approach, obstacles key terrain, cover and concealment	RCC	rescue coordination center
		RDF	radio direction finding
OB	order of battle	RFA	restricted fire area
obj	objective (graphics only)	RFI	request for information
OE	operational environment	RFL	restrictive fire line
OIC	officer in charge	RH-53	heavy assault, airlift, or minesweeping helicopter
OP	observation post		
OPCON	operational control	RII	request for intelligence information
OPFOR	opposing force		
OPLAN	operation plan	ROE	rules of engagement
opn	operation (graphics only)	RP	release point or rally point
OPORD	operation order	RSLC	Reconnaissance and Surveillance Leader Course
OPSEC	operations security		
ORP	objective rally point	RTO	radio telephone operator
	P	rv	rendezvous (graphics only)
pax	people (graphics only)		**S**
PEO	peace enforcement operation	S-2	Intelligence Officer
PIR	priority intelligence requirements	S-3	Operations and Training Officer
PKO	peacekeeping operation	S-6	C2 Information Officer
PL	platoon leader	SAFE	selected area for evasion
PLGR	precision lightweight GPS receiver	SAID	SAFE area intelligence description
plt	platoon (graphics only)	SALUTER	Size, Activity, Location, Unit, Time, Equipment, and Remarks
PMCS	preventive maintenance checks and services		
		SAM	surface-to-air missile
PO	peace operation	SARDOT	search and rescue point

Glossary

SCI	sensitive, compartmented information	TOE	table of organization and equipment
SEAD	suppression of enemy air defenses	TTP	tactics, techniques, and procedures
SEO	sniper employment officer		**U**
SERER	survival, evasion, resistance, escape, recovery	UAS	unmanned aircraft system
		UHF	ultra high frequency
SFG(A)	Special Forces Group (Airborne)	UNDER	proword used to describe a cache report
SIGINT	signals intelligence		
SINCGARS	Single-Channel Ground and Airborne Radio Subsystem	USAF	United States Air Force
		USAR	United States Army Reserve
SIR	specific information requirements	USSOCOM	United States Special Operations Command
SITEMP	situation template		**V**
SITMAP	situation map	VHF	very high frequency
SITREP	situation report	VoIP	voice over Internet protocol
SOCOORD	special operations coordinator	VSWR	voltage standing wave radio
SOF	special operations force		**W**
SOP	standing operating procedure	WAN	wide area network
SPIES	Special Patrol Insertion and Extraction System	WARNO	warning order
		WFF	warfighting function
SQI	special qualifications identifiers	WWII	World War II
SSB	single sideband	wx	weather (graphics only)
SSM	surface-to-surface missile		**XYZ**
surv	surveillance (graphics only)	XO	Executive Officer
	T		
TACAIR	tactical air		
TACCP	tactical command post		
TACP	tactical air-control party		
TACSAT	tactical satellite		
TAI	target area of interest		
TC	training circular		
tgt	target (graphics only)		
THFRS	Transformation High-Frequency Radio System		
TLP	troop leading procedures		
tm	team (graphics only)		
TOC	tactical operations center		

23 June 2009   FM 3-55.93   Glossary-5

Section II. TERMS

B

backbrief — A briefing by subordinates to the commander to review how subordinates intend to accomplish their mission

C

critical frequency — the highest frequency bent back to earth

H

handshake — the exchange of informtion between two electronic devices

N

net — a group of communications stations operating under unified control

O

outstation — a remote or outlying station

ordnance — explosives, chemicals, pyrotechnics, and similar stores, such as bombs, guns and ammunition, flares, smoke, or napalm

U

unilateral — undertaken by one person or party; one-sided

W

waveform — graphic representation of a wave shape that represents the relationship between two variables such as amplitude and frequency

References

REQUIRED PUBLICATIONS

These documents must be available to the intended users of this publication.

ARMY REGULATIONS

AR 40-400. *Patient Administration.* 06 February 2008.

AR 350-1. *Army Training and Leader Development.* 03 August 2007.

AR 385-63. *Range Safety.* 19 May 2003.

AR 600-9. *The Army Weight Control Program.* 27 November 2006.

FIELD MANUALS

FM 3-0. *Operations.* 27 February 2008.

FM 3-04.15. *Multi-service Tactics, Techniques, and Procedures for the Tactical Employment of Unmanned Aircraft Systems.* 03 August 2006.

FM 3-04.113. *Utility and Cargo Helicopter Operations.* 07 December 2007.

FM 3-05.211. *Special Forces Military Free-fall Operations.* 06 April 2005.

FM 3-05.212. *Special Forces Waterborne Operations.* 31 August 2004.

FM 3-06. *Urban Operations.* 26 October 2006.

FM 3-06.11. *Combined Arms Operations in Urban Terrain.* 28 February 2002.

FM 3-07. *Stability Operations.* 06 October 2008.

FM 3-09.32. *(JFIRE) Multi-service Tactics, Techniques, and Procedures for the Joint Application of Firepower.* 20 December 2007.

FM 3-11. *Multiservice Tactics, Techniques, and Procedures for Nuclear, Biological, and Chemical Defense Operations.* 10 March 2003.

FM 3-11.3. *Multiservice Tactics, Techniques, and Procedures for Chemical, Biological, Radiological and Nuclear Contamination Avoidance.* 02 February 2006.

FM 3-11.4. *Multiservice Tactics, Techniques, and Procedures for Nuclear, Biological, and Nuclear (NBC) Protection.* 02 June 2003.

FM 3-11.5. *Multiservice Tactics, Techniques, and Procedures for Chemical, Biological, Radiological, and Nuclear Decontamination.* 04 April 2006.

FM 3-11.19. *Multiservice Tactics, Techniques, and Procedures for Nuclear, Biological, and Chemical Reconnaissance.* 30 July 2004.

FM 3-19.30. *Physical Security.* 08 January 2001.

FM 3-21.8. *Infantry Rifle Platoon and Squad.* 28 March 2007.

FM 3-21.10. *The Infantry Rifle Company.* 27 July 2006.

FM 3-21.20. *The Infantry Battalion.* 13 December 2006.

FM 3-21.38, *Pathfinder Operations.* 25 April 2006.

FM 3-21.220. *Static Line Parachuting Techniques and Tactics.* 23 September 2003.

FM 3-24 . *Counterinsurgency.* 15 December 2006.

References

FM 3-34.210. *Explosives Hazards Operations.* 27 March 2007.

FM 3-97.6. *Mountain Operations.* 28 November 2000.

FM 4-01.011. *Unit Movement Operations.* 31 October 2002.

FM 4-02. *Force Health Protection in a Global Environment.* 13 February 2003.

FM 4-25.11. *First Aid.* 23 December 2002.

FM 4-25.12. *Unit Field Sanitation Team.* 25 January 2002.

FM 4-30.31. *Recovery and Battle Damage Assessment and Repair.* 19 September 2006.

FM 5-103. *Survivability.* 10 June 1985.

FM 6-0. *Mission Command: Command and Control of Army Forces.* 11 August 2003.

FM 6-22.5. *Combat and Operational Stress Control Manual for Leaders and Soldiers.* 18 March 2009.

FM 6-30. *Tactics, Techniques, and Procedures for Observed Fire.* 16 July 1991

FM 7-0. *Training for Full-spectrum Operations.* 12 December 2008.

FM 7-1. *Battle-Focused Training.* 15 September 2003.

FM 7-15. *The Army Universal Task List.* 31 August 2003.

FM 7-85. *Ranger Unit Operations.* 09 June 1987.

FM 7-92. *The Infantry Reconnaissance Platoon and Squad (Airborne, Air Assault, Light Infantry).* 23 December 1992.

FM 8-42. *Combat Health Support in Stability Operations and Support Operations.* 27 October 1997.

FM 21-10. *Field Hygiene and Sanitation.* 21 June 2000.

FM 22-6. *Guard Duty.* 17 September 1971.

FM 27-10. *The Law of Land Warfare.* 18 July 1956.

GRAPHIC TRAINING AIDS

GTA 01-14-001. *Battle Damage Assessment & Repair (BDAR) Smart Book.* 01 March 2007.

HANDBOOK

SH 21-76. *Ranger Handbook.* 2006.

JOINT PUBLICATIONS

JP 0-2. *Unified Action Armed Forces (UNAAF).* 10 July 2001.

JP 3-07.1. *Joint Tactics, Techniques, and Procedures for Foreign Internal Defense (FID).* 30 April 2004.

OTHER

Harris Corporation. *Operator's Manuals for the AN/PRC-117, AN/PRC-150, and AN/PRC-152, 2005 through 2007.* http://premier.harris.com/rfcomm.

Thales Corporation. *Operators Manual for the AN/PRC-148.* October 2004. https://secure.thalescomminc.com/customer_care_V2.asp.

USSOCOM Reg 350-6. *Special Operations Infiltration/Exfiltration Operations.* 25 August 2004.

Soldier Training Publications

STP 21-1-SMCT. *Soldier's Manual of Common Tasks Skill Level 1.* 18 June 2009.

STP 21-24-SMCT. *Soldier's Manual of Common Tasks, Warrior Leader Skill Level 2, 3, and 4.* 09 Sept 2008.

Technical Manuals

TM 11-5820-467-15. *Operator's, Organizational, Direct Support, General Support, and Depot Maintenance Manual for Antenna Group, AN/GRA-50.* 19 July 1961.

TM 11-5820-887-10. *Operator's Manual for Digital Message Device Group, OA-8990/P.* 20 August 1982.

TM 11-5820-919-12. *Operator's and Organizational Maintenance Manual for Radio Set, AN/PRC-104A.* 15 January 1986.

TM 11-5820-923-12. *Operator's and Organizational Maintenance Manual Radio Set, AN/GRC-213.* 14 February 1986.

TM 11-5820-924-13. *Operator's, Organizational and Direct Support Maintenance Manual for Radio Set, AN/GRC-193A.* 14 February 1986.

TM 11-5820-1025-10. *Operator's Manual for Radio Set AN/PRC-126.* 01 February 1988.

Training Circulars

TC 21-24. *Rappelling.* 09 January 2008.

RELATED PUBLICATIONS

These documents are quoted or paraphrased in this publication.

Department of the Army Pamphlet

DA Pamphlet 350-38. *Standards in Training Commission.* 13 May 2009.

Field Manuals

FM 1-02. *Operational Terms and Graphics.* 21 September 2004.

FM 2-0. *Intelligence.* 17 May 2004.

FM 3-05.70. *Survival.* 17 May 2002.

FM 3-25.26. *Map Reading and Land Navigation.* 18 January 2005.

FM 3-50.1. *Army Personnel Recovery.* 10 August 2005.

FM 3-50.3. *Multi-service Tactics, Techniques, and Procedures for Survival, Evasion and Recovery.* 20 March 2007.

FM 3-90.15. *Sensitive Site Operations.* 25 April 2007.

FM 5-0. *Army Planning and Orders Production.* 20 January 2005.

FM 5-19. *Composite Risk Management.* 21 August 2006.

FM 5-33. *Terrain Analysis.* 11 July 1990.

FM 6-02.72. *Tactical Radios Multiservice Communications Procedures for Tactical Radios in a Joint Environment.* 14 June 2002.

FM 6-02.74. *Multi-Service Tactics, Techniques, and Procedures for the High Frequency-Automatic Link Establishment (HFALE) Radios.* 20 November 2007.

References

FM 6-02.90. *Multi-Service Tactics, Techniques, and Procedures for Ultra High Frequency Tactical Satellite and Demand Assigned Multiple Access Operations.* 31 August 2004.

FM 6-99.2. *US Army Report and Message Formats.* 30 April 2007.

FM 7-92. *The Infantry Reconnaissance Platoon and Squad (Airborne, Air Assault, Light Infantry).* 23 Dec 1992.

FM 17-95. *Cavalry Operations.* 24 December 1996.

FM 23-10. *Sniper Training.* 17 August 1994.

FM 34-130. *Intelligence Preparation of the Battlefield.* 08 July 1994.

FM 90-4. *Air Assault Operations.* 16 March 1987.

JOINT PUBLICATIONS

JP 1. *Doctrine for the Armed Forces of the United States.* 02 May 2007.

JP 1-02. *Department of Defense Dictionary of Military and Associated Terms.* 12 April 2001

JP 3-50. *Personnel Recovery.* 05 January 2007.

OTHER

Combined Arms Center Publication *Operational and Organizational Concept for the Battlefield Surveillance Brigade (BFSB).* 14 November 2000.

Technical Memo 5-87. *Modern Experience in City Combat*, US Army Human Engineering Laboratory, March 1987.

REFERENCED FORMS

DA forms are available on the APD Web site (www.apd.army.mil); DD forms are available on the OSD Web site (www.dtic.mil/whs/directives/infomgt/forms/formsprogram.htm).

DEPARTMENT OF THE ARMY FORMS

DA Form 1594. *Daily Staff Journal or Duty Officer's Log.*

DA Form 2028. *Recommended Changes to Publications and Blank Forms.*

DA Form 5752-R. *Rope Log (Usage and History (LRA)).*

DEPARTMENT OF DEFENSE FORMS

DD Form 1833 TEST (V2). *Isolated Personnel Report (ISOPREP).*

INTERNET

Some of the documents listed elsewhere in the References, as well as all of the individual and collective tasks referred to in this publication, may be accessed at one the following Army websites:

Air Force Pubs	http://afpubs.hq.af.mil/
Army Forms	http://www.apd.army.mil
Army Knowledge Online	https://akocomm.us.army.mil/usapa/doctrine/index.html
Digital Training Management System	https://dtms.army.mil/DTMS (individual and collective tasks)
DoD forms	http://www.dtic.mil/whs/directives/infomgt/forms/formsprogram.htm
NATO ISAs	http://www.nato.int/docu/standard.htm
Reimer Digital Library	http://www.train.army.mil

Index

A

actions
  and responsibilities, 4-4 (*illus*)
  on objective, 4-17
activities matrix, 7-14 (*illus*)
air
  assault, 5-24
  infiltration/exfiltration annex, B-46 (*illus*)
AN/PRC-150
  compatible radios, 6-3 (*illus*), 6-6 (*illus*)
  in vehicular AN/VRC-104(V)3 configuration, 6-7 (*illus*)
analysis
  assumptions, B-14
  facts, B-13
  mission, B-11, B-15 (*illus*)
  of threat factors, 7-9
ANGUS (Initial Entry) Report, 6-13 (*illus*)
antennas, 6-18
  bidirectional pattern, 6-24 (*illus*)
  expedient insulators, 6-31 (*illus*)
  formula for calculating length of half wave dipole antenna, 6-27 (*illus*)
  half wave dipole, 6-26 (*illus*)
  high frequency antenna "V" type, 6-34 (*illus*)
    half rhombic type, 6-33 (*illus*)
    long-wire type, 6-28 (*illus*), 6-33 (*illus*)
    inverted "V" type, 6-27 (*illus*)
  omni-directional pattern, 6-25 (*illus*)
  sloping antenna, "V" type, 6-34 (*illus*)
  sloping wire antenna, 6-29 (*illus*)
  terminated sloping "V" antenna, 6-29 (*illus*)

antennas (*continued*)
  unidirectional pattern, 6-23 (*illus*)
  whip antenna, 6-30 (*illus*)
architecture management, 6-1
area
  of influence, 7-3
  of interest, 7-3
  of operations, 7-3
  reconnaissance, 4-21
Army aviation, 5-24
assessment, A-2
association matrix, 7-12 (*illus*)
assumptions, B-14 (*illus*)
attack, electronic, 6-16
automatic link sequence, 6-4 (*illus*)

B

backbrief, B-16, B-17 (*illus*)
base radio station, 6-2, 6-6
battle drills, H-1
battlefield surveillance brigade (BFSB) staff, 2-2
beyond line-of-sight (BLOS) equipment, 6-4
bidirectional antenna pattern, 6-24 (*illus*)
boat, rubber, 5-2 (*illus*)
BORIS (Intelligence) Report, 6-13 (*illus*)
box method, I-4 (*illus*)
branches and sequels, E-1
bridges, 4-25, 4-32
brief types, B-8 (*illus*), B-9 (*illus*)
buildings, 4-33
bunkers, 4-34

C

cargo straps, securing of, 5-18 (*illus*)
chain of command, 8-2
chemical, biological, radiological, and nuclear (CBRN), 4-66
classification, 6-22
  of evasion, 8-3
  of urban area by size, 7-2 (*illus*)

cold weather operations, 6-35, 6-36, D-2
combat
  assessment, 4-30
  patrol, 4-44
  rubber raiding reconnaissance craft, 5-1 (See also rubber boat.)
command and control
  Army operations, four types, 1-1
  employment schematic LRSC, 2-9 (*illus*)
    LRSD in an MSS, 2-10 (*illus*)
  LRSD and LRS team, 2-1
  nonstandard, 2-7
command post, 2-7
common types of antennas, 6-26
communications, 6-1
  annex, B-41 (*illus*)
  base, 6-6
  computers and intelligence, 2-14, 2-15
  data wire diagram, 6-10 (*illus*)
  security (COMSEC), 6-15
company operations base, 2-8 (*illus*)
computers, 6-2
confirmation brief, B-10
construction and selection, 6-25
contingencies, 4-29
contingency
  matrix, E-2 (*illus*)
  plan, E-1
converging routes method, 4-24 (*illus*)
coordination
  checklist, 5-28 (*illus*)
  for army aviation, F-1
count, 5-4 (*illus*)
countertracking procedures, I-1
courses of action, 4-12 (*illus*)
CRACK (Battle Damage Assessment) Report, 6-15 (*illus*)

Index

CYRIL (Situation) Report, 6-14 (*illus*)

D

damage types and levels, 4-31
dams and locks, 4-35
data wire diagram, 6-10 (*illus*)
databases
 debrief, B-33, B-34 (*illus*)
 decision brief, B-16
 decision support template, 7-20
desert operations, 6-35
desert operations, D-1
destruction priority, communications devices, 6-17 (*illus*)
disguises, 8-6
displacement, I-2 (*illus*)
dissemination of information, 4-29
distillation towers, 4-35
drawing technique
 common shapes, 4-59
 hatching, 4-60
 perspective, 4-59
 vanishing points, 4-59
 whole to part, 4-58 (*illus*)
duress codes, 6-11 (*illus*)

E

electronic warfare, 6-16
elements, reconnaissance and surveillance, 4-22 (*illus*)
employment, tactical, 6-7
environmental characteristics, 7-1
environmental
 effects on operations, 7-4
 analysis, 7-4
 environmental effects, 7-9
 threat evaluation
environments, unusual, 6-35
evaluation of threat, 7-9
evasion and recovery, 8-1
evasion and recovery, K-1
evasion plan of action, K-2 (*illus*)
evasion
 aids, 8-7
 areas, 8-9
evasion
 classifications, 8-3
example link diagram, 7-11 (*illus*)

example subsurface site, G-6 (*illus*)
execution (actions on objective), 4-17
exfiltration phase and extraction method, 4-17
expedient insulators, 6-31 (*illus*)
extraction method, 4-17

F

facilities, C-1
facts, B-13 (*illus*)
fan method, 4-23 (*illus*)
fast rope insertion/extraction system (FRIES), 5-20
fatigue, J-5
field expedient antennas, 6-30
field site, C-1
finished subsurface site, G-4
fire
 plans, 4-46
 support, 4-45
 support annex, B-42 (*illus*), F-4 (*illus*)
fishhook and dog leg methods, G-8 (*illus*)
fixed site, C-1, C-2 (*illus*)
footprints, I-3 (*illus*)
forcible occupation of site, G-8 (*illus*)
Forms
 DA Form 1594, 2-12
 DA Form 2028, xiii
 DA Form 5752-R, 5-20
 DD Form 1833 TEST (V2), K-5 (*illus*), K-6 (*illus*)
formula for calculating length of half wave dipole antenna applied to example, 6-27 (*illus*)
fragmentary order, B-8 (*illus*)
frequency management, 6-1

G

geographic environments, D-1
ground force personnel, 4-38
ground wave, 6-20 (*illus*)

H

half wave dipole antenna, 6-26 (*illus*)
hasty subsurface sites, G-2
hearing, J-4

heavy team and platoon operations, 4-29
helicopter operations, 5-9
helocasting operations, 5-7
HF, VHF, and UHF radios, 6-2
hide site, 4-28 (*illus*)
 actions in, G-9
high-frequency
 ranges in ionosphere, 6-20 (*illus*)
 skip zone and distance, 6-22 (*illus*)
high frequency antennas, directional, field-expedient, 6-33
 half rhombic type, 6-33 (*illus*)
 long-wire type, 6-33 (*illus*)
 "V" type, 6-34 (*illus*)
HMMWVs
 camouflaged, 5-42 (*illus*)
 procedures for loading into CH-47 for infiltration, 5-37 (*illus*)
 wedge formation, 5-40 (*illus*)

I

identification of gaps in existing databases, 7-4 (*illus*)
image-gathering equipment, 4-53
imagery, 4-52, 4-53 (*illus*)
infiltration phase
 and insertion method, 4-16
insertion method, 4-16
intelligence, 1-2
 annex, F-5 (*illus*)
 estimate annex, B-39 (*illus*)
 gaps, 7-4
 preparation of the battlefield, 7-1, 7-4
intelligence, surveillance, and reconnaissance, 1-10 (*illus*), 3-2 (*illus*)
 operations, 3-1
 requirements, 7-20
intent, B-11 (*illus*)
intermediate staging area
 use in planning, C-4 (*illus*)
inverted "V" antenna, 6-27 (*illus*)
ionosphere, 6-20 (*illus*), 6-21 (*illus*)

issues (*illus*)

J

jungle operations, 6-35, D-1

L

landing point diameters, 5-26 (*illus*)
landing zones, 5-25
laser designators, 4-45
leader reconnaissance, 4-25, G-7
liaison
　duties, employment, and coordination, 2-13
light sources and distances, J-3 (*illus*)
linkup, 4-29
　and dissemination of information, 4-29
　annex, B-43 (*illus*)
loading sequence, UH-60, 5-31 (*illus*)
long-range surveillance company (LRSC), 1-4
long-wire antenna, 6-28 (*illus*)
LRS tasks by operation, 4-10 (*illus*)
LRSD and LRS teams, 1-7
　capabilities and limitations, 1-8
　command and control, 2-1

M

marking procedures, 5-26 (*illus*)
message
　and report formats, 6-9
　header, 6-11 (*illus*)
method of extraction, 4-17
METT-TC, 4-7 (*illus*)
MIJI Report, 6-18 (*illus*)
military equipment, 4-36
mission
　analysis brief, B-10
　classification, 3-11
　development process, 3-1, 3-4 (*illus*)
　orders, 3-1, 3-3
　planning factors, 4-44
　planning folder, 3-5
　　contents, 3-5
　　development, 3-5
　　format, 3-9
　roadblocks (*illus*)

mission concept brief (MICON), B-19
　slides, B-20 through B-33 (*illus*)
mission, intent, and priority intelligence requirements, B-11 (*illus*)
missions, 1-3, 1-4
　special, 4-66
mountain operations, 6-35, 6-37, D-2
movement, 8-4, J-9

N

named areas of interest, 2-13
networks, 6-1
night
　vision, J-1
　walking, J-6

O

objective sketchpad, 4-56 (*illus*)
occupation of hide site, G-7
odor, J-5 (*illus*)
off-center viewing technique, J-2 (*illus*)
omni-directional antenna pattern, 6-25 (*illus*)
operation order, B-4, B-5 (*illus*)
operational environment, 7-1
operations, 5-45, 6-7
　Airborne, 5-46
　bases, 6-1
　capabilities and limitations, 4-65
　cold weather, 6-35
　desert, 6-35
　foot movement, 5-47
　heavy team and platoon, 4-29
　helicopter, 5-9
　helocasting, 5-7
　ISR, 3-1
　jungle, 6-35
　mobility platforms, 5-31
　mountain, 6-35
　other, 5-45
　reconnaissance, 4-21
　security, 3-11
　stability, 4-60
　stay behind, 5-46

surveillance, 4-26
　team, 4-1
　types, 4-61
　urban, 6-35

operations (*continued*)
　vehicle, 5-31
　waterborne, 5-1
OPORD, F-2 (*illus*)
orders
　and briefs, B-1
　fragmentary, B-8
　operation, B-4
　warning, B-1
organization
　of LRSC, 1-5 (*illus*)
　of tracking team, I-5

P

panoramic sketch, 4-55 (*illus*)
Pathfinder, 4-67
pattern analysis plot sheet, 7-18 (*illus*)
personnel
　recovery, 4-67
　security, 3-11
petroleum, oil, lubricants (POL), 4-38
phase
　exfiltration, 4-17
　planning, 4-1
pickup zones, 5-25
plan of action for an evasion, K-2 (*illus*)
planning area facilities and sites, C-1, C-3 (*illus*)
plans, 4-47, 8-2
polarization, 6-19
polyvinyl chloride site, G-3 (*illus*)
power plant turbines and generators, 4-39
primary missions, 1-4
primary, alternate, and contingency radios, 6-3
priority intelligence requirements, B-11 (*illus*)
priority of actions for rehearsal, 4-14 (*illus*)
priority of destruction, communications devices, 6-17 (*illus*)
priority of work, G-9
protection, electronic, 6-16

Prusik knot, recovery line with, 5-19 (*illus*)

R

radio wave propagation, 6-19
radios, 6-2
rail lines and rail yards, 4-39
reassignment, A-4
reconnaissance and surveillance
 elements, 4-22 (*illus*)
 squadron, 2-6
reconnaissance operations, 4-21 through 4-25, 4-47
recovery, 4-18
 line with Prusik knot, 5-19 (*illus*)
 of personnel, 4-67
 types, 8-2
recruitment, A-1
rehearsal
 area annex, F-6 (*illus*)
 priority of actions, 4-14 (*illus*)
relationship matrix, 7-13 (*illus*)
repair procedure, whip antenna, 6-30 (*illus*)
report formats, 6-11 (*illus*), 6-12 (*illus*)
reports, 6-9
resonance, 6-18
restated mission, B-15 (*illus*)
rigging
 fast rope
  other aircraft, 5-22 (*illus*)
  UH-60, 5-21 (*illus*), 5-22 (*illus*)
 of snap links, 5-17 (*illus*)
 of SPIES rope on UH-60, 5-17 (*illus*)
 of wood block, 5-18 (*illus*)
 SPIES, for CH-46 or CH-47, 5-20 (*illus*)
roads, 4-40
route
 reconnaissance, 4-25
 selection, J-5
rubber boat, 5-2 (*illus*)
ruggedized COTS laptop, 6-6
runways and taxiways, 4-40

S

satellite
 dishes, 4-41
 tactical, 6-9
scanning patterns, J-2 (*illus*)
scout swimmers, 5-5
secondary missions, 1-4
security, 3-11
 and reports, 4-27
selection, A-4
 and occupation of sites, 4-26
 of route, J-5
separation, 3-12
ships, 4-41
short count, long count, 5-4 (*illus*)
signals, J-7
site
 field and fixed, C-1
 hide, 4-28 (*illus*)
 selection, 4-26, 6-8, G-6
 sterilization, G-9
 surface, G-1
slides, mission concept brief, B-20 through b-33 (*illus*) (*illus*)
sloping antenna, "V" type, 6-34 (*illus*)
sloping wire antenna, 6-29 (*illus*)
smell, J-4
sounds, J-4 (*illus*)
special missions, 4-66
special operations forces
 comparison of LRSU operations to, 1-4
 organization, 1-4
 sustainment, 1-5
 subordinate organizations and key personnel, 1-5
Special Patrol Insertion/Extraction System (SPIES), 5-9
specified and implied tasks, B-12 (*illus*)
SPIES rope rigging on UH-60, 5-17 (*illus*)
steel towers, 4-42
storage tanks for petroleum, oil, lubricants, 4-38
success, elements of, 6-2
successive sector method, 4-24 (*illus*)

surface sites, G-1
surveillance
 operations, 4-26
 site, 4-27 (*illus*)
suspension line weave site, G-3 (*illus*)

T

tactical
 employment, 6-7
 operations center, 2-10
 tactical satellite, 6-9
target
 acquisition, 4-44
 detection, J-8
 folder, 3-9 (*illus*)
task organization
 outside named areas of interest, 2-13
tasks
 break contact from hide or surveillance site, H-8 (*illus*)
 break contact front (diamond or file), H-2 (*illus*)
 break contact front, left and right (australian peel), H-3 (*illus*)
 break contact left, right (diamond or file), H-4 (*illus*)
 react to enemy air attack, H-5 (*illus*)
 react to indirect fire or air attack, H-6 (*illus*)
team operations, 4-1
teams, 6-2
 LRSD and LRS, 1-7
terminated sloping "V" antenna, 6-29 (*illus*)
terrain, urban, 4-46
threat
 capability, 7-15
 courses of action, 7-16, 7-17
 factors, 7-9
time event chart, 7-15 (*illus*)
topographic sketch, 4-55 (*illus*)
tracker and dog team, I-5
tracking concepts and procedures, I-1
transformers, 4-42

Index

troop leading procedures and METT-TC, 4-7 (*illus*)
tunnels, 4-43
two-man surface site, with ghillie suits, G-1 (*illus*)
types of recovery, 8-2, 8-3 (*illus*)

U

UH-60
   loading sequence, 5-31 (*illus*)
   SPIES rope rigging on, 5-17 (*illus*)

UNDER (Cache) Report, 6-14 (*illus*)
unidirectional antenna pattern, 6-23 (*illus*)
units
   reconnaissance and surveillance, 1-2
unusual environments, 6-35
urban
   operations, 6-35, 6-37
   terrain, 4-46

V

vehicle
   load, 5-35 (*illus*)
   movement annex, B-44 (*illus*)
   movement coordination annex, F-7 (*illus*)

W

warfare support, electronic, 6-16
warning order, B-1, B-3 (*illus*)
waterborne operations, 5-1
wavelength, 6-18 (*illus*)

XYZ

zone reconnaissance, 4-23

Printed by Libri Plureos GmbH in Hamburg, Germany